GALÁPAGOS

Page 1: Lonesome George, last surviving Pinta tortoise.
Pages 2–3: Marine iguanas and volcanic eruption, Cape Hammond, Fernandina.
These pages: Galápagos hawk over steam vent, Alcedo Volcano, Isabela.
Pages 6–7: Blue-footed booby courtship display, Punta Cevallos, Española.

Published by Firefly Books Ltd. 2009

First published in 2009 in New Zealand by David Bateman Ltd.,
30 Tarndale Grove, Albany, Auckland, New Zealand

Publisher Cataloging-in-Publication Data (U.S.)

De Roy, Tui, 1953
Galapagos : preserving Darwin's legacy / Tui De Roy, editor.
[240] p. : col. photos. ; cm.
Includes index.
Summary: Essays and photographs describing scientific research and conservation efforts, past and present, on the Galapagos Islands.
ISBN-13: 978-1-55407-484-6
ISBN-10: 1-55407-484-3
1. Nature conservation -- Galapagos. 2. Biodiversity conservation-- Galapagos. 3. Natural history -- Galapagos Islands. I. Title.
333.951609866 dc22 GH77.E2.G353 2009

Library and Archives Canada Cataloguing in Publication

Galapagos : preserving Darwin's legacy / edited by Tui de Roy.
Includes index.
ISBN-13: 978-1-55407-484-6
ISBN-10: 1-55407-484-3
1. Natural history--Galapagos Islands.
2. Galapagos Islands. I. De Roy, Tui
QH77.C5G35 2009 508.866'5 C2009-900488-7

Published in the United States by Firefly Books (U.S.) Inc.
P.O. Box 1338, Ellicott Station
Buffalo, New York
14205

Published in Canada by Firefly Books Ltd.
66 Leek Crescent
Richmond Hill, Ontario L4B 1H1

Book design: Trevor Newman
Map: Black Ant Design
Printed in China through Colorcraft Ltd, Hong Kong

This book is recognized by the Galápagos National Park and the Charles Darwin Foundation as an official publication celebrating their joint 50th anniversary year.

GALÁPAGOS

PRESERVING DARWIN'S LEGACY

Tui De Roy
Editor and Principal Photographer

With contributions by experts and major researchers
highlighting new knowledge, recent discoveries
and breakthroughs in applied conservation science.

FIREFLY BOOKS

Contents

The Galápagos Islands

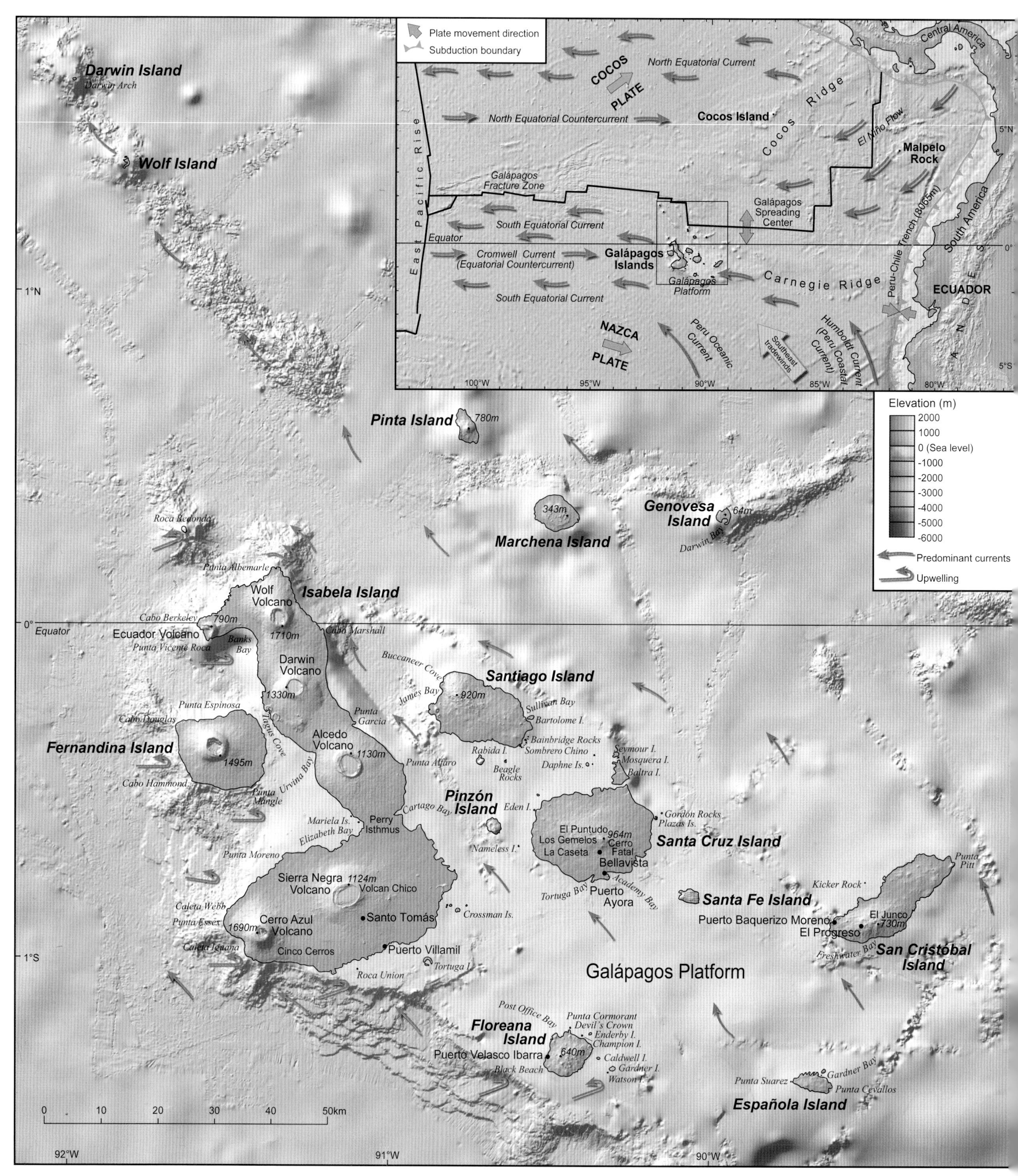

Plate movement direction
Subduction boundary
COCOS PLATE
North Equatorial Current
North Equatorial Countercurrent
Cocos Island
Cocos Ridge
Central America
El Niño Flow
Malpelo Rock
East Pacific Rise
Galápagos Fracture Zone
Galápagos Spreading Center
South Equatorial Current
Equator
Cromwell Current (Equatorial Countercurrent)
Galápagos Islands
Galápagos Platform
Carnegie Ridge
Peru-Chile Trench (8065m)
South America
ECUADOR
ANDES
NAZCA PLATE
Peru Oceanic Current
Southeast tradewinds
Humboldt Current (Peru Coastal Current)
100°W
95°W
90°W
85°W
80°W
5°N
0°
5°S
Elevation (m)
2000
1000
0 (Sea level)
-1000
-2000
-3000
-4000
-5000
-6000
Predominant currents
Upwelling
Darwin Island
Darwin Arch
Wolf Island
1°N
Pinta Island
780m
Marchena Island
343m
Genovesa Island
64m
Darwin Bay
Roca Redonda
Punta Albemarle
Isabela Island
Wolf Volcano
1710m
Cabo Berkeley
790m
Ecuador Volcano
Equator
Punta Vicente Roca
Banks Bay
Cabo Marshall
Darwin Volcano
1330m
Buccaneer Cove
Santiago Island
920m
James Bay
Sullivan Bay
Bartolome I.
Punta Espinosa
Cabo Douglas
Tagus Cove
Punta Garcia
Alcedo Volcano
1130m
Bainbridge Rocks
Sombrero Chino
Rabida I.
Beagle Rocks
Daphne Is.
Seymour I.
Mosquera I.
Baltra I.
Fernandina Island
1495m
Cabo Hammond
Urvina Bay
Punta Mangle
Punta Alfaro
Pinzón Island
Eden I.
Gordon Rocks
Plazas Is.
Cartago Bay
Mariela Is.
Perry Isthmus
Elizabeth Bay
Nameless I.
El Puntudo
964m
Los Gemelos
Cerro Fatal
La Caseta
Bellavista
Santa Cruz Island
Punta Moreno
Sierra Negra Volcano
1124m
Volcan Chico
Tortuga Bay
Academy Bay
Puerto Ayora
Santa Fe Island
Kicker Rock
Punta Pitt
Caleta Webb
Punta Essex
1690m
Cerro Azul Volcano
Santo Tomás
Crossman Is.
Puerto Baquerizo Moreno
El Junco
730m
El Progreso
Caleta Iguana
Cinco Cerros
Puerto Villamil
Tortuga I.
Freshwater Bay
San Cristóbal Island
1°S
Roca Union
Galápagos Platform
Post Office Bay
Punta Cormorant
Devil's Crown
Enderby I.
Champion I.
Floreana Island
Puerto Velasco Ibarra
640m
Caldwell I.
Gardner I.
Watson I.
Black Beach
Gardner Bay
Punta Suarez
Punta Cevallos
Española Island
0
10
20
30
40
50km
92°W
91°W
90°W

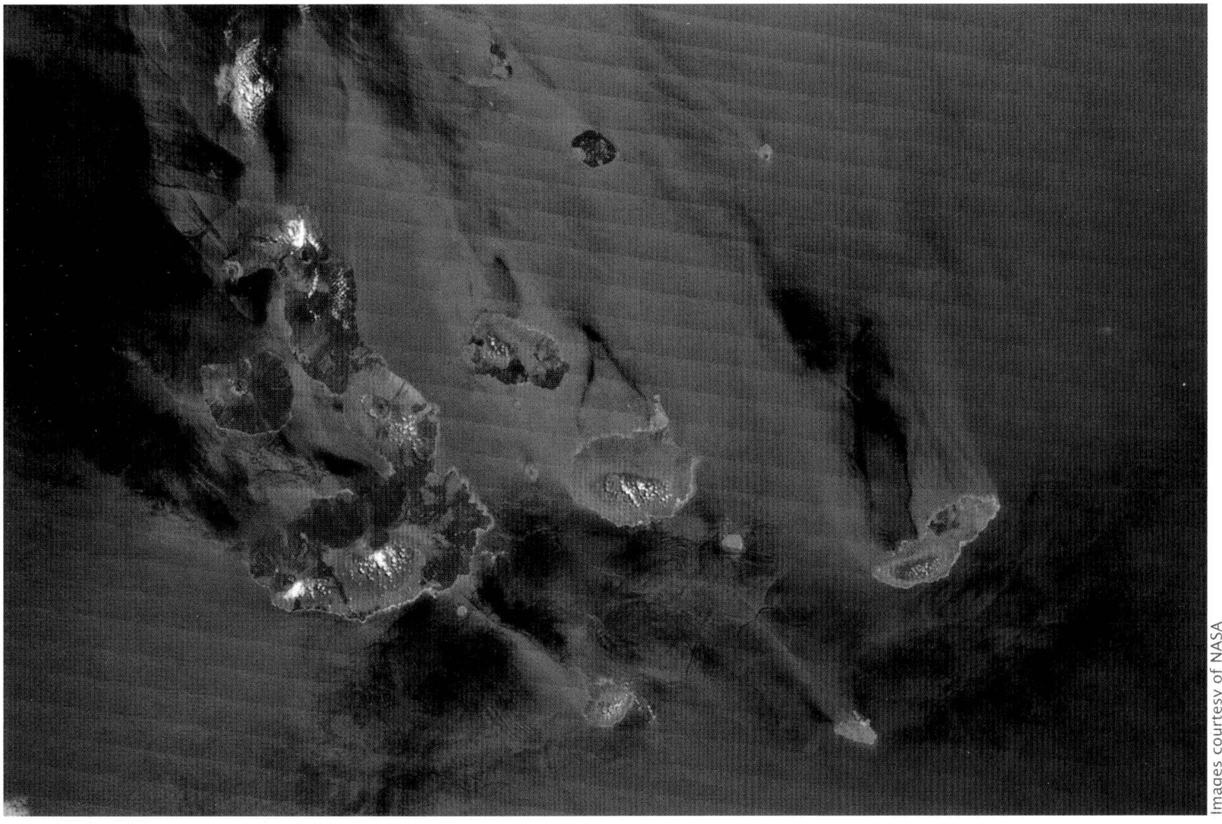

Images courtesy of NASA

LEFT: A NASA MODIS image shows nutrient plumes and strong eddies flowing through the archipelago, as well as the islands'lush vegetation areas. (The striping corresponds to the 40-pixel wide sweep of the satellite detectors as light from the Earth's surface hits them at different angles.)

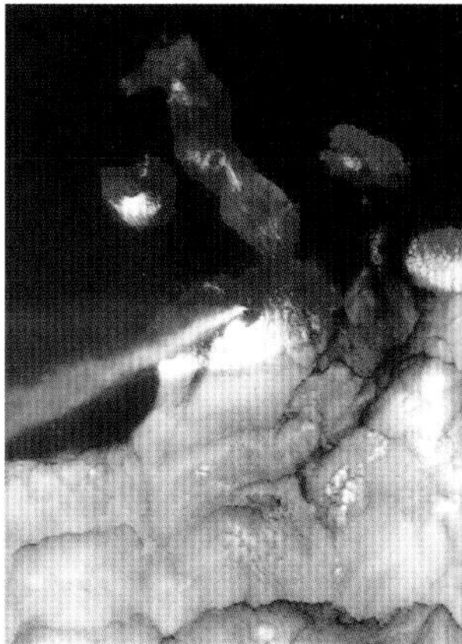

ABOVE: The gas plume from Sierra Negra's 2005 eruption (chapter 1) drifts high above a layer of cold-season inversion cloud (chapters 2, 3 and 4).

LEFT: The landforms below the sea in this relief map clearly show the extensive Galápagos Platform upon which the main south and central islands are built, and especially the abrupt drop-off along its southern and western edge. Also visible are the unique lobed terraces south of Isabela, which are described in chapter 1. The northern islands, though not rising high above sea level, emerge from the deeper ocean floor, forming substantial submarine features, along with several seamounts, some of which may be the remnants of past islands, such as the prominent one north of San Cristóbal. The inset regional map places the Galápagos Islands in the context of tectonic plates and fracture zones, also discussed in chapter 1. Arrows indicating major oceanic currents highlight the complexity of their movements, and the exceptional climatic setting described in chapters 2, 3, 6, 7 and 12. The confluence of several water masses, and particularly their cooling influence, is what nurtures many of the keystone species in the marine ecosystem, such as the fishes, marine iguanas, pinnipeds and seabirds, covered in chapters 9, 16, 17, 18, 19 and 20. The distribution of many of the terrestrial species is also reflected by the wind and current patterns as shown, which affected how species were transported from island to island over time (e.g., chapters 11 and 13), making those relegated on the older, southeastern islands the most genetically isolated.

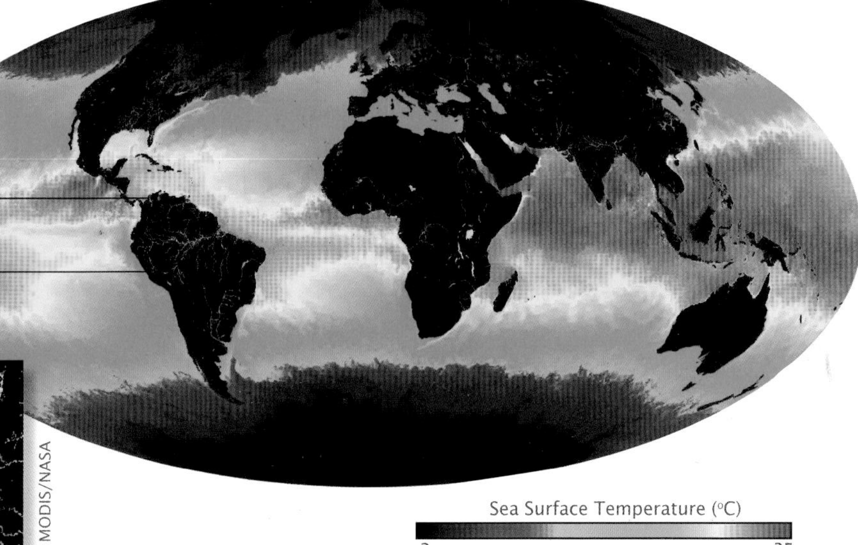

MODIS/NASA

ABOVE: Averaged over one month (May), daily global sea surface temperatures recorded by NASA show the remarkable pattern of cold waters around Galápagos, which drives the islands' exceptional climate and unique ecosystems.

BLUE-FOOTED BOOBY, PUNTA VICENTE ROCA, ISABELA. The southern hemisphere winter marks the Galápagos seasons. In May–June, cold waters return, and nutrient-laden upwellings rise to the surface around western Isabela and Fernandina islands, promoting massive plankton blooms. This is the engine driving the vibrant marine ecosystem, while cool sea fog laps the parched volcanic shores. The onset of the cool season is also when many seabirds begin to nest, such as this blue-footed booby prospecting a colony site at the base of Ecuador Volcano, bisected by the equator. For detailed discussions of the exceptional climatic setting and the unique wildlife of this region of the archipelago, refer to chapters 3, 9, 12, 18, 19 and 20.

Prologue: A World Flagship of Inspiration

Tui De Roy

The Roving Tortoise,
Worldwide Nature Photography,
14 Burnside Road,
Takaka, New Zealand.
<photos@rovingtortoise.co.nz>

HALF A CENTURY AGO a bold vision was born: to celebrate 100 years since the publication of Charles Darwin's revolutionary book *On the Origin of Species by Means of Natural Selection*, the world would join forces to preserve the islands that had helped formulate his ideas, and which in turn had changed scientific thinking forever. In 1959, the government of Ecuador declared the Galápagos Archipelago its first national park, while scientists and naturalists from around the world created the international Charles Darwin Foundation (CDF). Thus, 97% of all land areas, excluding small enclaves already colonized by a few hundred settlers, would be preserved forever, and a permanent research station would gather the knowledge needed to support the government in the wise administration of the islands.

Thanks to this unique collaboration between Ecuador and the world, many aspects of the Galápagos ecosystem are in better condition today than they were back then. Endemic species on the brink of extinction have been bred back to healthy numbers, and destructive introduced animals have been removed from many of the islands, returning them to a near-pristine state in several cases. Numerous scientific discoveries have also come to light — and continue to do so at an astonishing pace — including a bevy of new species and the rediscovery of some plants and animals once believed extinct. Together, these accomplishments have earned the Galápagos Islands the reputation as a world flagship of conservation.

BELOW: First sighted high on Wolf Volcano a decade ago, in 2009 the enigmatic pink iguana was formally described as a species new to science, a fitting tribute to this important anniversary year.

Nowadays the pace of the work has vastly accelerated, with ever more ambitious projects undertaken, and formidable new challenges emerging and multiplying. As the CDF and the Galápagos National Park (GNP) pull together to tackle increasingly demanding tasks, it is time to pause and celebrate. On the occasion of four combined anniversaries — the 50th year of both the GNP and the CDF, 150 years since the publication of Darwin's tome on evolution, together with Darwin's 200th birthday — this book aims to illuminate many of those groundbreaking accomplishments, along with sobering lessons applicable to the future.

Surprisingly perhaps, much of the initial international impetus to conserve Galápagos first emanated from faraway Belgium, which has hosted the permanent legal home of the CDF ever since. From the outset, the Belgian government, with its long tradition of scientific exploration in Africa, Antarctica and beyond, through the CDF established a Belgian scientific mission in Galápagos. A half-century later, it has again provided generous support to make the publication of this book possible.

Coincidentally, my own roots, too, originate in Belgium. In part inspired by many of those grandfathers of Galápagos conservation, in 1955 my parents left my birth country, before I was two years old, to join a handful of pioneers already living in the islands. So it is perhaps fitting that my profession has followed a path in conservation photography and writing, much of it focused on Galápagos.

When I began to approach long-standing researchers and conservation experts for contributions to this book, I was unprepared for the overwhelmingly positive responses I would receive. These were people who have dedicated their lives to Galápagos — unraveling its natural mysteries as well as mapping out solutions to the ever-mounting threats looming ahead. Little did I realize that in the subsequent months of editing I would find myself immersed in a wondrous world of discovery about a place I thought I knew so well. Each contributor has written a special essay, revealing in his or her own words their most outstanding contributions to our understanding of — and ability to care for — these magical islands.

My intimate familiarity with Galápagos enabled me to work very closely with the authors, the resulting email correspondence adding up to more than 2000 messages, many of them fascinating, in-depth exchanges of ideas. In some cases, our daily flurries of communications took on a breathless pace as we bounced questions and answers back and forth, both sides admitting to finding the process itself to be a stimulating and thought-provoking experience. I even had the satisfaction of learning that some of my reasoning helped generate new directions for future research. Chapter after chapter, a myriad facets of Galápagos research and analysis emerged that had not previously been publicized to the lay audience, beginning with Sarah Darwin's reflections upon a family history of Galápagos research — from her great-great-grandfather Charles Darwin to her own work on threats to the endemic Galápagos tomatoes.

On these pages the reader will travel from the deep inner workings of 'hotspot' volcanoes to how the convolutions of our atmosphere have molded an ecosystem designed around feast-or-famine conditions. We peek into past species' compositions by exploring fossil records in caves, and decipher rainfall records encrypted in lake-floor sediments. We learn of baffling extinctions, whether natural or man-induced, and are enthralled by the recent discovery of hundreds of species new to science, from lichens and mollusks to the enigmatic pink iguana, whose genetic lineage places it at the dawn of Galápagos time scales.

For the first time, we can follow the oceanic wanderings of hammerhead sharks, revealing an island-hopping 'golden triangle' across the eastern tropical Pacific, and begin to appreciate the reasons why the majestic waved albatross has recently slid into the International Union for the Conservation of Nature's (IUCN's) 'critically endangered' category. We see the evolutionary process remolding species of Darwin's famous finches, even as we mourn the disappearance of an equally diverse product of adaptive radiation in the bulimulid land snails, both processes happening right before our eyes. On the behavioral front, we confront the harsh world of fur seal pups and booby chicks, sacrificing their siblings in their bid for survival, while females rule the roost among both endemic hawks and flightless cormorants. Some mysteries endure, like why albatrosses take their eggs for a wander while incubating, or the storm petrels whose nests have never been found. Inevitably, we encounter some contradictory conclusions between authors, whose very different studies lead to divergent interpretations, for example, climate predictions and whether lost forms of Darwin's finches represent species extinctions or just lost island varieties.

These accounts also remind us again and again of the extreme fragility of Galápagos. There are heartrending details about the ravages wrought by alien diseases, climate change and the first-ever oil spill, counterbalanced by the exhilarating story of successfully applied biological warfare. The re-emergence of an endemic mammal believed extinct for nearly a century, and likewise the presence of tortoise hybrids long gone from their original islands, fill us with renewed hope for 'lost' species. The tide turns on ecological devastation when we read about the world's largest successful feral animal eradication project, or the restoration of the Española tortoise, from a mere few survivors to a wild population that is now self-expanding after more than 30 years of captive breeding and repatriation. And finally, we ponder the future, listening to the advice of experts on how to avoid a very plausible biological holocaust due to ever-increasing contact with the rest of the world.

This unique collection is an evocative, authoritative anthology to nurture the curious mind and to stretch our imaginations. Above all, it is a glorious celebration of what we have learned about Darwin's famed 'natural laboratory of evolution,' and what it will take to preserve it.

Tui De Roy

ABOVE: With dusk settling over the rumbling volcano, lava flows roil the sea at Cape Hammond during the 1995 Fernandina eruption, while a juvenile brown pelican rests after a day feeding on fish killed by the heat.

GIANT TORTOISES, ALCEDO VOLCANO, ISABELA
Giant tortoises begin to stir as the sun breaks over the rim of the shallow, slumbering caldera of Alcedo, a scene no doubt repeated countless times since these ancient reptiles first colonized the young islands millions of years ago. The largest males compete for wallowing space in fast-shrinking ponds at the end of the rainy season. Many of them have spent the night here, sometimes ramming and shoving each other until all available puddles are filled to capacity, although the exact reason for this predilection is not clear. Perhaps the thick mud retains the sun's warmth, or more likely it helps rid them of bothersome ticks. As the day warms, they will walk away in search of greener pastures to graze, and shade trees to avoid overheating, retracing their steps unerringly before sunset. For detailed information on tortoise research and conservation, climate cycles, and the dynamics of their volcanic home, refer to chapters 1, 2, 3, 4, 11 and 24.

Foreword: Galápagos Research
A Family Tradition
Sarah Darwin

Natural History Museum,
Cromwell Road,
London SW7 5BD,
United Kingdom.

ABOVE RIGHT: Ripe endemic tomatoes, Isabela.
BELOW LEFT: Feeding on *Opuntia* flower pollen, the cactus finch on Santa Cruz is one of 13 species that, as a group, bear Charles Darwin's name for their classic role in the study of evolutionary processes.
BELOW MIDDLE: A Galápagos mockingbird on the summit of Fernandina Island finds moisture in the acrid fruit of wild tomatoes.
BELOW RIGHT: Endemic tomato, *Solanum galapagense*, Cape Douglas, Fernandina.

I FELT VERY HONORED when Tui De Roy approached me to write the foreword to this important and remarkable book. We share a fascination for the Galápagos Islands, but Tui brings a unique perspective of the archipelago, as she spent nearly 40 years — and indeed was raised — in Galápagos. This rich immersion in the flora, fauna and culture of the islands, as well as her dedication to this enchanted place, shows throughout Tui's work.

In 1831 a young and enthusiastic Charles Darwin boarded the HMS *Beagle* as companion to the captain. The voyage was sponsored by the British Navy in order to map the southern oceans. No one could have guessed that, in time, this journey would be instrumental in forever changing the world view of life on earth, while making the small islands of Galápagos famous.

It was while he was in the Galápagos Archipelago that Darwin noted small differences in the mocking-birds from different islands. Also pointed out to him by the local governor were the island variations among the giant tortoises. Later, the finches that were eventually named after him became one of the other triggers for his thoughts on evolution.

When Darwin made his Galápagos field collection he did not separate the finches according to their original islands, but when the ornithologist John Gould classified them in March 1837 he found there were 13 separate species with different beak shapes and other features apparently corresponding to differing life habits. It was later that year that Darwin wrote in his notebook that he believed evolution had taken place in the Galápagos birds, and indeed in all living things, including humans (Browne, 2006).

In the 170 years since that momentous turning point, the ensuing fame of the Galápagos Islands has become a double-edged sword. On the one hand, this has attracted funding and researchers from around the world in efforts to document, understand and preserve the natural history of the islands and their surrounding waters. On the other, tourists in their thousands have been drawn to this unique and beautiful archipelago, along with a steady flow of entrepreneurial settlers. Sadly, this influx has come to exert huge pressures on the fragile island ecosystems, and threatens to damage the very situation the visitors seek to experience.

Even so, the Galápagos Islands remain, to a large degree, nearly intact. Unlike many other island groups, they have experienced few species extinctions. As such, they represent a global example of science and conservation working together to stem — and even reverse — a tidal wave of threats to their world's biodiversity. Lessons are being drawn on how

to seek ways to minimize the impact of tourism and associated industries. Meanwhile, world-class programs and innovative methods to eradicate invasive species are proving extremely successful.

The first large collection of Galápagos plants was gathered by my great-great-grandfather Charles Darwin. Among these were two species of endemic Galápagos tomatoes. During the past few years I have been working as a botanist on these taxa at the Natural History Museum in London and University College London, and I have been studying Charles Darwin's very specimens. Working on Galápagos plants has given me insights into the fragility of the islands' biota, and the many threats the flora is facing. Even though the islands may look relatively untouched, this process has been underway since the first sailors began hunting tortoises on these wild shores centuries ago, setting fires and leaving behind a trail of destructive alien species. Charles Darwin could not have known that among the new species described from his Galápagos collections there are some endemic plants that would never again been seen alive.

In my conversations with people interested in Galápagos, I often find that there is confusion about the true nature of the archipelago. This is, I think, in part due to the many television documentaries covering just one of two extreme views of the islands: either the remarkable beauty and near-pristine nature of the islands or the devastation from human impact. I hope that this book will redress this balance and reaffirm to the world the importance of the Galápagos Islands and why they need to be conserved — now more than ever.

Charles Darwin's account in *The Voyage of the* Beagle was my first literary impression of Galápagos. However, it was Tui's photographs that provided my first visual experience of the islands. What Tui has managed to do here so expertly is to marry harmoniously the current work of the scientific and conservation community with fantastic photographs. The scientific community and indeed the world have been waiting for a book such as this for a long time.

An acclaimed list of authors have each contributed their own story, some having lived in the islands for many years and all having worked there to become experts in their chosen fields. From internationally distinguished scientists to local natural historians, all the authors have a unique perspective to share. Their accounts are written in such a way as to be accessible to all and a joy to read.

Once you have experienced Galápagos, your life may never be quite the same. For many people, Tui's photographs will be their only encounter with this enchanted place, and this is no bad substitute as she truly captures the spirit of the islands and their wild inhabitants in this exceptional book.

Sarah Darwin

PLIGHT OF THE GALÁPAGOS TOMATOES

The two endemic species of Galápagos tomato (*Solanum cheesmaniae* and *S. galapagense*) have evolved into exceptionally hardy plants, often flourishing on sun-baked, salt-laced lava shores. Their ability to grow in these harsh conditions has made them valuable in plant breeding programs to enhance salt tolerance (and other characteristics) in cultivated tomatoes. Thus it is possible we may have all unwittingly eaten some part of a Galápagos tomato in far-flung corners of the world.

These plants have also developed some remarkable relationships with resident animals, their acrid orange to yellow fruits being very attractive to land birds and giant tortoises. Not only does this offer the plants free transport and readily available fertilizer packets aiding their dispersal, but the seeds actually germinate better after passage through a tortoise's digestive tract. Today, not only have these animal partners been much reduced in abundance and distribution, but a new threat has emerged with the introduction of two species of cultivated tomatoes (*S. lycopersicum* and *S. pimpinellifolium*). The latter has spread along roads and via garbage dumps on the north slope of Santa Cruz Island, becoming a problem plant and a classic invasive species. One of the threats that the introduction of closely related species poses is the potential of hybridization with the rarer Galápagos tomatoes, leading to the loss of the genetic integrity.

The survival of the Galápagos tomatoes is important not simply because they are unique in the world. Our crop plant relatives also need preserving in the wild for future generations of tomato eaters — both humans and the Galápagos wildlife with whom they evolved.

Even during the dry season, sprawling clumps of endemic *Solanum cheesmaniae* flourish on the arid, salt-laced lava along the north coast of Isabela.

GALÁPAGOS HAWK
OVER ALCEDO VOLCANO
CALDERA, ISABELA.

Using the warm breeze over steaming sulfur fumaroles, a juvenile hawk pauses briefly on an old snag where a forest once grew before shifting volcanic heat altered this dynamic landscape. Historically Alcedo has been the least active of the young shield volcanoes of the western islands, yet 100,000 years ago it was the scene of an exceptionally violent, explosive eruption. A thick layer of pumice still blankets much of the outer slopes, and genetic studies infer that giant tortoises were nearly exterminated by the event. Galápagos hawks hold year-round hunting and breeding territories when mature, but mottled young birds like this one often join together as an errant band that roams the island freely to find prey, overwhelming the aggressive defences of the all-dark resident adults. For more on volcanism and the fascinating social life of these endemic hawks, refer to chapters 1 and 15.

Director, Galápagos National Park Service,
Isla Santa Cruz,
Gálapagos, Ecuador.
www.galapagospark.org

The Galápagos National Park

Half a Century Dedicated to Conservation

Edgar Muñoz

FEW CONSERVATION AREAS IN Latin America have attained the high management standards of the Galápagos National Park (GNP). Over 8000 square kilometers (3100 square miles) in total area, the protection of its natural ecosystems and the development of sustainable-use concepts have presented extraordinary challenges. In the five decades since its creation, the GNP is distinguished by its innovative planning methodology and management concepts, establishing itself as a pioneer in the administration of protected areas. This work has been guided by the design and implementation of four consecutive management plans, the GNP being the first in all of South America to develop such an operational tool (1974), applied and updated on a continuous basis and replicated as a model in many other protected island systems. The most recent of these working documents, formulated in 2005, represents a strategic interdisciplinary process with an ecosystem focus that strives to manage the causes of the problems ahead of their effects.

RIGHT: An endemic Galápagos sea lion performs playful underwater pirouettes through shafts of sunlight entering a sea grotto along the north coast of Seymour.

The GNP also stands out on a world scale with the development of a remarkable array of solid, powerful national and international partnerships, both governmental and nongovernmental, to serve its conservation interests. While these relationships, many of them enduring across the decades, number far too many to enumerate in detail, a few examples are helpful to give an idea of their range and scope. Locally, these include the Environmental Police and Ecuadorian Navy to help law enforcement on land and sea, the INGALA (Instituto Nacional Galápagos) Council and other local governments on planning issues, and the Participatory Management Board to seek consensus on resource extraction from the Galápagos Marine Reserve, to name but a few. Internationally, the GNP's most long-standing partner is the Charles Darwin Foundation (CDF), with whom many ecological restoration activities have been co-managed. Historically, the GNP has also maintained very strong links with many foreign government agencies and the United Nations, enabling it to implement a number of its most ambitious undertakings, such as the famed Isabela Project with the Global Environment Facility's backing. Other durable relationships exist with conservation organizations around the world, such as the 'Friends of Galápagos' organizations in Europe, the United States, New Zealand and Japan, several ecotourism enterprises, universities or research organizations, and many others too numerous to mention here, but all of whom can be called upon by the GNP for technical and other support.

Some of the GNP's proudest achievements and important milestones include:

- Major contributions in the captive breeding and repatriation of seven species of giant tortoises to Española, Pinzón, San Cristóbal, Santa Cruz, Santiago and southern Isabela (two types), with operating facilities in the three main islands: the *'Fausto Llerena'* Tortoise Center on Santa Cruz, the seminatural *'Cerro Colorado'* enclosure with hatching facility on San Cristóbal and the *'Arnaldo Tupiza'* center at Villamil, Isabela.

• Restoration of endangered land iguana populations of Cartago Bay (Isabela), Conway Bay (Santa Cruz) and Baltra Island, successfully combining captive breeding with control of introduced predators. Initiated in 1976, the program surpassed expectations and could therefore be wound down for Santa Cruz in the late 1990s. Likewise for Baltra in 2008, after the successful reestablishment of the species on its island of origin, with the release of some 420 captive-bred young and the eradication of cats.

• The creation of the Galápagos Marine Reserve covering 1.33 million square kilometers (over half a million square miles), placed under GNP management in 1998 through the formulation of the Galápagos Special Law, and administered by means of a participatory management system involving the key stakeholders in the islands.

• Biological control of the cottony cushion scale, a devastating foreign insect pest, by means of the introduction of its natural predator, the Australian ladybird beetle, in close cooperation with the CDF.

• The rigorous management system for all tourism activities within the GNP, including vessel licensing, fixed itineraries, strictly defined visitor sites and guides who must accompany every group. These principles have ensured that in almost four decades of tourism no measurable direct impact has been noted and visitors today can still have the same experience as in the early days. This model has served as an example in other parts of the world.

• Eradication of introduced feral animals from all islands in the attached table, projects which in many cases required long years of sustained effort.

Other outstanding achievements have included the first successful international lawsuit to settle in favor of environmental damage reparation in the case of the *Jessica* oil spill of 2001, and the creation of an internationally backed endowment fund for the permanent control of invasive species in Galápagos.

ABOVE: Sally-lightfoot crabs and marine iguanas share a wave-washed boulder during exceptionally high tides.

All of these examples are part of the lengthy suite of actions required in today's world in order to fulfill the GNP's entrusted mission: to conserve the natural capital of the archipelago, a task made possible thanks to the dedicated support and commitment to cooperation of all of our partners, particularly the local governments, the Galápagos community as a whole, and above all the park rangers present and past, who all have in their time given of themselves — and even in some cases given their lives — for the conservation of Galápagos. Each and every one of them has contributed, in one way or another, to all of our successes of the last 50 years, and thus must always be remembered.

SPECIES	ISLAND													
	Isabela	Floreana	Santa Cruz	San Cristóbal	Santiago	Bainbridge	Marielas	Rábida	Española	Baltra	Pinta	Santa Fe	Marchena	Seymour
Goat	†*	†**			†			†	†	†	†	†	†	
Pig					†									
Donkey	†*	†			†									
Cat										†				
Dog	†	†	†											
Black rat						†	†							†**
Red fire ant												†	†	
Rock pigeon	†		†	†										
Tilapia (fish)				†										

*complete eradication on the northern half only, with a few remaining on the southern two volcanoes. ** awaiting final confirmation through follow-up monitoring.

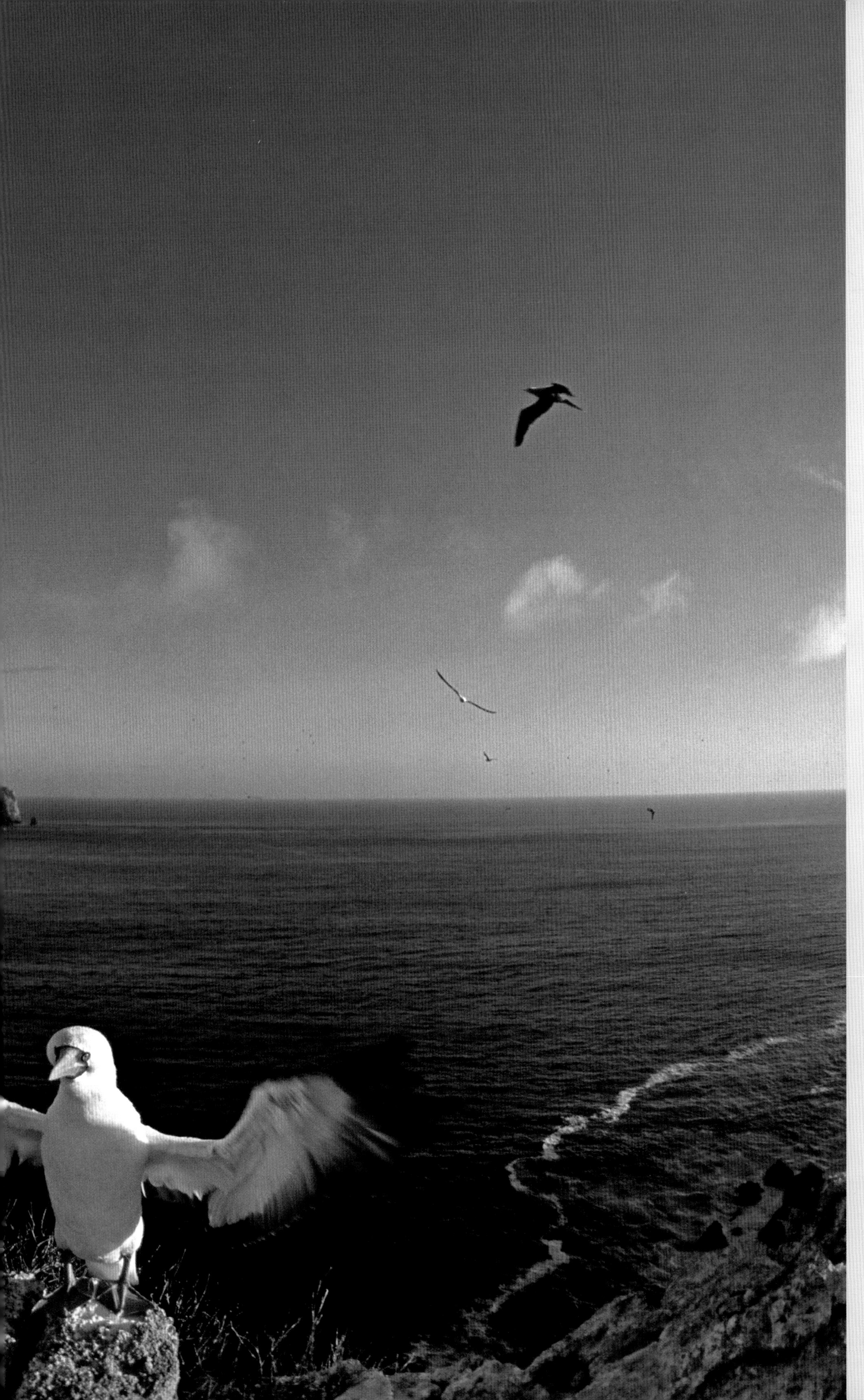

NAZCA BOOBY, WOLF ISLAND.

Up until 2002 the Nazca booby was considered the same species as the somewhat heftier masked booby ranging in all tropical oceans of the world. Yet the birds found nesting primarily in Galápagos and on the small rock of Malpelo near Colombia are distinctive both genetically and taxonomically, with more slender, brightly colored beaks among other features. The most extensive nesting colonies are found on the small outpost islands of Wolf and Darwin, from where large flocks join offshore feeding frenzies that often include dolphins, sharks, tuna and other marine predators. Long-term studies of their unique breeding system are revealing startling survival strategies among both chicks and adults. To learn more about remarkable Galápagos seabird behavior, as well as different aspects of the northern islands, read chapters 10, 16, 17, 18, 19 and 26.

The Charles Darwin Foundation

History of a Science and Conservation Vision

Peter Kramer

President, Charles Darwin Foundation (AISBL), Puerto Ayora, Santa Cruz, Galápagos, Ecuador. www.darwinfoundation.org

WHEN I FIRST CAME to Galápagos in 1962 conservation was not a widespread concern. Policymakers had only recently begun serious efforts to save the biodiversity of Galápagos. In those times, it was still common for visitors from the continent to take home tortoises as souvenirs. They were simply following the example of travelers and collectors from all over the world who had removed thousands of plants and animals of numerous species — for food, for commerce or as samples for museums and zoos. In fact, the visionary scientists and public servants who founded the Galápagos National Park (GNP) and the Charles Darwin Foundation (CDF) 50 years ago were inspired into action by looking back on centuries of destruction. Throughout the 19th and first half of the 20th centuries notable scientists and authors like Rollo Beck and William Beebe were already describing the situation as bleak. Giant tortoises had been driven close to — and on some islands, all the way to — extinction. Galápagos fur seals had barely escaped total extermination. Whales, once ruthlessly exploited by whalers from both sides of the Atlantic, had become scarce. At the same time, it seemed clear that this wave of annihilation would be accelerated by the great number of invasive animals and plants that people had introduced, and were still introducing, either on purpose or involuntarily, into the Galápagos ecosystem.

This sort of drama was not new. Sadly, all the world's oceanic islands and archipelagos have suffered the same fate: humans have exploited indigenous wildlife, very often to extinction, and have introduced many species that act as predators and competitors capable of dramatically altering the delicate insular fauna and flora composition.

During the early 1900s, many scientists believed that this was simply the normal course of history, and that it would be impossible to halt the rapid loss of Galápagos species. Many expeditions thus redoubled their efforts to collect as many specimens for science as possible, because their extinction seemed unavoidable.

The founders of the CDF and the GNP took a much more daring and optimistic view: they wanted to seize the last opportunity anywhere in the world to preserve an entire oceanic archipelago and its ecosystems in a state that was still very close to pristine, or that at least had the potential of being restored. Nowhere else was it possible to experience and study the processes of natural selection and biological evolution so directly and intimately. To protect and manage the Galápagos Islands would bring great scientific and economic benefits to humanity and to the country of Ecuador, the islands' owner and guardian.

In spite of many potentially destructive ideas and enterprises that have beset the islands since that time, the vision of those pioneers has prevailed. Conservationists note that, in many ways, the Galápagos environment is in better condition today than it was 100 years ago, and is often hailed as one of the very few true conservation success stories in the tropics. Not only have relatively robust conservation and research infrastructures been operating for half a century, but real on-the-ground progress is in evidence everywhere: the looming threat of extinction has been averted from a number of iconic Galápagos species, with tortoise and land iguana populations recovering in their restored native habitats; invasive herbivores have been removed from most islands where they once ran rampant, and work is accelerating toward their complete eradication from all protected areas, which represent 97% of the total surface of the islands. Meanwhile, although a number of marine species continue to be seriously overexploited, the Galápagos Marine Reserve provides a basis for both conservation and sustainable use of the abundant resources of the archipelago's surrounding seas.

I believe a large part of this success is the fruit of years of public education and continuous professional training, two of the underpinnings of the CDF's charter. Half a century ago knowledge of Galápagos was still sketchy: scientists saw them as a special place for evolutionary and geological research, while the Ecuadorian public knew the islands only as a prison camp and a frontier for adventurous settlers. The new message transformed those images totally: the Galápagos Islands and their wildlife are unique

BELOW: A Galápagos penguin perched on the lava shoreline of Bartolomé personifies the fragility of a unique marine ecosystem, where tropical sea urchins have become more numerous with prolonged El Niño events and declining numbers of urchin-eating slipper lobsters due to fishing pressures.

and spectacular, they must be preserved, and they offer outstanding opportunities for nature tourism. This enlightened perception has on the one hand turned Galápagos into a conservation and research hotspot, and on the other into a prime destination for visitors, generating a steady flow of cash which now amounts to many millions of dollars per year. This development, however, has created new and serious challenges.

Many of today's immediate Galápagos conservation quandaries come as a consequence of this uncontrolled growth. Tourism has provided a very significant stimulus to the local economy, resulting in a chain reaction of massive immigration spurring increased natural resource extraction and continuous movement of goods and people between the continent and the islands, as well as within the archipelago. This traffic, in turn, leads to the breakdown of the biological barriers that were the essential basis for all that makes Galápagos so special and unique. Up until relatively recent times, the islands had had no human inhabitants, while plant and animal colonization was limited by 1000 kilometers (about 600 miles) of ocean. Such rapid change is now dangerously stressing the delicate original ecosystems.

Clearly, we scientists and policymakers, while concerned about resolving species survival problems, did not always pay enough attention to our own species. In a relatively short time, humans have become a significant and even determining factor in the Galápagos ecosystem. Henceforth we will have to focus on the question of how the human species can be made to cohabit within the specific Galápagos ecology without causing the destruction of the entire system. How can we alien immigrants fit in without pushing out the other species that are the essence of the islands and which make this such a valuable place? How must we behave and organize ourselves socially, economically and culturally?

Consequently, the future portfolio of research and education efforts on Galápagos will include social and economic studies, as well as the essential conservation-oriented biological baseline work. At the same time, it is important to facilitate the many fundamental studies in evolution, ecology and geology that serve to expand our horizon of knowledge of this extraordinary archipelago and the entire natural world.

Celebrating our first half-century together with the GNP, at the same time as Darwin's 200th birthday, the CDF looks back on a remarkable history, steeped in international cooperation, which knows no equal anywhere in the world. It came into existence under the aegis of UNESCO (the United Nations Education, Science and Conservation Organization), simultaneously with the creation of Ecuador's first protected area, to mark the centennial of Charles Darwin's landmark treatise on evolution, *On the Origin of Species*. The CDF's mission would be to generate knowledge, garner international support and provide advice to assist the Ecuadorian government in preserving the Galápagos Islands for posterity. To this end the Charles Darwin Research Station — the Foundation's operative arm — was constructed on Santa Cruz Island, where it has pursued conservation science without interruption since its 1964 inauguration. The CDF's legal seat is registered in Belgium where, ever since the times of our founding president Victor Van Straelen, government scientific institutions have been steadfast supporters.

Through a binding agreement, the Foundation continues to enjoy a unique status as guest of the Ecuadorian government, still charged with the same responsibilities today. With its focus on generating and broadcasting knowledge, it is thus positioned to play a central role in the ongoing efforts to integrate humans with their environment. To achieve this work the organization benefits from a remarkable structure: international outreach and support, exclusive dedication to Galápagos, and strong institutional roots in Ecuador.

Due to the loss of its protective isolation, Galápagos is in peril, but it will survive if we humans can succeed in adapting to an island way of life that is different from what we know elsewhere, just as the endemic species have done before us. This will require the adoption of several fundamental concepts: islanders respecting the need for biological isolation; an economy adapted to the special island situation; and Galápagos visitors who impact the islands as little as the migratory birds coming from the Far North and leaving again every year.

Peter Kramer

ABOVE: A nocturnal swallow-tailed gull departs the breeding colony at dusk to feed far out at sea, using highly sensitive eyes to snatch squid and fish that rise from the depths under the cover of darkness, Genovesa.

Freda Chapman

BELOW: Visitors from around the world come to the Charles Darwin Research Station on Santa Cruz, where the organization's founding president, Professor Victor Van Straelen of Belgium, is remembered.

GIANT OPUNTIA CACTI, SANTA FE.

The tallest cacti in the world, reaching up to 15 m (50 ft), stand against time on the arid coast of Santa Fe, where tradewinds waft salty moisture yet rain rarely falls. These slow-growing trees may take 150 years to attain this size, having survived El Niño devastation near the cliff edge, where their root system escaped deadly waterlogging. It is possible this stand may have already been developing when the giant tortoises went extinct on these shores, and witnessed the introduction of goats that once ravaged the vegetation. But endemic rice-rats and land iguanas, who depend on the cacti for food and water, did survive to see the goats exterminated and their island restored. For information on this island and its unique species, and an in-depth study of the cactus forest, turn to chapters 7, 11, 13, 15 and 25.

Islands on the Move
Significance of Hotspot Volcanoes
Dennis Geist

Department of Geological Sciences, University of Idaho, Moscow, ID 83844, United States. <dgeist@uidaho.edu>

Dr. Dennis Geist has been working in Galápagos since his Ph.D. fieldwork during the 1982 El Niño. He earned his Ph.D. under the direction of Alexander McBirney, who pioneered the geology of the islands. He has been a professor at the University of Idaho since 1990, where he won the Research Excellence award. He also works on volcanoes in the Andes, Hawaii, Iceland, Antarctica and Yellowstone, mostly with his partner in science, Karen Harpp. His daughter Beryl is now carrying on the family tradition of fieldwork in Galápagos. He is a member of the General Assembly of the Darwin Foundation and a Fellow of the Geological Society of America. His Galápagos research has been funded by the U.S. National Science Foundation and the National Geographic Society.

ALTHOUGH BETTER KNOWN for its role in our understanding of the origin of life and evolutionary mechanisms on oceanic islands, the Galápagos Archipelago is also one of the world's premier natural laboratories for the study of the earth, including tectonic processes and the origin of volcanoes. Just as many geographic conditions conspire to produce a unique setting for colonization and adaptation of life, Galápagos is also in a unique geologic environment, one that has yielded insights ranging from the internal flow of rocks deep inside the planet to how volcanoes are born, grow and change.

My own interest in Galápagos volcanoes began in 1982, when I circumnavigated San Cristóbal Island on foot, in the first detailed geological study of that island. Some of the research discussed here results from the 22 field expeditions I completed since, but much can also be credited to my former students and other collaborators, especially Terry Naumann, Bill Chadwick and Mark Kurz.

The Galápagos Spreading Center

To understand the importance of the geologic origin and development of Galápagos it is best to start with the big picture, with the causes of island formation as the result of movements of rocks deep within the earth. Much of the geologic uniqueness of the archipelago is its tectonic setting: a hotspot adjacent to a mid-ocean ridge. This combination of the two tectonic regimes is found nowhere else on earth. Hotspots in the middle of plates are common and include Hawaii, Yellowstone, the Canary Islands and

BELOW LEFT: A fast-flowing river of lava, typical of the basaltic volcanoes of Galápagos, overflows its banks as it contours an older, vegetated cone on the west slope of Fernandina, charring a small palo santo, *Bursera graveolens*, forest.
BELOW RIGHT: The author's team approaches the fountaining lava of the Sierra Negra eruption of 2005 along the caldera rim.

Réunion. Iceland is a hotspot directly superimposed on a mid-ocean ridge. The Galápagos Islands are neither in the middle of a plate nor directly at a plate boundary. Instead, they lie within several hundred kilometers of the Galápagos Spreading Center, a submarine volcanic ridge that is notable for its deep-sea hydrothermal vents.

The easiest way to think of the Galápagos Spreading Center is as a giant tear, where the earth's rigid outer shell has given way because it is being pulled in two different directions. The Cocos Plate is sinking to the northeast beneath Central America, whereas the Nazca Plate is being pulled straight east by the Andean system. The ridge is offset in several places by faults, and there is a large one at 91 degrees west longitude, a complex region of heavily scarred crust, which profoundly affects the northern islands. The current proximity of the spreading center and the Galápagos hotspot may be a coincidence, although most believe that the hotspot 'captures' the spreading center by creating a weakness in the crust. The term 'hotspot' refers to a fixed volcanic region set away from a plate boundary. Most hotspots form from rocks that rise buoyantly from deep within the earth's mantle, in a thermal plume. (A common misconception is that the deep earth is molten; instead, it is made of solid rocks, which are plastic and relatively soft when they are hot.) Doug Toomey, Emilie Hooft and their colleagues have used seismic waves to make images of the earth's deep interior in the Galápagos region. They have measured a hot mass of rocks extending down to at least 700 km (430 miles). Like most things that are more difficult to define the farther away they are from the observer, the bottom of this mantle plume has not yet been detected — it could well extend into the lower mantle, to a depth of 3100 km (1900 miles). In contrast, the hot zone beneath the spreading center extends only to about a depth of 200 km (120 miles),

ABOVE AND BELOW: In 1995, a radial fissure running down the western flank of the Fernandina shield volcano poured copious new lava into the sea, while flightless cormorants nearby remained unperturbed.

LAVA TYPES

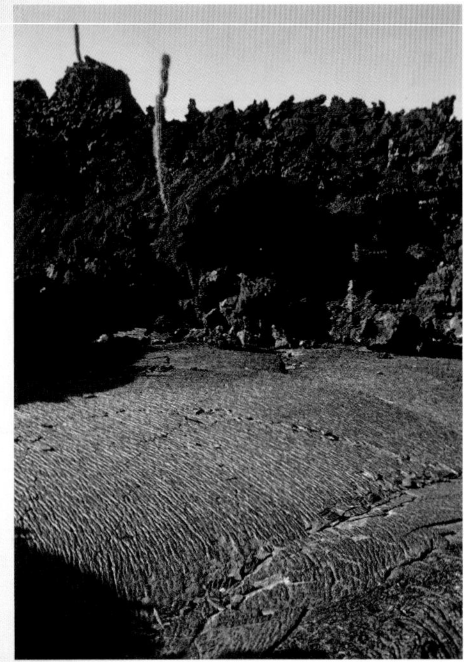

An a'a' flow partially covers older pahoehoe lava on eastern Santiago Island.

Lava flows can take on very different surface textures depending on how they flowed and cooled, even though their composition may be identical. Pahoehoe is the name given to ropy-textured lava that results when very fluid lava cools with a stable crust preserving its liquid shapes, whereas a'a' refers to jagged, broken lava that forms when the solidified surface crust tumbles and breaks as it is conveyed on a still-molten layer below. Both terms are borrowed from Hawaiian words, where these lava types are commonplace.

Tui De Roy

RIGHT: Few places are as deeply eroded as the sea cliffs of Ecuador Volcano, exposing innumerable layers of tuff, lava and scoria, as well as vertical lava dikes where magma rose to the surface through narrow fissures.
FAR RIGHT: In a land often reshaped by volcanic activity, a mockingbird sings atop a cactus singed by hot scoria fallout on Fernandina.

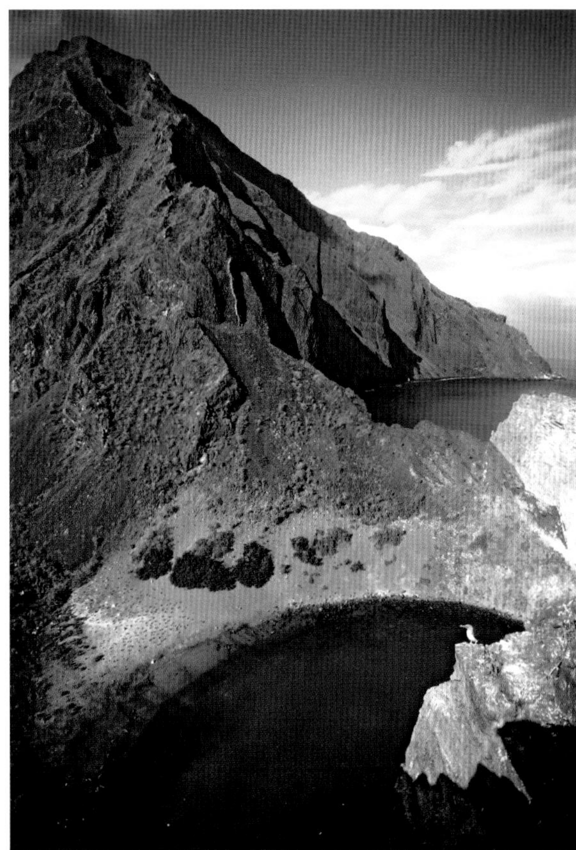

indicating that its magmas are supplied from the upper mantle.

The chemical compositions of Galápagos lavas also provide a clue to their origins, as they are richer in potassium, uranium and thorium than lavas from the Galápagos Spreading Center. Bill White has shown that they are likewise higher in the radiogenic daughters of uranium and thorium (isotopes of lead), indicating that the magma source of the Galápagos hotspot has been relatively enriched in radioactive elements for at least hundreds of millions of years. This radioactive enrichment is probably why the hotspot is hot and thus explains the Galápagos mantle plume: the radioactive heat provides thermal buoyancy to the rocks, which causes parts of the lower mantle to ascend as a plastic plume. The leading theory for the origin of this enrichment in radioactive elements is that the mantle plume originated from ancient crust that foundered beneath a mountain chain like the Andes and sank to the top of the earth's core hundreds of millions of years ago. Another clue to the origin of the Galápagos plume is the large amount of the isotope helium-3 in some Galápagos magmas. Helium-3 has taken a one-way path from the mantle to the atmosphere to space since the earth accreted. Lavas from Fernandina, in particular, are rich in helium-3, indicating a contribution from a primordial part of the earth's interior, an area that has only partly degased over the earth's entire history.

Thus, the prevailing theory is that the Galápagos Islands originate from a plume, which ascends from the deep mantle. The plume itself is solid but plastic, and rises because it is thermally buoyant. It is hot because of a concentration of radioactive elements, a combination of primordial material and recycled materials that foundered from the surface into the deep mantle hundreds of millions of years ago.

As the Galápagos plume nears the surface (within about 150 km, or 90 miles), it begins to melt owing to decompression. Unlike ice, which melts at lower temperatures the higher the pressure, rocks melt at higher temperatures the deeper you go into the earth, so rocks at the extreme pressures of the deep mantle are probably never hot enough to be molten. As the mantle plume ascends, pressure is lessened concurrently, but the plume remains hot because it takes a long time for heat to be conducted out of it. So the rocks begin to melt not because they are heated any further, but because the pressure on them is lowered.

A map of the seafloor (see page 8) in the Galápagos region shows the results of these processes. The spreading center runs east–west as an underwater ridge to the north of the islands. The central part of the archipelago is built on a volcanic platform that extends from Floreana in the south to Roca Redonda in the north. The northern islands (Wolf, Darwin, Pinta, Marchena and Genovesa) are separate from the platform and probably not formed as a direct result of the mantle plume. If you look at the seafloor around these islands, you can see that the islands are merely the emergent parts of linear ridges. These elongate volcanoes suggest some form of cracking. In fact, the orientation of the ridges is exactly what one would predict from the stresses that build up from the offset of the ridge at 91 degrees west. These stresses are common everywhere along the earth's ridge systems (where such offsets are also common), but near Galápagos, the stresses on the Nazca Plate are revealed by the Galápagos mantle plume.

A consequence of the eastward motion of the Nazca Plate is that new volcanoes are continually built over the Galápagos plume, and older ones are carried away to the east. Radiometric age determinations support this theory. The oldest rocks in the archipelago are two million to three million years old and are found on San Cristóbal, Española and Santa Fe. In contrast, most historic volcanic activity is on the western islands of Fernandina and Isabela. Lavas from the central islands (Pinzón, Santiago, Santa Cruz and Floreana) are intermediate in age and range to about 1.5 million years old.

Despite its importance for biological studies, determining the age of emergence of each Galápagos island is an unresolved problem because the volcanoes are frosted over with a veneer of young lavas, and the near absence of erosion makes sampling through that veneer impossible. Thus, the oldest lavas on each island are never exposed. It is possible that each island formed 'downstream' from the present focus of the hotspot, close to its present position, and consequently is no older than its oldest sampled lava. Another equally viable possibility is that each island originally emerged at the apex of the hotspot; hence, its age could be determined by taking its distance from the hotspot divided by the velocity of the Nazca Plate's eastward travel. Although one estimate of that velocity is 22 km (14 miles) every million years, others have calculated speeds nearly three times faster.

Galápagos volcanoes

Galápagos volcanoes come in two morphologic styles: the classic gentle-sloped shields of the central and eastern archipelago and the impressive, dome-shaped 'Galápagos shield' type of the western archipelago. Galápagos shields are typified by their steep upper slopes, enormous calderas and curious arrangements of fissure vents, with circumferential fissures near the

ABOVE: Erupting every few years, Fernandina is one of the most active volcanoes in the world. In 1978, lava flowing down from a rim vent caused the lake to boil, whereas 10 years before the caldera floor collapsed from a depth of 300 m (1000 ft) to around 900 m (3000 ft). In the 1980s, the lake vanished, then later reformed.

PARASITIC CONES

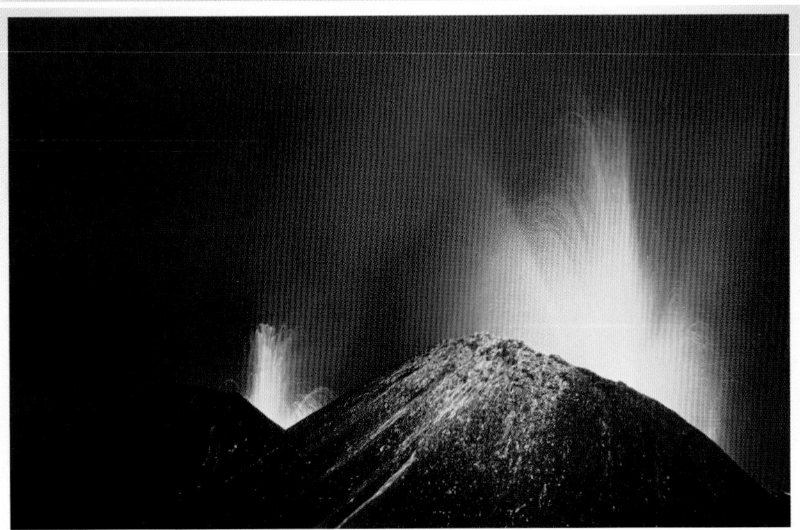

Spatter cones forming on the eastern flank of Cerro Azul volcano in 1979, the arced trajectories of lava bombs outlined by this dusk time-exposure.

Thousands of small volcanic cones punctuate the Galápagos landscape. Usually aligned along either radial or circumferential fissure systems that riddle the large shield volcanoes, there are three distinct forms, depending on the consistency and location of the erupted lavas and gases. Spatter cones are the most common, so termed because they are formed by the splattering of molten 'lava bombs' that build up around the vents, whereas scoria cones consist of smaller, lighter fragments that cool in midair when lava fountains are propelled by escaping gases. Tuff cones are made of consolidated volcanic ash resulting from fallout when water enters the eruptive fissure, causing a steam explosion upon meeting the rising magma and sending fine pulverized dust high into the atmosphere. They are collectively called parasitic. The results of innumerable past eruptions, they predominate on the older islands.
Tui De Roy

caldera rim and radial fissures on the lower flanks. The steep slopes are caused by the eruptive styles of the volcanoes. Lavas erupted from the summit tend to be small-volume pahoehoe, because of the decreased head from getting the magma up to a higher elevation. These small-volume lavas pile up to build the upper part of the edifice. Lavas erupted from the lower radial fissures are higher volume a'a', with greater eruption rates due to the higher pressure when lava escapes from the base of the volcano.

Galápagos calderas form by collapse, when the magma reservoir beneath it is partly evacuated by eruptions, creating a void into which the caldera floor can slump. The calderas are highly dynamic: they swell elastically between eruptions (sometimes breaking and causing earthquakes) and are formed sequentially, over many successive eruptions. For at least 13 years before the 2005 eruption, Sierra Negra's caldera floor had inflated by over 5 m (16 ft), then it subsided an equivalent amount during the week of the eruption. These volcanoes effectively breathe, inhaling and expanding as magma intrudes into their insides, and exhaling and contracting when they erupt.

From the magnitude and geometry of the volcanoes' swelling, we know that the tops of the magma reservoirs are only about 2 km (1.25 miles) beneath many of the Galápagos shields. These reservoirs are staging areas, where the magma cools. As its temperature drops from about 1200°C to about 1150°C (from 2200 to 2100°F), the magma slowly crystallizes (partially, not completely) in its reservoir, forming large crystals (1 mm to 20 mm, or up to three-quarters of an inch, long) of green olivine, white plagioclase, and black pyroxene. Most of the crystals are deposited at the base of the magma chamber, but others are carried in suspension in the magma and

RIGHT: A classic shield volcano, and one of the youngest in the group, at 1710 m (5600 ft) Wolf, on northern Isabela, marks the highest point in Galápagos. Its domed summit holds a 600 m (2100-ft) deep caldera and its steep eastern flank is streaked with bare a'a' lava flows cutting through scrubby vegetation.

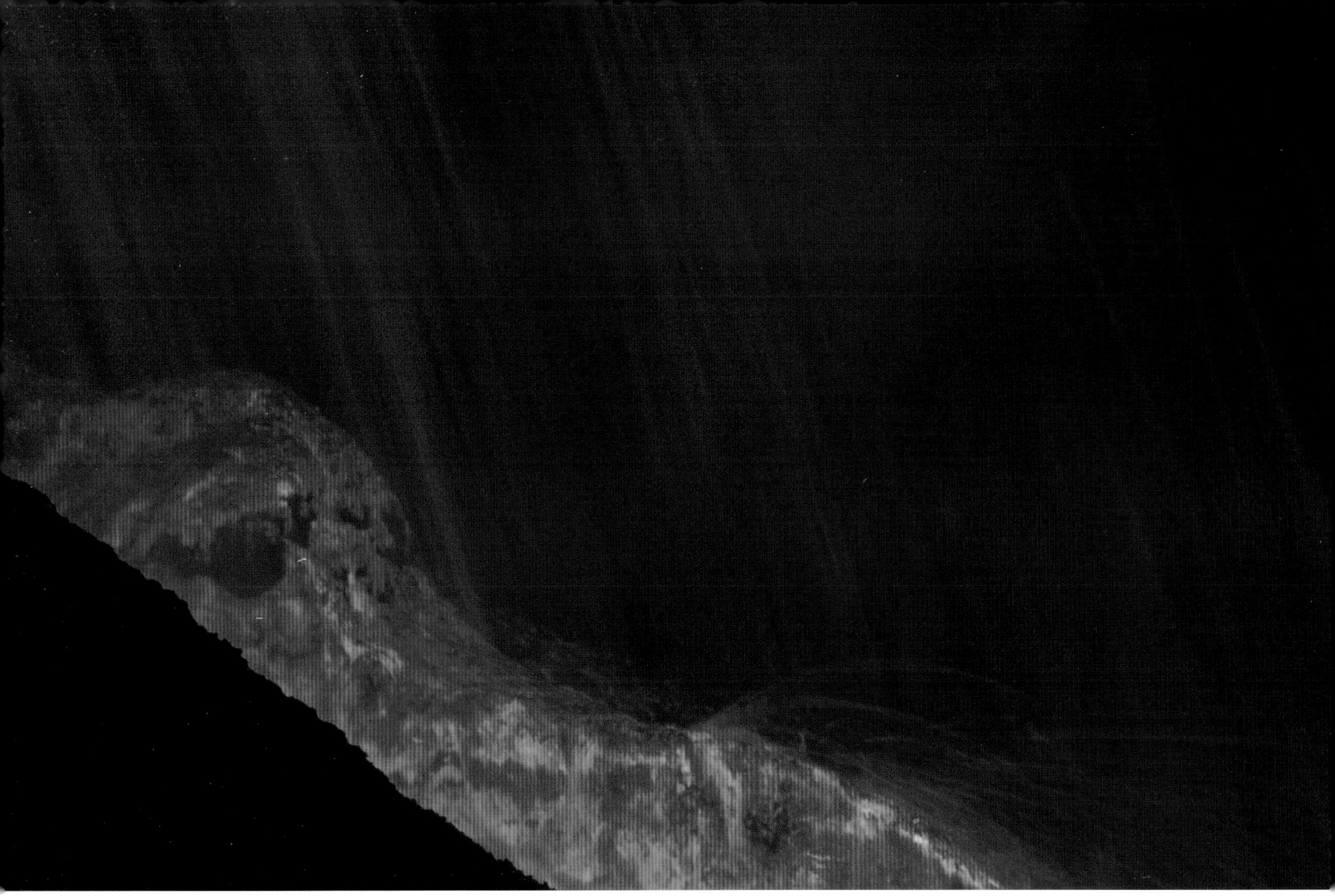

incorporated in the flowing lava, forming the crystals one sees in most Galápagos rocks. If you were to drill a hole into the center of Sierra Negra or Fernandina volcanoes you would go through about 2000 m (6500 ft) of solid rock, then several hundred feet of mostly liquid magma (with a few crystals), then several miles of crystal-rich 'mush' that is still at magmatic temperatures. Finally, you would reach the Galápagos plume in the earth's mantle.

Alcedo is a unique volcano. Its shallow magma body and mush column have cooled more than the other western volcanoes, to about 900°C (1650°F). Consequently, extensive crystallization produced an abnormally large volume of rhyolite, a magma rich in silica, which is more normally associated with continental volcanoes such as those in the Andes. Alcedo's rhyolite erupted as pumice and obsidian in an extraordinarily explosive event about 100,000 years ago.

The plumbing systems of the older volcanoes to the east (for example, Santiago, San Cristóbal and Santa Cruz) are more transient, owing to lower magma productivity. Here the magmas cool and crystallize at depths up to 15 km (9 miles), and no stable chamber forms. Nonetheless, these volcanoes are still capable of producing historic eruptions, such as Santiago's extensive pahoehoe flow in Sullivan Bay dating from 1904 or Marchena's 1991 eruption.

Because they are islands, the Galápagos volcanoes are profoundly affected by the interaction between magma and water. Before they emerge, all eruptions are underwater, and most produce a characteristic form of lava known as 'pillow' flows. An oceanographic expedition in 2001 sampled over 100 different submarine lava flows from the deep ocean floor surrounding the archipelago and most were pillow lavas. Some of the submarine lava fields cover much greater areas than their subaerial (or above

ABOVE: Fountaining lava from degasing magma during the Sierra Negra 2005 eruption rains down continuously, with molten lava bombs splattering around the vents gradually building a string of new parasitic cones.

LEFT: After the 1988 eruption, once-lush *Scalesia* trees and land iguanas lay scorched under a blanket of scoria fallout on the rim of Fernandina. As the islands grow, such destructive cycles may define the evolutionary course of endemic species.

ABOVE: The panoramic view from the top of Bartolomé reveals a volcanic landscape of young lava flows dotted with spatter, scoria and pale tuff cones eroding into olivine-rich beaches.

BELOW: Tree ferns and other plants thinly cloak the volcanic contours of older islands, such as El Puntudo in the Santa Cruz highlands.

water) counterparts, particularly those to the west of Fernandina and Cerro Azul volcanoes. On the southern margin of the Galápagos Platform (a large plateau on the tectonic plate, averaging 500–700 m, or 1650–2300 ft, beneath the sea surface, which forms the pedestal upon which the individual volcanoes are built), large lava flows are stacked on top of one another, forming a remarkable sequence of lobate terraces that, so far as we know, are unique morphological forms on this planet. Submarine volcanic rifts crosscut the terraces at Fernandina, Genovesa and Floreana, which is odd, because these volcanoes lack focused rifts.

When magma erupts through shallow water, violent explosions occur because the lack of confining pressure permits water to flash instantaneously to steam. The violent expansion fragments the magma into fine ash, which is carried in dusty tephra clouds that travel at hundreds of miles per hour, to be deposited and eventually consolidated as tuff.

Recent efforts have attempted to establish the extent to which the geological history of the islands controls the colonization, establishment and evolution of the life forms that settled on them. One of the most important controls is the sinking of islands. Volcanic islands subside on our planet because plate motion carries them away from their source of magma, and in the process the plate cools and contracts. This effect is readily observable in Galápagos: when you line up the mountains west to east, their elevations decrease, from Fernandina to Alcedo to Santa Cruz to San Cristóbal. This shrinkage of the plate continues to the east of San Cristóbal, where there are seamounts that were islands millions of years ago (at least 14 million years ago, and perhaps as much as 90 million). The implication is that, although they are now reduced to seamounts, many of these past islands might have been colonized by terrestrial organisms tens of millions of years before the current islands emerged above the sea. Moreover, it is important to understand that life has evolved in Galápagos on a constantly changing template: new islands emerge and old islands sink, and all are moving by plate tectonics. Over geologic time scales (even a few million years), the current map of Galápagos becomes irrelevant. Also, in just the past 20,000 years, the rise and fall of ice-age sea levels has fluctuated up to 130 m (425 ft), changing the biogeographic template further.

Innumerable studies of the biodiversity of individual archipelagos have shown that the main limiting factor is island area, and work by Howard Snell and Alan

Tye has demonstrated that this relationship holds true for the plants and vertebrates in Galápagos. In an interdisciplinary study, we have shown that the ages of almost 100 islands and islets also play a small but measurable role in island species diversity, whereas the nature of the substrate does not (see also chapters: Parent & Coppois; Grant; Parker; Tye).

Volcanism on Alcedo Volcano has played a rather unique role in that island's genetic diversity: the explosive eruption of rhyolite nearly wiped out the Alcedo race of tortoise 100,000 years ago. There is substantive genetic evidence that only one female, or her clutch of eggs, survived the eruption — a graphic demonstration of how volcanoes continually reshape the biology of Galápagos.

ABOVE: Lava cacti, *Brachycereus nesioticus*, are among the first plants to grow on fresh lava, but their molds show this can be a tenuous existence.
ABOVE AND BELOW LEFT: Around 100,000 years ago, a huge, explosive eruption covered Alcedo Volcano in a thick pumice layer that nearly exterminated the giant tortoises. Their descendants can be seen here, meandering among sulfur fumaroles inside the caldera.

Living Water
Investigating an Elusive Element
Noémi d'Ozouville

Coordinator, Galápagos Islands Integrated Water Studies, Barrio Estrada, Puerto Ayora, Isla Santa Cruz, Galápagos, Ecuador. <pajarobrujo@gmail.com>

Dr. Noémi d'Ozouville worked for six months as a volunteer at the Charles Darwin Research Station between 2000 and 2001, an experience that left her with a compelling desire to pursue the study of the hydrological functioning of the Galápagos Islands, a field that seemed to have long remained an absolute unknown. In 2007, she successfully defended her Ph.D. on the hydrogeology of the Galápagos Islands. Originally from France, Dr. d'Ozouville now lives permanently in Galápagos.

THE DAY THE FIRST MAN set foot on the parched earth of the Galápagos Islands, thirst was on his mind. Prior to his arrival, incredible life forms had developed on the islands, adapting to the availability of water in various forms and the capricious nature of its occurrence. But man doesn't adapt to the environment ... man needs water, and as time goes by, more and more of it.

The winds and oceanic currents long maintained the isolation of the Galápagos Islands. They also shaped the climate of the archipelago and had a direct impact on water availability, with weathering, seasons, vegetation and evolutionary processes all playing their part. Although situated on the equator, the archipelago has a relatively cool and dry climate due to the influence of the southeast tradewinds that shunt along the cold Humboldt Current and the upwelling of the Cromwell Current (also known as the Equatorial Countercurrent) along its western edge. A few months of hot, rainy weather occur when the Intertropical Convergence Zone (ITCZ) moves south toward the equator and thus closer to the islands (see also chapters: Sachs; McMullen; Grove).

RIGHT: The young Galápagos geology, typically riddled with volcanic cones and fissures such as the interior of Santiago Island, is highly porous and not conducive to surface water runoff.

The engravings above and below are reproduced with permission from John van Wyhe ed., The Complete Work of Charles Darwin Online (http://darwin-online.org.uk)

Photo courtesy Noémi d'Ozouville

Photo courtesy Noémi d'Ozouville

A historical perspective

In 1535, Tomás de Berlanga, becalmed and stranded in the Galápagos Islands, miraculously found water on the fifth day of his search. He wrote to the King of Spain both of his ordeals and of the discovery of the new land, but gave the islands no name. Throughout the 17th, 18th and 19th centuries the fascinating tales of pirates and whalers abound with references to the search for the precious resource on Galápagos. The availability of fresh water affected itineraries, the duration of stays in the islands, and the capture of enemy vessels.

The year 1835 saw HMS *Beagle* sailing through the waters of the archipelago, recently claimed by Ecuador (see also chapter: Bungartz). By then, a few settlements had been established on two islands that had small freshwater sources, though due to social chaos and harsh living conditions these colonies were eventually abandoned. After three weeks in the Galápagos Islands, the *Beagle*, like many preceding vessels, was in dire need of fresh water. So Captain FitzRoy took off to the south coast of San Cristóbal Island, to where a fur seal hunter had mentioned there was abundant fresh water flowing into the sea. This source is not mentioned in the literature before this date, yet it remains permanent and plentiful to this day. The fresh water of San Cristóbal later allowed Manuel Cobos to set up a sugar cane industry in

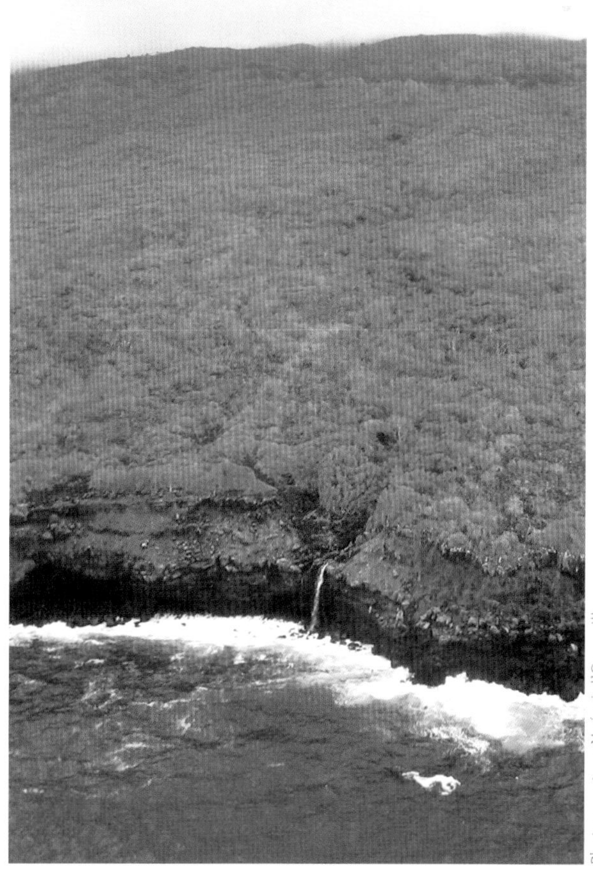

Photo courtesy Noémi d'Ozouville

ABOVE AND LEFT: The south coast of San Cristóbal, one of the older islands, is the only place in Galápagos where streams reach the sea year round. Engravings titled 'Watering Place' from the 1835 voyage of the HMS *Beagle* attest to the importance of these rare features to early navigators, while comparisons with today's photographs show little change over time, even though the interior of the island has been heavily transformed by farming and introduced plant species.

ABOVE: Although a very arid region today, heavily eroded contours at the eastern end of San Cristóbal denote the work of water runoff on old volcanic tuff cones, where blue-footed boobies nest. BELOW: The south-facing slope of the Santa Cruz highlands is bathed in mist during six months of the year, yet pools and rivulets soon disappear into the permeable ground.

the highlands of that island (1879–1904), the first successful settlement and enterprise in Galápagos.

The famed author, researcher and explorer William Beebe, confronted by the harsh reality of Galápagos in 1924, remarks, 'had fresh water been the sole object of our explorations, we would have been indeed empty handed,' and subsequently that 'fresh water was now the chief topic of conversation.' Much happier, he later states: 'we reached Freshwater Bay (south coast of San Cristóbal), and there at last we saw what we had almost come to believe was non-existent anywhere in the world — wild fresh water. Two waterfalls, one quite small, the other of good size, plunged over the edge of the cliffs into the sea. Niagara or Kaieteur never looked so wonderful.' Notwithstanding the rarity of this precious resource, Beebe's book inspired others to visit and settle in Galápagos!

Between 1941 and 1946, an American Air Base was set up on the island of Baltra, and fresh water was brought over from San Cristóbal in barges. A pipeline carried it from a dam in the highlands (where Cobos had obtained his supply in the previous century) to the coast and into storage reservoirs. Throughout the rest of the archipelago, the availability of fresh water has always been a concern for all, including field scientists, as expressed by Duffy in *Water, water anywhere: the search for the potable in Galápagos* (1981).

The islands now sustain more than 30,000 permanent inhabitants, plus 170,000 tourists visiting each

year. Technology (desalinization, water treatment, electrical pumps) has overridden the natural limitation of fresh water availability, the constraining factor for human settlement and population increase. Yet even today, little attention is paid to understanding the dynamics of such a vital part of life.

Living water

The relationship between mankind and water versus the adaptability of the fauna and flora of Galápagos to its presence or absence initiated my desire to investigate the hydrological functioning of the islands from a scientific perspective. It was clear from the start that establishing a conceptual model would require not only observation and monitoring of hydrological phenomena such as evaporation, runoff and infiltration, but also knowledge of geology, the internal structure of the islands, soil and weathering, and climate data. The ongoing multidisciplinary study, which I initiated in 2003 through the University Pierre and Marie Curie in Paris, is revealing fascinating and surprising facets of the Galápagos Islands.

To begin, two islands were chosen for investigation: the 'waterless' island of Santa Cruz and the island of 'permanent rivers,' San Cristóbal. From the outset, I was confronted by the lack of precise and accurate topographical data and the ensuing difficulty in delineating watersheds within which the hydrological water budget can be calculated. Luckily, as of July 2003, I was able to use the recently available Shuttle Radar Topography Mission 90 m data for South America, including Galápagos. Integrating the data with radar images, it was possible to generate a remarkably detailed 20 m (65 foot) grid with elevation accuracy better than 15 m (50 feet). From this data, large watersheds running from summit to coast in a radial manner were identified, as were tiny catchments between the exit points of the larger ones. Scrutinizing the landforms it occurred to me that water catchments are important considerations in ecosystem and management studies: not only do they encompass traditional habitat delineations, such as coastal, arid, transition and humid zones but all water flow is related within and across these features.

The next step was to relate the occurrence of recharge (runoff and/or infiltration) to the hydrological system, which is determined by rainfall and evapotranspiration. This in turn required the analysis of climatic data: long-term information from the meteorological recordings at the Charles Darwin Research Station and at the inland village of Bellavista, on Santa Cruz, plus short-term records from automatic rain gauges set up during the project. I found that recharge occurs in the highlands above 400 m (1300 ft) elevation during the 'garúa' season (June–

December) and occasionally at all altitudes during heavy rainstorms in the so-called 'invierno' or hot season (January–May). On an annual basis, evaporation in the highlands accounts for more than 50% of the rainfall. Of the remaining precipitation, infiltration into the substrate is the dominant recharge process, representing 75 to 90% of the available total. Based on this data, I was able to define the 'hydrological year' in Galápagos as beginning in the month of June, with the onset of the garúa season, and finishing in May after two of the statistically driest months of the year.

One of the most fascinating aspects of my fieldwork was discovering the numerous permanent streams on the south-central mountainside of San Cristóbal that are fed by small springs attributed to outcropping perched aquifers. The water-filled crater of El Junco, whose waters are mythically reported to arrive from the snowmelt of the Andes, is commonly thought to be the source of the permanent rivers. This is also mythical because, although it has an overspill that sporadically feeds the easternmost river it is a closed system whose level changes in relation to rainfall and evaporation (see also chapter: Sachs). Filtrations through the bottom of the lake are possible but not sufficient to account for the volume of water flowing through the rivers, or for the presence of rivers further to the west, below Cerro San Joaquin. Based on field observations, it appears that the aquifers form

ABOVE: Deep fractures in the basaltic layers on Santa Cruz provide access to the water table, where sea water and fresh water are mixed in the basal aquifer.

LEFT: Fog-shrouded El Junco lake on San Cristóbal is often credited as the source of flowing water on that island, yet in reality it is a sealed system.

LAND AND SEA DISCOVERIES

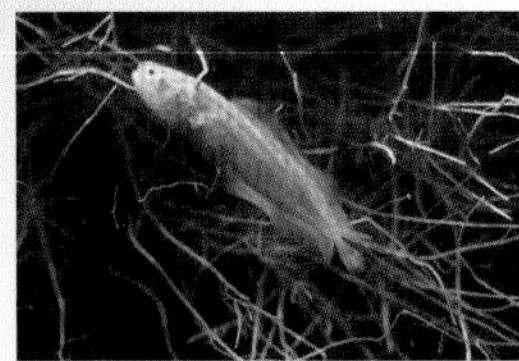

In the 1950s and 1960s my father, André De Roy, nurtured a dream of finding a hidden freshwater oasis in the sunbaked hinterlands of our home island, Santa Cruz. While hunting goats (part of the pioneering life on Galápagos in those days), he spent many hours in this quest, scrambling down narrow basaltic fissures and crawling under thickets of thorny scrub. Gradually, he discovered an underworld of clear brackish water connected to the sea by tidal movements. Resting in one of these dim caverns after quenching his thirst one day, he noticed the faintest of movements among a filigree mat of floating plant roots. It turned out to be a tiny fish, virtually blind (vestigial eyes could be seen embedded in its transparent tissue) and almost devoid of any pigment, leading a secretive and extremely sedentary life in almost total darkness. It was eventually described as a new species in 1965: *Ogilbia galapagosensis*, the pink brotula (top), evolved in the underground aquifers described in this chapter. Ever inquisitive, André later noticed a similar shaped little fish slithering into the narrowest crevices of the lava reef just in front of the Charles Darwin Research Station, a 10-minute walk from the center of Puerto Ayora, the largest settlement in Galápagos. This fish was bright red and sported pin-prick eyes. It was another new species, the orange brotula, named *Ogilbia deroyi* (above) in my father's honor the following year.

Tui De Roy

within the highly fissured basalts, and flow occurs at stratigraphical contacts between permeable and impermeable layers.

My conviction that the key to understanding Galápagos hydrogeology lay inside the island itself led to a helicopter-borne geophysical investigation carried out in collaboration with a Danish team of scientists. The method involves interpreting the response of subsurface strata to an induced electromagnetic field in order to detect layers of different properties and thicknesses. The resistivity of each layer is dependent on the nature of the rock, water content, mineralization of water and clay content.

The resulting three-dimensional scan of the internal structure of both islands went far beyond all our expectations. For both islands, we saw a continuous and homogeneous deeply penetrating saltwater intrusion around the whole coastal fringe. Above the salt water lies the freshwater lens that forms the basal aquifer. The thickness of the freshwater lens is greater on the windward mountainsides. I also gathered ground-based data on the basal aquifer of Santa Cruz from the open coastal fractures named 'grietas', and the recently drilled deep bore 5 km (3 miles) inland from the township of Puerto Ayora, both of which give direct access to this groundwater system. The water flow within this aquifer was thus characterized: water is present in the large-, medium- and small-scale fissures of the unweathered basalts. Possibly a perched aquifer also occurs on the southern mountainside of Santa Cruz, but with no outcrop to generate springs and rivers.

This new research lies at the crossroads between the geological sciences, which reveal the volcanic nature of the archipelago, and the biological sciences, which seek to understand the ecosystems and their living organisms. Water is a key factor by which the two are closely interrelated. Further work will be carried out to deepen our understanding of the subterranean water flow, especially the presence of perched aquifers, and to extend the study to the islands of Floreana and Isabela. The active volcanoes on Isabela will bring new information about the behavior of water cycles in the proximity of magma chambers. Beyond the interest of earth scientists alone, this research opens new fields of investigation such as hydroecology, and can help provide viable solutions toward a more sustainable relationship between human development and the protection of unique ecosystems.

RIGHT AND FAR RIGHT: The same view from the rim of Cerro Azul on Isabela, taken a decade apart, reveals a bone-dry caldera in 1973, but an extensive lake had appeared in the wake of a major El Niño in 1983. Today, only the cinder cone in the middle holds water.

Paleoclimate and the Future
A Knife-edge Balance
Julian P. Sachs

Dr. Julian Sachs has been working on climate reconstruction from lake sediment cores in Galápagos since 2004, with similar investigations in Palau, the Marshall Islands, the Gilbert Islands, the Northern Line Islands, and Clipperton Atoll. He is an Associate Professor of Chemical Oceanography in the School of Oceanography at the University of Washington in Seattle, Washington.

School of Oceanography,
University of Washington,
Seattle, WA 98195,
United States.
<jsachs@u.washington.edu>

STRADDLING 0 DEGREES LATITUDE in the eastern Pacific Ocean, the Galápagos Islands are extremely sensitive to El Niño-Southern Oscillation (ENSO) events. Normally, the climate in this region is arid and remarkably cool compared to most tropical settings. A combination of local and remote oceanic and atmospheric processes accounts for the cold seawater that bathes the Galápagos Archipelago, causing this unusual situation in the eastern Pacific Ocean. Tradewinds that blow along the equator from east to west drive surface waters into the northern and southern hemispheres at right angles to the wind. These displaced waters are replaced by subsurface water that is upwelled, or 'pulled' up, from deeper and colder layers, contributing to the locally cool conditions. The Humboldt Current, with its origins in subpolar waters around Antarctica, is kept cold by upwellings along the Peruvian coast, from where it flows northwest to Galápagos and combines with the cool and westward-flowing South Equatorial Current. Yet another source of cold water originates from the opposite direction via the Equatorial Countercurrent (also known as the Cromwell Current), a deep stream moving west-to-east, contrary to the surface flow, and rising up where it encounters the undersea rampart of the Galápagos Platform.

Together, these cold waters are responsible for an inversion of air temperatures such that cold air near the ocean surface sits trapped beneath buoyant air warmed by the tropical sun, preventing the development of atmospheric convection cells that would otherwise bring abundant rainfall to tropical islands. A fog layer, locally called 'garúa,' envelops the highlands of Galápagos, resulting from the condensation of moisture out of that warmer air where it comes into contact with the colder layer below. This situation prevails during the latter half of the year, usually lasting from June to December, in tandem with the southern (austral) winter.

During the southern summer, in mild years, the rain

BELOW: The strong presence of the cold upwelling Cromwell Current, or Equatorial Countercurrent, around the western part of the archipelago frequently brings thick sea fog lapping the arid shores of Fernandina Island.

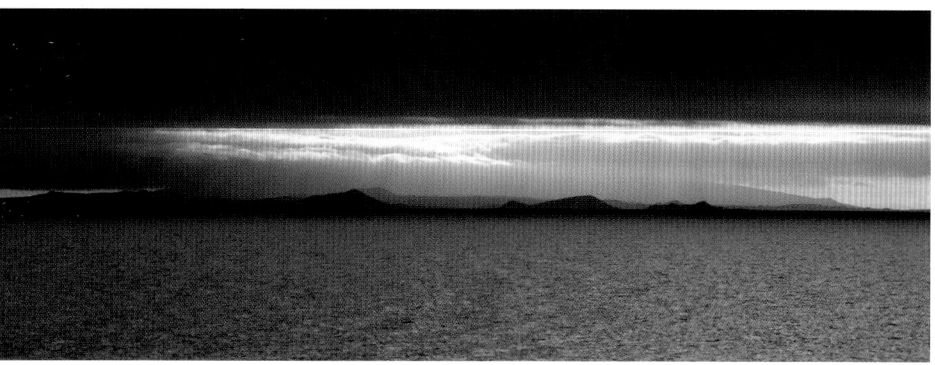

ABOVE: An atmospheric inversion layer caused by cold sea temperatures drapes a low cloud layer over the islands, breached only on their leeward, northern sides. Known locally as the garúa season, mist shrouds the highlands from June to December, yet no rain falls on the lowlands.

BELOW: Used by the locals to harvest salt, a briny pan near the Santa Cruz shoreline may harbor the records of prior climate in layered sediments, while dead barnacles on the surrounding rocks reflect high sea levels during recent El Niño conditions.

band known as the Intertropical Convergence Zone (ITCZ) that spans the tropical Pacific Ocean, moves south from its normal position well north of the equator. When it reaches its most southern seasonal migration, this convergence of southeasterly and northeasterly winds causes the inversion to collapse and convection to occur, bringing some heavy precipitation to Galápagos, usually in February-March.

What differentiates a true El Niño event, which generally occurs once every two to seven years, is when the ITCZ effectively remains stuck at its southern extent for an unusually long period, sometimes for many months on end. During particularly strong El Niño years, extensive depression of the subsurface cold ocean water causes sea surface temperatures to rise in the entire eastern equatorial Pacific Ocean (EEP) region, and the atmospheric inversion around Galápagos disappears completely, intensifying tropical convection cells further. During such periods, these usually arid Islands receive several times their normal rainfall, while global precipitation patterns are altered, with droughts in normally wet locations and torrential rains in desert environments, especially coastal South America south of the equator. At the same time, as the upwelling of cool, nutrient-rich water wanes, fish stocks plummet and marine birds and mammals perish, causing widespread ecologic and economic disruptions throughout the region.

ENSO and Galápagos

My interest in the Galápagos began when, in the summer of 2004, I set out to determine how the largest global climate anomaly on the interannual time scale, ENSO, behaved before the beginning of human perturbations to the world's climate systems, and consequently how it might change in the future. The Galápagos Islands are perfectly situated for this line of inquiry because they sit on the equator in the eastern tropical Pacific region most strongly impacted by El Niño. They have the added advantage of being home to many different types of fresh, brackish and hypersaline water bodies likely to contain the sediment layers we need to reconstruct rainfall patterns from the past.

The two other locations I first targeted for this work were Christmas and Washington islands in the Northern Line Islands of the central equatorial Pacific, and Palau in the western tropical Pacific.

These three study sites span the entire tropical Pacific and are situated within regions that have a very strong rainfall response to El Niño. Like Galápagos, Christmas and Washington islands become very wet during El Niño events, while Palau experiences drought.

A fundamental question in climate dynamics concerns the sign and magnitude of change in the ENSO system under different global climate conditions than those of today. Will El Niño cycles become more intense and/or frequent in response to global warming? Presently, we do not know. Predictions range from a significant strengthening of the phenomenon to no effect, or even weakening. Climate data that predate the latter part of the 20th century, a primary requirement in order to evaluate these possibilities, are scarce and, as a result, extrapolations from them are often contradictory.

Some climate-modeling studies indicate that El Niños will become more pronounced and/or more frequent in a warmer climate resulting from a buildup of anthropogenic (human-induced) greenhouse gases in the atmosphere. Different models, and the paleoclimate data that supports them, however, suggest exactly the opposite. That we are not able to confidently predict the sign of change in the ENSO system, never mind its magnitude, is worrisome. It suggests that there may be important physics in the climate system that we do not understand, or that our translation of such physics into mathematical models is flawed. The best hope of rectifying this situation is to generate detailed paleoclimate records in an attempt to validate these divergent climate hypotheses and models.

It is toward this end that I became involved in producing the longest, most detailed and most unambiguous reconstructions of ENSO variability yet obtained from the EEP. In order to do that my colleagues and I are measuring the isotopes of hydrogen in the molecular remains of plankton in sediment cores that we collected in 2004 and 2008 from lakes, lagoons and bogs on six islands in the Galápagos Archipelago, deposits generally spanning the last 10,000 years.

Owing to the higher vapor pressure of water (H_2O) relative to deuterated water (DHO) — that is, water that has one deuterium atom, which is just a hydrogen atom with a neutron in its nucleus, and therefore a mass of two rather than one, and one hydrogen atom as opposed to two hydrogen atoms — evaporation enriches lakes and oceans in deuterium (D), while precipitation enriches them in hydrogen (H). During El Niño events, when torrential rains occur in the EEP, we expect the D/H ratio of lakes to decrease. Conversely, during La Niña episodes (when opposite conditions to El Niño prevail), the rain-starved lowlands of the Galápagos become tinder-dry and the normally rain-soaked volcanic peaks are moistened only by fog-drip from the enveloping stratus clouds. Thus, with a single source (rain) and sink (evaporation), the D/H ratio of water in closed lake systems is sensitive to the balance between precipitation and evaporation. During El Niño events their level tends to rise and the D/H ratio decreases. Conversely, during periods of weak ENSO activity the levels fall and the D/H ratio of the water increases. (See photographs of Cerro Azul crater on page 40.)

Since changes in the hydrogen isotopic ratio of water must accompany the hydrologic changes, we are reconstructing water D/H variations in lakes and lagoons by measuring D/H values of lipid molecular fossils from plants and algae whose sole source of hydrogen is the lake water in which they live. Our studies with cultures of seven species of phytoplankton, and likewise in the field, have shown that the D/H ratio in algal lipids tracks the D/H ratio of the water in which they live with near perfection. These variables are recorded in the remains of such

ABOVE: Roiling thunderstorms over Santa Cruz are a daily event during an El Niño year, when nutrient-poor seas tend to be exceptionally clear.
BELOW LEFT AND RIGHT: Contrasting undersea conditions show barren rocks and bleached coral heads during prolonged El Niño conditions (left), compared to lush seaweed beds and abundant invertebrate life when nutrient-rich upwellings are profuse.

UNIQUE FLAMINGO ENCLAVE IN THE PACIFIC

The Caribbean flamingo, *Phoenicopterus ruber* (above), closely related to the old world greater flamingo, is the only member of this ancient bird family with a foothold in the Pacific Ocean. Living in hypersaline pools tucked behind beaches along the lava shorelines, small flocks feed on minute crustaceans, drawing their intense color — the brightest of any flamingo — from the pigments of microscopic red algae that represent the base of this remarkable salt-loving food chain. Though permanent residents, they lead a precarious existence and rarely breed. Flamingo habitat is very susceptible to slight sea-level changes and heavy rains, particularly during El Niño events, which may flood nesting areas or allow fish to enter the lagoons from the sea, decimating their food supplies. Periodic archipelago-wide censuses show the Galápagos population hovers below 500 individuals.
Tui De Roy

organisms deposited long ago in undisturbed lake sediment layers. The Galápagos Islands present an ideal opportunity to record these variations through time, with lakes serving as fixed sampling stations of the eastern Pacific marine temperature variations.

The story in the sediment

In September 2004 we cored the sediments of El Junco Lake on San Cristóbal Island. This lake occupies a maar (explosion crater) in the cone forming the island's 750 m (2460 ft) summit, and became the focus of our primary investigation. It is enveloped for most of the year by the stratus clouds characteristic of the EEP climate. The lake is endorheic (a closed basin) and fed only by rain falling directly onto its surface and on the narrow annulus of the crater rim (see also chapter: d'Ozouville).

After flying to San Cristóbal, our team set up in Hotel Mar Azul, where we were able to use the restaurant area to spread out our equipment and sediment cores. The first day was spent hauling our gear over the crater rim to the lake edge. Fortunately, there is a road that meanders most of the way up the volcano, so we only had to carry our equipment uphill for quarter of a mile or so. Though aware that El Junco Lake was cool and enveloped by clouds most of the time, one still has 'tropical' in mind when being less than 100 km (60 miles) from the equator. Not so. Except for a few sunny hours, we ended up working in many layers of clothing and in full raingear for the two to three weeks we conducted our sediment coring.

Samples were extracted from El Junco using a piston corer operated from a raft constructed from two rubber dinghies and a plywood platform. The upper

RIGHT: With its impermeable clay floor allowing sediments to accumulate over millennia, the author's main study site in misty El Junco lake on San Cristóbal is revealing startling insights into past climate patterns.

3 m (10 ft) of sediment was soft mud and easy to penetrate. Below that came a 9-m (30 ft) thick layer of very stiff red clay that was extremely difficult to bore into. After twice losing a core barrel at around 4 m (13 ft) below the lake floor, we managed to fine-tune our method, but it was slow going, usually hammering the core head into the substrate inch by inch, then extracting the full core barrels, again inch by inch, with a giant corkscrew-like apparatus. On several days we recovered just one or two sections. Several times we abandoned the coring platform to let its buoyancy do the work pulling out buried core barrels for us. After three weeks, however, we obtained close to 40 m (130 ft) of sediment from seven different sites in the lake, with the deepest hole reaching about 12 m (40 ft) and quite possibly the last warm (interglacial) period 120,000 years ago. After this, we were able to enjoy the beauty and unique flora and fauna of the Galápagos Islands for a couple of days, before returning to the laboratory to start the real work.

Our measurements of hydrogen isotope ratios in molecular fossils from green algae in the El Junco Lake sediment, though still ongoing, have produced some surprising results. First, we discovered that there was just one other period during the last 5000 years that was as persistently dry — presumably the result of strong and/or frequent La Niña events — as the late 19th and early 20th centuries, and this happened in the 10th century A.D., during a period known as the 'Medieval Warm Period.' Second, the wettest episode of the entire last 5000 years of Galápagos climate — presumably the result of strong and/or frequent El Niño events — was from 1200 to 1800 A.D., a period known as the 'Little Ice Age.' If we extrapolate these findings into the future, when tailpipe and smokestack emissions will likely double the amount of carbon dioxide and other greenhouse gases in the atmosphere, warming the global climate dramatically, we would predict that the Galápagos Islands will become even dryer than they are today. If this scenario is correct, such a shift toward aridity would likely have severe consequences for plant and animal species that are already at or near their tolerance level for water deprivation.

Our findings are at odds with other studies on the frequency and intensity of El Niño events during the last millennium, which mostly suggest strong and/or frequent El Niño prevalence during the Medieval Warm Period and weak and/or infrequent events during the Little Ice Age. As additional data from other locations are generated, and computer-based simulations of the climate are performed, we will be able to confirm whether our interpretation of the data from El Junco is correct. Unfortunately, most current computer simulations of the climate system are not even able to reproduce the ENSO cycle accurately, and they incorrectly place an ITCZ in both hemispheres. In addition, there are very few sediment or coral archives in regions most sensitive to El Niño with the length and temporal resolution required to address this question in detail. So the controversy over El Niño predictions will not be solved in the immediate future without recourse to deciphering the detailed scripts we have begun to uncover in these lakes. Extrapolating from the unusually strong El Niño events of the last 30–40 years into the future is probably too simplistic, given the relatively short period of time that this covers. It may be that such recent, multi-decadal periods of strong ENSO occurred every few centuries and thus may be independent of the buildup of greenhouse gases that began 150 years ago. Thought-provoking as these initial findings may be, we will unfortunately need a lot of additional data before we can draw definitive conclusions.

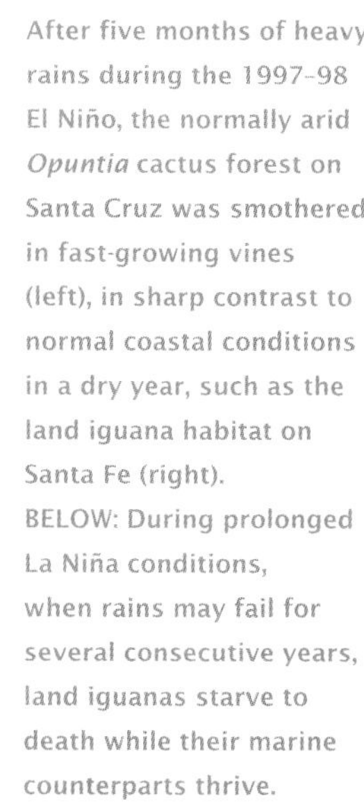

ABOVE LEFT AND RIGHT: After five months of heavy rains during the 1997–98 El Niño, the normally arid *Opuntia* cactus forest on Santa Cruz was smothered in fast-growing vines (left), in sharp contrast to normal coastal conditions in a dry year, such as the land iguana habitat on Santa Fe (right).
BELOW: During prolonged La Niña conditions, when rains may fail for several consecutive years, land iguanas starve to death while their marine counterparts thrive.

Vertebrate Diversity

The Long View

David W. Steadman

Florida Museum of Natural History, University of Florida, P.O. Box 117800, Gainesville, FL 32611, United States. <dws@flmnh.ufl.edu>

Dr. David Steadman is Curator of Ornithology at the Florida Museum of Natural History, University of Florida. He has worked in the Galápagos Islands since 1978, just before beginning his Ph.D. program at the University of Arizona under Paul Martin. Dave has studied living and extinct birds (and other vertebrates) on nearly 200 tropical islands in the Pacific Ocean and Caribbean Sea, as well as a number of places in the continental Neotropics. His primary passion is to use prehistoric information to help understand natural versus anthropogenic processes of colonization and extinction.

BELOW: Like other endemic species, Galápagos hawks and giant tortoises, such as these on Alcedo Volcano, Isabela, have disappeared from several of their ancestral islands due to human activity, but their past distribution can be reconstructed from bones preserved in lava caves and crevices.

LUCK CAN PLAY AS important a role in science as in any other part of life. And luck was certainly on my side that day in 1977 when, while sorting fossils from Blackbone Cave (Puerto Rico) at the Smithsonian Institution, I learned of some electrifying news from my friend and colleague Storrs Olson. S. Dillon Ripley, Secretary of the Smithsonian and a member of the Charles Darwin Foundation, had just returned from the Galápagos Islands and expressed disappointment to find no Smithsonian scientists doing research there in spite of the considerable financial and logistical support supplied.

Why not, Storrs asked me, extend my fossil research to this archipelago? Most vertebrate fossils from tropical islands came from limestone caves, but while no such features existed in Galápagos, fossils of birds were beginning to surface in the Hawaiian Islands, another group of oceanic volcanoes with considerable evolutionary notoriety. My nesophily still in its infancy, I jumped at the chance and at 8:20 a.m. on 29 December, 1977, I found myself staring dumbfounded at the magnificent Andes as I landed in Quito, Ecuador. Dr. Minard ('Pete') Hall, Professor of Geology at the Escuela Politécnica Nacional, coached me on the geology of Galápagos. Five days later I flew to the islands, utterly in awe, and eager to explore them from a fresh perspective.

Much of the young volcanic topography making up the Galápagos landscape contains caves that formed during eruptions. Basaltic flows, while cooling, often continue to advance beneath a hardened crust, leaving long hollowed chambers called lava tubes. These can run for miles in many parts of the archipelago, such as the southern slope of Santa Cruz and the Post Office Bay region of northern Floreana. Through the millennia, these caves might accumulate sediment and bones. But the locations of only a few lava tubes were known to the scientific community when I arrived.

After several days in the islands, it became clear that the best way to find and evaluate as many sites as possible during a two-week reconnaissance would be to talk to Galápagos residents, thus taking advantage of their past years of explorations. I promptly met the De Roy (André, Jacqueline, Tui and Gil) and Devine families (Bud, Doris and Steve). I also had informative chats with Miguel Cifuentes, head of the Galápagos

National Park (GNP), and Craig Macfarland, director of the Charles Darwin Research Station (CDRS). Soon I was investigating lava tubes on Santa Cruz, at times accompanied by two bright, energetic boys who lived in Puerto Ayora, Daniel Fitter and Jason Gallardo. I was shown Cueva de Kübler, a large lava tube that runs beneath the road between Puerto Ayora and Bellavista. Fossils were lying all over the floor of this spectacular cave, which since has been visited (and trampled) by countless thousands of tourists. I also found fossils of reptiles, birds and mammals on ledges in the massive crevice behind the tortoise pens at CDRS. It was clear that a bony gold mine awaited me in the Galápagos Islands.

The fossils

On subsequent trips, totalling about a year in the field (1978, 1980, 1982–85, 1991, 1995, 1998), I discovered fossils from 15 sites on the islands of San Cristóbal, Santa Cruz, Floreana, Rábida and Isabela. Five of these locations were shown to me by others; 10 were found by me or my field companions, among the many of whom I must single out María José Campos, Gayle Davis, Jacinto Gordillo, James Hill, Paul Martin, Godfrey Merlen, Mary Kay O'Rourke, Miguel Pozo, Arnaldo Tupiza, Winter Vera, and my brothers Ed and Lee, for their extraordinary efforts. Fossils were much better preserved in the arid lowlands than in the humid highlands. Santa Cruz had the richest fossil record, whether in terms of the number of sites, total fossil specimens or the species represented. In order to recover bones of even the smallest taxa, such as geckos (*Phyllodactylus* spp.), all excavated sediment was sieved through screens of 12.5, 6.4 and 1.6 mm mesh (½, ¼ and 1/16 in mesh), and sometimes subsampled with 0.80 mm (1/32 in) mesh. The remains of native rodents were discovered on each of the five islands except Floreana. Using radiocarbon dating methods, the fossils were dated mainly to the Holocene (the past 10,000 years); the oldest of 26 radiocarbon dates across the archipelago were 21,570 ± 280 and 21,820 ± 300 years before present (YBP) for fossils of the wedge-tailed shearwater (*Puffinus pacificus* undescribed subspecies) from Grieta de Gordillo in the arid lowlands near Punta Pitt, San Cristóbal.

Most fossils were deposited in caves as the prey remains of Galápagos barn owls, who regurgitate

ABOVE: Only a half-hour's walk from the town of Puerto Ayora, a large lava tube known as Cueva de Kübler turned into a fossil hunter's treasure trove. BELOW: Nesting barn owls are responsible for depositing the vast majority of plentiful bone remains that are well preserved deep in dry cave recesses.

FRAGILE WORLD OF THE RICE-RATS

Bold and inquisitive, a large Fernandina rice-rat shares a visitor's dinner.

Rodents and bats are the only native land mammals in Galápagos. The small mouselike rice-rats diverged into different species, much as did Darwin's finches, but their fate has been precarious at best since the arrival of humans and their retinue of introduced species, particularly nefarious black and ship rats, cats and mice. Of the six named and currently recognized species of rice-rats, three are extinct, one was rediscovered in 1997 after not being sighted for 60 years, and another was seen alive for the first time in 1995.

GALÁPAGOS RICE-RAT (*Aegialomys galapagoensis*)
San Cristóbal & Santa Fe Islands
First collected by Charles Darwin in 1835 on San Cristóbal, where it has never been seen alive again, although prehistoric bones occur abundantly in lava tubes. Its extinction was likely due to predation and competition from introduced species, especially with a large human settlement beginning in the mid-1800s. The Santa Fe island population was considered a separate species, *Aegialomys* (formerly *Oryzomys*) *bauri*, until it was amalgamated with *A. galapagoensis* in 2006. In the absence of introduced mammals, it still thrives here though the island is only 24 km^2 (9.3 square miles), 259 m (850 ft) high, and extremely arid.

DARWIN'S RICE-RAT (*Nesoryzomys darwini*) — Extinct
Santa Cruz Island
First collected in 1906 and last seen in the 1930s, its extinction coincides with the introduction of black rats. Nothing is known about how this small species shared habitats with *N. indefessus*, a larger rice-rat of the same genus.

SANTA CRUZ RICE-RAT (*Nesoryzomys indefessus*) — Extinct
Santa Cruz and Baltra Islands
Once believed to be the same species as the large Fernandina rice-rat but now considered to be separate, *N. indefessus* was last collected in 1934, so its life history remains unknown. It is by far the most commonly occurring species in vertebrate fossil sites on Santa Cruz.

SANTIAGO RICE-RAT (*Nesoryzomys swarthi*)
Santiago Island
First collected in 1906, this rice-rat was believed to be extinct by the time it was described in 1938. Extensive searches led to its rediscovery in 1997 in spite of the presence of aggressive black rats, against which it seems to hold a tenuous competitive advantage in very dry areas, being able to survive on a meager diet of cacti.

LARGE FERNANDINA RICE-RAT (*Nesoryzomys narboroughi*)
Fernandina Island
Abundant on all parts of the island thanks to the absence of introduced species, it is fearless and highly exploratory in its behavior, able to make a living even around clusters of cacti or mangroves.

SMALL FERNANDINA RICE-RAT (*Nesoryzomys fernandinae*)
Fernandina Island
Smaller than its more common cousin on the same island, this species was discovered from bones in owl pellets in 1979, and soon declared extinct for lack of live evidence. In 1995, it was found in vegetation patches mainly near the summit of the volcano, where it is vulnerable to potentially devastating eruptions.

Finally, we note that fossil remains of three forms of *Nesoryzomys* await formal description as new species: two are from lowlands of the Wolf Volcano, Isabela (large and small species in the general size ranges of those known from Santa Cruz and Fernandina), and a third from Rábida, apparently resembling the Santiago species.
David Steadman, with Robert Dowler

bone-filled pellets that accumulate beneath the ledges where they roost. These owls probably were responsible for depositing nearly all prehistoric bones in caves except for those of tortoises and land iguanas that had plunged through holes in the roofs of lava tubes.

While the fossil sites on Santa Cruz and Isabela revealed many small vertebrates, they were dominated by the remains of extinct native rodents, namely the rice rats *Nesoryzomys indefessus* and *N. darwini* and the giant rat *Megaoryzomys curioi* on Santa Cruz, and a similar trio on Isabela (the rice-rats *Nesoryzomys* undescribed spp. 1 and 2, and the giant rat *Megaoryzomys* undescribed sp.). On Santa Cruz the fossils also included two extinct undescribed species of Darwin's finches, one a large-nostriled form related to the small ground finch *Geospiza fuliginosa*, and the other being intermediate in bill shape and size between the medium ground finch *G. fortis* and vegetarian finch *Platyspiza crassirostris*.

One of the most remarkable fossil sites in the Galápagos is Barn Owl Cave on Floreana. My first excavation there (Ex-1), in October 1980, reached bedrock at a depth of 43 cm (17 in). Among pieces of wood found on its dry, sediment-covered floor, two were dated at 640 ± 50 YBP (cf. *Cordia* sp.) and 2420 ± 25 years YBP (*Acacia* sp.). That such old wood was lying on the surface in Barn Owl Cave in 1980 suggests that rates of sedimentation (and therefore precipitation) here had been low during recent centuries if not millennia. Also scattered all about the surface of this cave at the time were bones of the tortoise *Geochelone galapagoensis*, snake *Alsophis biserialis*, Floreana mockingbird *Mimus trifasciatus*, and large ground finch *Geospiza magnirostris magnirostris*, four species that have been gone from Floreana for more than a century.

On 24 May 1983, Gayle Davis, Godfrey Merlen and I watched as sediment flowed into Barn Owl Cave during the torrential rains that fell from November 1982 through July 1983. The cave's floor was covered in wet, sticky sediment that was washing into the main entrance and down the debris slope. It was impossible to continue our excavations under these conditions, not to mention that we were all sick, being bitten nonstop by fire ants, and being drenched by downpours of the infamous 1982–83 El Niño event, the most pronounced ever recorded in Galápagos. We left Floreana early, vowing to return someday under drier conditions.

On 5 November 1995, Winter Vera and I were back in Barn Owl Cave. We excavated a 1 x 1 m (3ft by 3 ft) test pit (Ex-2), but because of an inflexible boat schedule, we had to stop at a depth of 73 cm (29 in) without reaching the cave's bedrock floor. Layer I of Ex-2 consisted mainly of sediment deposited during the 1982–83 El Niño, its top made up of the dark soil that had washed into the cave 12 years prior. The lower half was clearly mixed with sediments that had been deposited in the cave before the mid-19th century, because it contained bones of four extinct species.

The porous soils and fractured lavas in Galápagos typically allow rainwater to infiltrate the substrate rather than to flow along its surface. In a site with a small catchment basin, such as Barn Owl Cave, sedimentation occurs only during times of extremely high rainfall. The floor of this cave had been dry in October 1980, when it was still littered with the exposed bones of extirpated species. That the sediment deposited here during the 1982–83 El Niño

Photo courtesy Heidi Snell/Visual-Escapes.smugmug.com

ABOVE: Oblivious to the dangers below, a giant tortoise walks over a bone-littered lava tube on Wolf Volcano, northern Isabela.

LEFT TOP: Oversize rodent bones littering the floor of many caves on Santa Cruz are all that is known of the extinct giant rat, *Megaoryzomys curioi*, a species that disappeared around the time of man's arrival, yet has never been seen alive. Smaller rice-rat bones and recently extinct Bulimulid land snails can also be seen.

LEFT BOTTOM: As in many other oceanic settings, introduced black rats, *Rattus rattus*, are responsible for many island extinctions.

ABOVE: Volcanic cones dotting the Floreana Island landscape, and the many lava tubes that radiate from them, became an ideal hunting ground for the author.
BELOW: A race of large ground finch, *Geospiza magnirostris*, with an even larger bill than those alive today, became extinct on Floreana not long after being collected by Darwin in 1835.
RIGHT: Floreana was the first island colonized by people, in 1832, the same year the archipelago became part of Ecuador. Yet the small settlement of Black Beach has grown little to this day.

buried the old wood and bones still exposed 12 years before, suggests that this climatic event had fueled the first major influx of sediment for centuries. The 1982–83 El Niño has been independently evaluated by other scientists to have represented the warmest and wettest event in the equatorial Pacific Ocean during this century, and perhaps for several centuries before that. I agree completely.

Layer II at Barn Owl Cave is well stratified with six to eight sets of alternating bands of light and dark sediment. About 70% of the total vertical extent of Layer II is dark, but never as dark or clayey as in Layer I. Layer III, which is radiocarbon dated (using tortoise bone) at 8290 ± 70 YBP, has 10–12 sets of alternating bands of light and dark sediment. On average, the sediment in Layer III is lighter in color and has larger, more angular pebbles and cobbles than that in Layer II. The lighter bands of sediment have a higher component of minerals that precipitate within caves, such as gypsum, fluorite, apatite and calcite. I interpret the lightly colored sediments to have been deposited during relatively dry times, whereas the darker sediments have a much larger fraction of material derived from soils washed into the cave during heavy, sustained rains, such as during strong El Niño events. All else being equal, one would expect a lower overall rate of sediment accumulation in Barn Owl Cave during periods of relatively dry climate. Based on changes in sediment type, pollen and spores from cores obtained at El Junco Lake on San Cristóbal, Paul Colinvaux and Eileen Schofield proposed that the Galápagos climate from ca. 8000 to 3000 YBP probably was drier than earlier or later in the Holocene. Layer III of Ex-2 may represent this relatively dry 5000-year interval.

Ex-2 of Barn Owl Cave was packed with fossils, about two-thirds of which represented reptiles (gecko *Phyllodactylus sp.*, lava lizard *Microlophis grayii*, snake *Alsophis* sp., tortoise *Geochelone galapagoensis*). Birds made up nearly all of the remaining fauna, with a few bones of the extirpated red bat *Lasiurus borealis*. The prehistoric birds from Barn Owl Cave are dominated by the Galápagos dove (*Zenaida galapagoensis*), the Floreana mockingbird, and various Darwin's finches, especially the

extinct and largest form of large ground finch (*G. m. magnirostris*). Less common species of birds from Ex-2 include the Galápagos hawk (*Buteo galapagoensis*), black rail (*Laterallus jamaicensis*), barn owl (*Tyto alba punctatissima*), and sharp-beaked ground finch (*Geospiza nebulosa/difficilis*), none of which still live on Floreana, an island that has sustained more than its fair share of extinction since people settled here 177 years ago.

I would love to do more excavations and analyses at Barn Owl Cave. To extract more biologically or climatologically informative data will require finer stratigraphic resolution than the 5 to 10 cm (2-4 in) levels used thus far, because most of the light or dark bands are only 1 to 3 cm (¼-1 in) thick. New excavations of the finely stratified bone deposit in this cave could also shed light on residency times, turnover rates, and the relative abundance of species under natural conditions.

So what else do the Galápagos fossils tell us?

Fossils have documented at least 24 species or populations lost from the five islands studied, namely tortoises on Floreana and Rábida, geckos on Rábida, land iguanas on Rábida, snakes and red bats on Floreana, various rice-rats on San Cristóbal, Santa Cruz, Rábida and Isabela, giant rats on Santa Cruz and Isabela, hawks, rails, barn owls and mockingbirds on Floreana, sharp-beaked ground finches on Santa Cruz and Floreana, two unnamed species of finches on Santa Cruz, and a large ground finch on San Cristóbal and Floreana. Evidence from radiocarbon dates, stratigraphy, historic documents and museum specimens suggests that all of these losses took place during the past few centuries of human influence, except perhaps those from Rábida, where four radiocarbon dates range from 5700 ± 70 to 8540 ± 100 YBP. Therefore, if modern conditions become suitable, the prehistoric data can be used to guide translocations of species to islands that once made up part of their natural range.

LEFT AND ABOVE: The endemic snake and hawk are both extinct on Floreana. BELOW LEFT: Rábida Island, where land iguana, gecko and giant tortoise bones were found, although DNA analysis suggests the latter may have been brought here from Santiago by early sailors.

TEETERING ON THE BRINK

The Floreana mockingbird, *Mimus trifasciatus*, is one of the rarest, most endangered birds in the world. Extinct on its island of origin since the 1880s, its existence hangs by a tenuous thread on two extremely arid satellite islets, Gardner and Champion, where the tiny surviving populations are quite sedentary, fluctuating around 60-80 and 20-45 individuals, respectively. Although capable of flying, the Floreana mockingbird does not like to travel far, which could stem from the fact that it evolved in the absence of native rodents, and, similarly to its cousin on Española, actually behaves much like one. Spending a lot of time scrabbling in the meager ground cover may explain its disappearance once rats and cats were introduced to the main island. The total numbers fluctuate between wet and dry years, and may double after good rains. Plans are underway to restore it to its home island, using a combination of methods that include the experimental release of surplus young birds after the rainy season, establishment of a captive breeding program, and control of introduced predators, combined with intensive monitoring of real and potential threats such as diseases. A very costly and long-term proposition, this project awaits a sponsor prepared to go the distance to save the species. No currently recognized Galápagos bird species has gone extinct to date — let us not allow the Floreana mockingbird to become the first!

Tui De Roy

ABOVE: An unidentified subspecies of wedge-tailed shearwater disappeared on San Cristóbal some 21,000 years ago.
BELOW: Three species of as-yet unnamed rodents were discovered on the west slope of Wolf Volcano, Isabela.

Prehistoric information about vertebrates is also valuable because it can be used to estimate the composition and rate of change of communities in the absence of human impact. Even though the size of vertebrate populations in Galápagos can vary considerably on annual or decadal scales because of fluctuations in climate and innumerable biotic factors, the fossil record would argue that extinction was a rare event before humans colonized the archipelago. In fact, after examining more than a half-million fossils of reptiles, birds and mammals, I believe that the rate of vertebrate extinction following the arrival of people has been at least 100 times greater than in prehuman times. This is consistent with the long-term vegetation record of the Galápagos, where Paul Colinvaux and Eileen Schofield found little evidence of non-anthropogenic extinction and colonization in the archipelago during the past 10,000 years.

Scientists have long recognized that indigenous species of vertebrates may not be able to survive the environmental changes that accompany human activities on oceanic islands (see also chapters: Watkins, Merlen). This realization is especially evident to the west of Galápagos, on the hundreds of remote islands in Oceania, where a huge wave of extinction took place in the wake of human colonization over the last 3000 years.

The Galápagos Islands have been spared such prehistoric human impact, with people first arriving only a few centuries ago. But the species composition certainly has changed even in that short time, with the loss of a number of indigenous taxa or populations, and the establishment of many non-native species. Still, the biotic changes in Galápagos have been less drastic than in any other tropical island group. We owe it to ourselves and to this relatively intact insular biota to seize any opportunity to keep it that way.

Lichen Discoveries
Bright Bold Color Specks, Tiny and Overlooked
Frank Bungartz

Charles Darwin Foundation, Baseline Studies, Inventories & Collections, Puerto Ayora, Santa Cruz, Galápagos, Ecuador.

Dr. Frank Bungartz works at the Charles Darwin Research Station (CDRS) as a cryptogamic botanist investigating plantlike organisms that lack conspicuous flowers (from Greek 'crypto' meaning hidden and 'gameein' meaning to marry): lichens, fungi and bryophytes. He also manages the Station's natural history collections. Born in Germany, he completed his M.S. at Bonn University, spent some time working on lichens in Great Britain, and earned a Ph.D. at Arizona State University, where he was one of the principal editors of the three-volume *Lichen Flora of the Greater Sonoran Desert Region.*

THE PLANE TOUCHES DOWN as we approach a dry, austere landscape. Looking through an oval window I see cacti, sparse palo verde trees (*Parkinsonia aculeata*) with green photosynthetic trunks, and rocks of bare, weathered red lava — almost a desert. I had not anticipated this. Not on the equator, not in a tropical archipelago (see also chapter: McMullen).

The Galápagos Islands lie at the crossroads of subtropical and tropical climates. The Humboldt Current brings cold seawater from the Peruvian and Chilean coast. Stronger than the warm, tropical flow from Panama, the cool temperate sea dominates Galápagos weather patterns, and for most of the year pushes rainfall away, northward. Therefore, organisms of very different ancestries have made it here: famously, penguins now live on the equator, alongside cacti, in a desertlike landscape.

Today, of course, the islands are renowned for their unique organisms — their giant tortoises, their marine iguanas, even Darwin's finches, though inconspicuous and drab. Year after year, tourists and scientists alike flock to this archipelago to admire and study Darwin's Enchanted Islands. I came here for less spectacular elements, equally unique, though often overlooked. In 2005 I was hired by the CDRS for an unusual job, as a lichenologist to study the archipelago's rich and diverse lichen flora.

Arriving at the airport, on the flat island of Baltra,

LEFT: Damp sea breezes nurture abundant lichen growth on low-lying Española, covering both the arid vegetation and rocky ground. Among the mosaic of species are epiphytic *Roccella* dangling from palo santo trees, *Bursera graveolens*, and yellowish endemic *Lecanora pseudopinguis* dominating the encrusting assemblages on the lava boulders.

ABOVE: Lichen-nurturing fog banks hug the ground as clouds stream into the caldera of Darwin Volcano on Isabela.
BELOW: Long strands of *Usnea mexicana* hang like curtains in the upper transition zone of many larger islands, where mist blows frequently but rain is rare.

I could have been fooled. Though lichens easily survive long periods without water in a state of physiological dormancy, they eventually will not endure in the most hostile, arid landscapes. On Baltra, even the toughest lichens seemed to be having a hard time, a few hiding from scorching sun in shaded cracks of otherwise bare lava.

From here, we take the ferry across to Santa Cruz. An old bus then conveys us up a straight road into the highlands, and down on the other side to Puerto Ayora. During this journey, the vegetation drastically changes. Suddenly trees are covered in long, drooping curtains, *Ramalina usnea*, *R. anceps* and *Usnea mexicana* to name but a few of the many lichens that form the natural pale green drapes.

'The Tortoises which live on these islands where there is no water ... live chiefly on the succulent Cactus: I have seen those which live in the higher parts, eating largely of a pale green filamentous Lichen, which hangs like presses from the boughs of the trees ...' [from *Charles Darwin's Zoological Notes*, edited by Richard Keynes, 2000].

Even the most famous naturalist ever to visit Galápagos might have erred: though curtains of lichens can be extremely abundant, they are an unlikely food even for reptiles. The nutrient content of lichens is low, and they are rich in organic insoluble salts, known as lichen acids, secondary metabolites that are difficult to digest and probably a major deterrent to protect the lichen against grazing.

But Galápagos lichens serve another important function. During the dry season when rains are scarce, the lush forest at higher elevation relies on other sources of fresh water. Garúa, a fine mist, hangs in the highlands as a dense fog. Pale lichen curtains trap this moisture as the wind blows through them, the water condenses and, when they are saturated, the excess drips to the ground. Shallow pools accumulate and fresh water seeps into the soil. Thus, on an archipelago with few natural wells and at least seven months of sparse rainfall, lichens contribute significantly to the natural water cycle. The tortoises like these puddles too. Darwin noticed them enjoying the shallow, muddy pools; where they drink and bathe, and thus rid themselves of ticks.

In 1535 the islands were discovered by accident, when Tomás de Berlanga was stranded here with his fleet. Sailing from Panama to Peru, his ships drifted to the archipelago because of missing trade winds. After helpless weeks at sea, discovering the barren lava flows, with monstrous marine iguanas, prickly pears and giant tortoises, must have been a dreadful shock. Berlanga described the land as 'dross, worthless, because it has not the power of raising a little grass, but only some thistles ...' Thus, for years Galápagos drew little interest; a few British pirates hiding from the Spanish fleet, or whalers hunting the seas for their prey, touching land only for tortoise meat.

Why then a sudden interest in the 1830s? Why were these remote, hostile islands suddenly claimed as a territory by the young state of Ecuador? Darwin's commentary on the unique Galápagos Islands biodiversity had not been written. Indeed, he had not even visited them yet. So what attracted the attention?

In 1831 General José Villamil discovered that *Orchilla* grew abundantly along the Galápagos coasts. He managed to convince the Ecuadorian government that this lichen, also known as dyer's moss (*Roccella gracilis*) had economic potential. Fermented in ammonia (or urine from livestock), the species turns a deep purple or blue, its hue dependent on the pH level of the bath. Litmus, another name for this vibrant natural dye, was once a highly prized resource. When stocks became depleted on other tropical coasts, its price suddenly seemed attractive enough to justify even the colonization of islands as remote as Galápagos.

But for botanists it still took years to notice the abundance of other lichens. In 1905–06 Alban Steward participated on an expedition mounted by the California Academy of Sciences to explore the biodiversity of the archipelago. He later wrote in his 'Notes on the lichens': 'When one lands for the first time on almost any of the islands, one is immediately struck by the great abundance of lichens ...' [A. Steward, 1912].

The year 1964 marked a turning point for Galápagos research. Five years after the islands had been declared Ecuador's first national park, and coinciding with the inauguration of the CDRS, the Galápagos International Scientific Project (GISP) culminated in a massive multidisciplinary expedition, when some 60 scientists, with unprecedented logistical support including helicopters, fanned out across the archipelago for a month. As a member of this expedition, Bill Weber was the first lichenologist to visit the islands, and recalls in his published account how he was overwhelmed with the lichen diversity: 'I felt that I could not possibly get on top of this fantastic lichen flora in the six weeks at my disposal.' [W. A. Weber, 1986].

A fellow expedition member later wrote: '[Bill Weber] came in from his first day of field work like a small boy from a candy store.' [Robert L. Usinger, 1972].

What Bill could not have known then is that in his first inventory he would discover many species that today are lost, never to be found again, for example *Usnea articulata*. In 45 years his 'candy store' has become depleted.

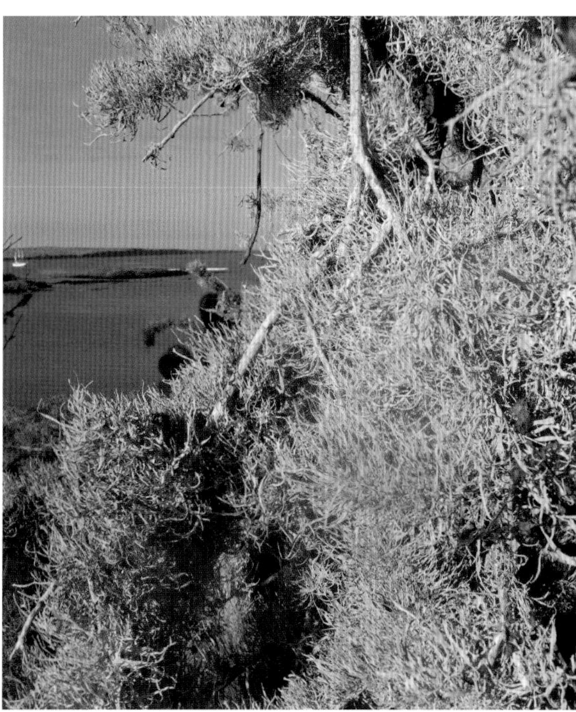

ABOVE: A young giant tortoise wanders amid lichen-clad shrubbery on the south slope of Alcedo Volcano, an area often bathed in rainless clouds. Several species of lush *Ramalina* were discovered during the author's survey, some of them new to science.

LEFT: Shaggy clumps of dyer's moss, *Roccella gracilis*, hang from coastal vegetation on Santa Cruz, where humid sea breezes are constant. Used for making deep purple dyes, these were much-prized by early entrepreneurs.

RIGHT, FROM TOP: An astonishing number of species new to science have emerged during the author's groundbreaking work, including: an intriguing form in the genus *Lecanora* (A), found in the transition zones of several islands, which he nicknamed 'popcorn lichen' while it awaits publication; delicate *Ramalina darwiniana* (B), described in 2007, from the Santiago lowlands; hieroglyph-shaped *Schismatomma spierii* (C) from the harsh, salt-encrusted coast of Bartolomé, described in 2009; an as yet undescribed *Caloplaca* species (D) from Pinta.
BELOW LEFT: Growing on cactus spines kept damp by sea breezes, a free-growing alga of the genus *Trentepohlia* very often occurs as a photobiont (the algal partner in charge of photosynthesis) in tropical lichens.
BELOW FAR RIGHT: A rare species usually restricted to primary tropical forests, *Usnea angulata* is a sensitive indicator of environmental change. This 1964 museum specimen was collected in the Santa Cruz highlands by Bill Weber, but the species can no longer be found in Galápagos today.

Frank Bungartz
A

Frank Bungartz
B

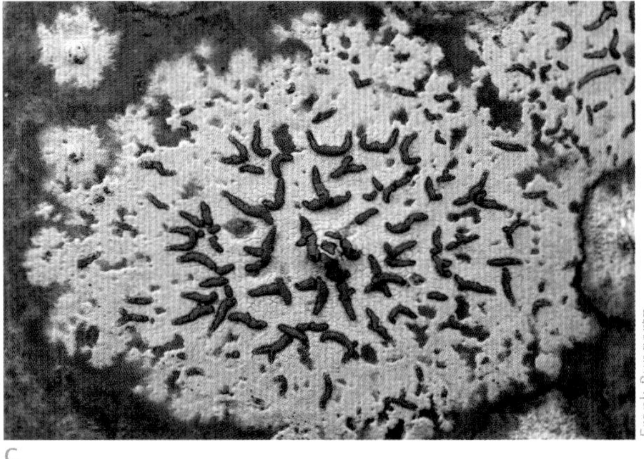
Frank Bungartz
C

Environmental indicators

Many lichen species are extremely sensitive indicators of intact environments. They grow unusually slowly, and their recovery rates are poor. When native ecosystems are disturbed, sensitive species soon disappear. And even when things appear back to normal, their absence bears witness to prior disturbances.

Slow growth rates are, in part, explained by their biology. Lichens are symbiotic organisms, composed of two different partners growing closely together as one. What we perceive as a unit is really a fungus hosting microscopic algae within. Unlike other fungi, lichens therefore function as 'plants'. Their algal symbionts produce sugars, the nutrient supply for the fungus. Production rates need to be sufficient for both partners, and therefore they take time to grow. Just as a farmer prefers a particular kind of cattle, lichen fungi domesticate their own particular breed of algae; and like farmers' cattle are adapted to our needs, lichen algae are uniquely adapted to the needs of lichen fungi. This relationship is fragile. When attempts are made to cultivate an intact symbiosis, the relationship typically loses its balance: the fungus aggressively feeds on its algae, and algae dissociate from the symbiosis. In nature, however, lichens are surprisingly adept. Some are rare and specialized; others are common and opportunistic species. Some even grow on artificial substrates like plastic bottles or abandoned cars. In Galápagos, lichens grow on tortoises! At first, this may seem an unlikely habitat, but the carapace is the perfect environment for many species. Tortoises are famously long-lived, sufficient time for a lichen to establish itself. Their carapace is exposed to sunlight and it is an otherwise unoccupied habitat. When at night a tortoise digs its back into an earth mound, nutrient-loving lichen species get their fair share of dust. But being on the move has disadvantages too. When they mate the shells of female tortoises get polished by the males, and only closely attached, flattened lichens survive this challenge.

Frank Bungartz
D

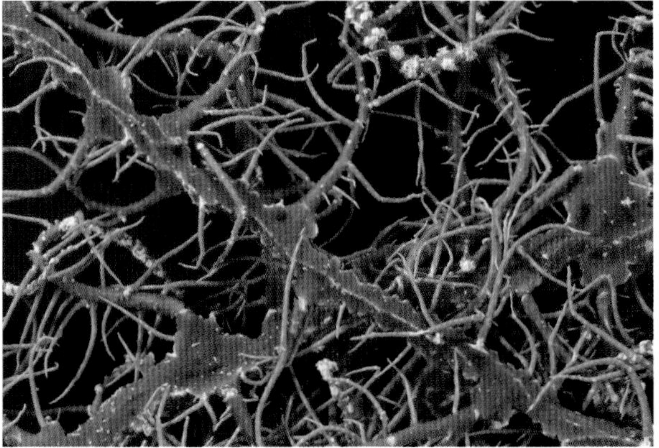
Frank Bungartz

Bill Weber worked on his first checklist over many years. Eventually, he reported more than 200 species, knowing that there were still many more to be identified. For decades, large numbers of Galápagos lichens remained entirely unknown. One of the most diverse groups of tropical encrusting lichens are the Graphidaceae. Because their fruiting bodies form lines, they are also known as script lichens. Of this diverse group, Bill recorded only eight species. Nearly half a century later, having searched the archipelago over the last three years, we can now report 39 different species in this group alone. The same pattern holds true for most other groups. In the common lichen genus *Ramalina*, eight of 15 species were not previously recorded, four of these being totally new discoveries previously unknown to science.

Today, compared to Weber's initial count of 200, our current inventory is approaching a total of almost 600 known species. Most exciting are an astounding number of totally new and yet unnamed lichens. We at the CDRS estimate that at least 30 Galápagos lichen species remain new to science. So far only a few have been named and we are busy describing the remainder that still have no official scientific names. Most of these new species are likely to be endemic, meaning they grow nowhere else on earth. Many of these species have evolved in the dry, lower parts of the archipelago. Perhaps here the evolutionary pressure has been more intense, with a stronger necessity to adapt to the hostile environment of the lava flows and scorching sun. However, a few unnamed species have also been found in the humid highlands, especially where the forests are least disturbed. Some of the new species come in bizarre, unusual shapes; we informally call one 'popcorn lichen' while it awaits its official binomial. The evolutionary pressures that favor these strange forms are unknown, yet islands generate such special tendencies, uniquely so in the case of Galápagos.

Not all new species were overlooked; Weber collected many of them and even mentioned this 'hidden' diversity awaiting discovery. But like the slow-growing objects of his studies, lichenology is a science with little resources. Even in Galápagos, where they are truly abundant, lichens receive a lot less attention than spectacular iguanas or giant tortoises. Their identification takes time and many species are still poorly known. The inconspicuous organisms require exhaustive studies at the microscope and an analysis of their diverse chemistry. Only recently, molecular techniques not available in Weber's time, are helping us to better understand their taxonomy and pry apart their poorly known diversity.

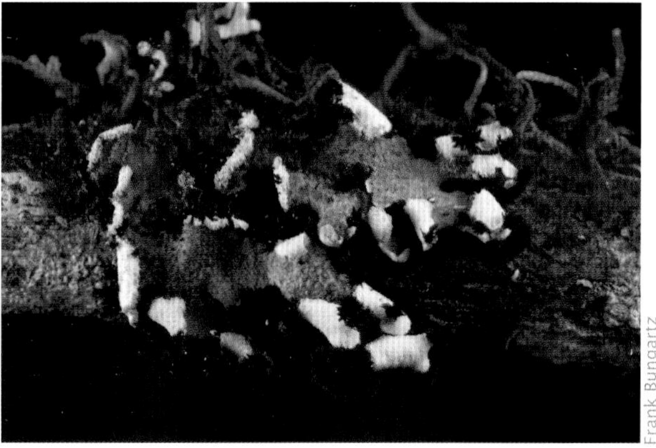

Frank Bungartz

A

Frank Bungartz

B

Frank Bungartz

C

Frank Bungartz

D

LEFT, FROM TOP: Exhaustive analysis is required to determine each new species or first Galápagos record: the rare velvety *Erioderma sorediatum* (A) and knobbly *Leptogium javanicum* (B) from the Santa Cruz highlands, both new records but found elsewhere in the tropics; looking blood-stained, *Cladonia didyma* (C) from the rain-drenched slopes of southern Isabela is a cosmopolitan species collected by the author for the first time; an oddly tentacular, possibly new, species of *Usnea* (D) from the transition zone of San Cristóbal awaits identification.

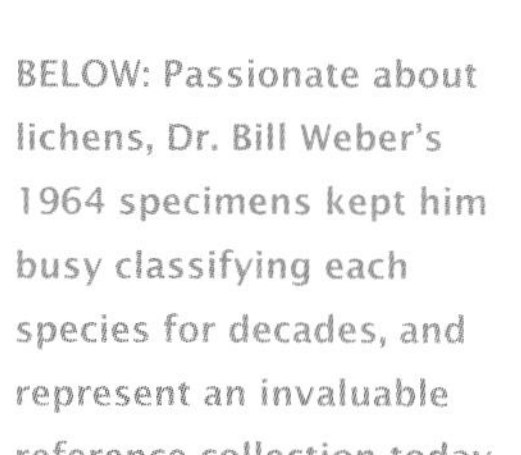

BELOW: Passionate about lichens, Dr. Bill Weber's 1964 specimens kept him busy classifying each species for decades, and represent an invaluable reference collection today.

Photo courtesy Allyn G. Smith

RIGHT, CLOCKWISE FROM TOP LEFT: All represented in Weber's original collections but also occurring elsewhere in the tropics, an amazing variety of shapes, colors and forms from Isabela and Santa Cruz include: the two-toned *Pseudocyphellaria aurata* (A); disklike *Dictyonema glabratum* (B), growing among rain-soaked epiphytes in *Scalesia* and *Miconia* forests; and closely related frilly white *Dictyonema sericeum* (C); strange, gelatinous *Leptogium phyllocarpum* (D) absorbs vast amounts of water to survive in the arid lowlands; blood-red *Cryptothecia rubrocincta* (E), clinging to tree trunks in transitional and wet zones; South American tropical reindeer lichen *Cladonia confusa* (F) grows in the Galápagos highlands on misty open ground, but is increasingly outcompeted by invasive introduced plants.

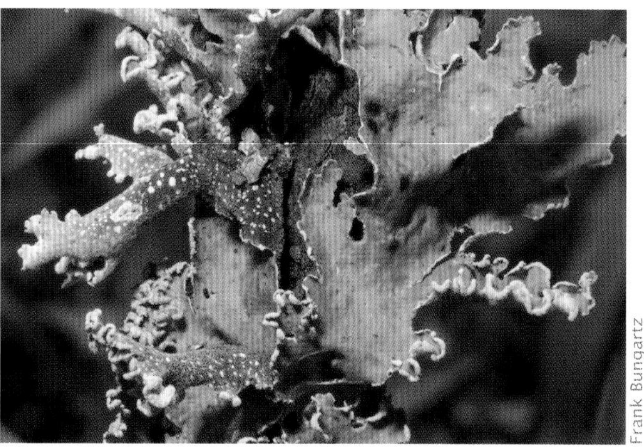

Frank Bungartz

A

Frank Bungartz

B

F

Frank Bungartz

C

Frank Bungartz

E

Frank Bungartz

D

Frank Bungartz

ABOVE: The author carefully chisels a small flake of rock harboring a precious specimen at Punta Pitt on San Cristóbal.

The situation today

I am standing in front of a boulder with my hand lens, looking at a small shard just chiseled off the rock. This thin crust has some tiny specks that receive my full attention. Have I seen it before? Perhaps not ... Carefully I wrap the piece in tissue paper, collecting my specimens in paper bags on which I note my observations. Where did I find this specimen? When did I collect it? Was it growing in full sun or in shade, on a vertical face of a boulder or on a flat rock?

Back at the CDRS my samples will help me to understand the ecology of each species. Sectioning the material, studying its anatomical characteristics under the microscope, entering its collection information into our database — slowly a picture emerges ... Where does this species grow? How is it distributed across the islands? Is it rare or common? What are the special characteristics of the sites where it grows? Broader questions also emerge. Could it be an indicator for its environment? How do the lichens fit in?

Collection methods have changed little since Weber's time, and his historic collections are now a most valuable information resource. Studying his specimens and comparing them with our own gives us clues to abundance and distribution, showing us which species were here in the past even though they might not all have been identified.

Ecology is a cumbersome business. Interactions between organisms are complex. To understand this intricate web, scientists tend to simplify. They establish models and are excited if their models approach reality. But there is more to ecosystem modeling than just powerful computers and complicated algorithms. 'Garbage in, garbage out' is a common saying among ecosystem modelers, meaning raw data fed into a simulation need to be of high quality. How can we begin to understand a system, if its elements remain unknown?

The Galápagos Islands have been called a natural laboratory of evolution. It is no accident that the islands stimulated evolutionary thinking. These remote specks of land are isolated; they require a long journey by sea, providing ample time to think for a young Charles Darwin during his long voyage on the *Beagle*.

Remote isolation. Few organisms survive their own journey across the sea. Then barren lava flows await with a scorching sun. This must be the most uninviting prospect to almost any newcomer. Species that survived had to adapt. They crossed an unbearable distance and became established despite a hostile environment. But for the few that survived there was now room — new, open, unconquered spaces, niches that could now be claimed.

With their microscopic propagules, with spores and tiny thallus fragments, lichens have developed effective long-distance distribution mechanisms. They can travel further than any heavy plant seed or fruit. Arriving on the barren lava cliffs along the coast, conditions were still a challenge. Lichens, however, are better adapted to these extreme environments than many other organisms. They survive on bare rock, where plants fail to claim territory. They dry out, lie dormant and rehydrate when fog drifts in from the sea. They are survivors; poor competitors, but survivors.

Conservation science must not ignore the tiny and obscure in favor of the big and spectacular. Tourism in the islands relies on the spectacular, but the spectacular will only survive if the ecosystem remains intact. While many other oceanic islands have lost most of their original biodiversity, the unique flora and fauna of Galápagos are still its main attraction. Understanding this fragile insular balance is the key to its conservation. Perhaps, after all, it did make sense that on these remote islands they now hired a lichenologist who tries to understand how these obscure organisms fit in.

Frank Bungartz

ABOVE: Capable of developing on the most inhospitable surfaces, the common *Pyxine cocoes*, widespread in arid tropical regions, clings to the base of a discarded plastic drink bottle in the lowlands of San Cristóbal.

Frank Bungartz

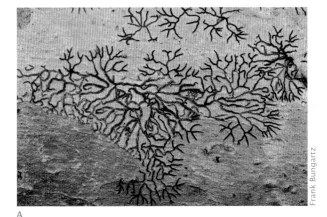

A

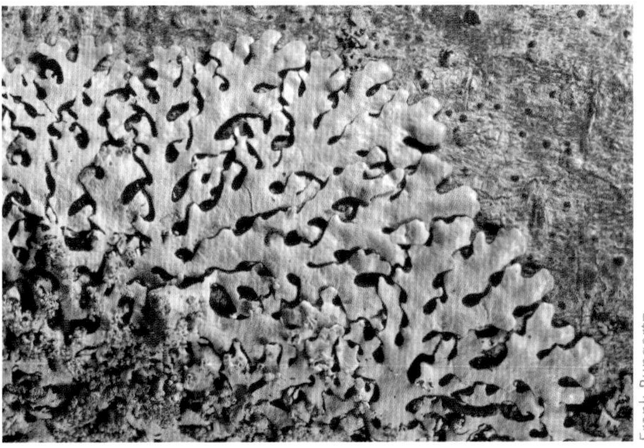

Frank Bungartz

B

Frank Bungartz

D

Frank Bungartz

C

LEFT, CLOCKWISE FROM TOP LEFT: Thanks to the atmospheric inversion layer that characterizes the Galápagos climate, high levels of humidity in the air allow for a profusion of intricate encrusting lichens attached to the trunks of trees from the arid lowlands to the misty summits, including: the filigree-like *Graphis elongata* (A) from Santa Cruz highlands; and the palmate *Hypotrachyna osseoalba* (B) in the transition zone of San Cristóbal, both new records for Galápagos but found in other tropical and subtropical parts of the world; bubbly *Ochrolechia africana* (C) from the Santiago lowlands, another tropical species; the colorful South American script lichen *Graphis subchrysocarpa* (D), known from only two specimens, the first collected in the highlands of Santa Cruz by Weber, the other on Pinta by the author.

Insular Flora

More than 'Wretched-looking Little Weeds'

Conley K. McMullen

President, Southern Appalachian Botanical Society, Department of Biology, MSC 7801, James Madison University, Harrisonburg, VA 22807, United States. <mcmullck@jmu.edu> http://csm.jmu.edu/biology/mcmullck/index.htm

Dr. Conley K. McMullen is Associate Professor of Biology at James Madison University in Harrisonburg, Virginia. His Galápagos research focuses on pollination biology, floristics, and plant systematics. He has published numerous articles, spoken widely and, in 1999, published *Flowering Plants of the Galápagos*. He is a member of the General Assembly of the Charles Darwin Foundation (CDF), the Darwin Scientific Advisory Council of Galápagos Conservancy and a Fellow of the Linnean Society.

CHARLES DARWIN WAS NOT favorably impressed after visiting the Galápagos Archipelago in the autumn of 1835. Having collected specimens of all flowers he encountered, Darwin described the plants of San Cristóbal Island as 'wretched-looking little weeds.' My impression almost 150 years later, in the fall of 1983, could not have been more different. Having just arrived as a visiting scientist, I found myself surrounded by a striking assemblage of herbs, vines, shrubs and trees, many in full flower.

Given Darwin's eventual renown as a naturalist, it would be folly to blame his assessment on a lack of interest in plants, or a general absence of zeal. How then can one explain such a marked difference in opinion of two scientists visiting the same archipelago during the same season of the year? The answer is two-fold. First, during Darwin's visit to Galápagos, his field observations were restricted primarily to the lowlands of the four islands he visited, as few opportunities for exploration of the highlands were available at that time. To fully understand why this is important, one must first know that there are two major seasons in the Galápagos. During the warm season (January through May), the skies are typically clear throughout the day, although interrupted by occasional downpours, some of them quite heavy. At this time, the vegetation of the lowlands reaches it peak productivity. The cool season lasts from June through December, when a persistent cloud cover produces a mixture of rain and mist in the highlands

BELOW: Warm season rains, normally brief and sharp, provide a spectacular burst of annual plants, which can grow many times their normal size when El Niño conditions take over.
RIGHT: Within a week of the first rains, the delicate *Croton scouleri* bursts into bloom.
FAR RIGHT: The lava cactus *Brachycereus nesioticus* is an endemic pioneering species growing on young lava but disappearing when plant succession diversifies.

ABOVE: Although fine clouds shroud the higher elevations, during a normal dry season the coastal lowlands lay dormant for up to 10 months of the year, when most trees and shrubs are leafless. Buccaneer Cove, Santiago. BELOW: The little Galápagos blue butterfly *Leptodes parrhasioides* is a dainty species well adapted to many of the tiny flowers of the arid coastal zone, such as *Cordia leucophlyctis.*

known as garúa, while the lowlands receive virtually no rainfall. Darwin's visit coincided with the cool season, when many of the lowland plants go dormant and lose their leaves as a survival strategy and are, in fact, fairly 'wretched-looking.'

The second reason that our experiences were so different is that my first visit happened to take place soon after the famous El Niño event of 1982–83. During such a climatic event, the archipelago experiences an unusually large amount of rainfall, and a consequent increase in vegetation that must be seen to be believed (see also chapter: Sachs). Thus, Darwin's experience might be thought of as a classic example of 'right place, wrong time.' And it demonstrates the danger of drawing conclusions based on a single visit to a particular locale.

Despite any shortcomings of his historic visit, Darwin was not unaware of the uniqueness of the Galápagos flora, especially following discussions with his close friend and botanical colleague, Sir Joseph Dalton Hooker, after returning to England. Indeed, Hooker's writings, based partially on Darwin's field notes and plant collections, inspired future generations of scientists to explore the discipline that became known as phytogeography (plant geography). Basically, this is the area of research that investigates the flora of a particular region and attempts to discover where the plants came from, how they got there, how they survived and reproduced, and how they may have adapted to their new surroundings. By the time I first visited Galápagos, many of these questions had already been answered, but other areas of research were still blank fields.

Plant evolution on Galápagos

All plants within the archipelago fall into one of three major categories: Endemics are found only in the islands; natives occur naturally in the archipelago, as well as in other parts of the world; exotics were brought to Galápagos by humans, either by accident or intentionally for cultivation.

It was well known, given the fact that the islands are volcanic in nature and arose from the ocean floor, that the natives and the ancestors of the endemics must have arrived through long-distance dispersal. Studies have determined that the majority of these species originated in western South America (Venezuela to Chile). Others have geographical relationships with other parts of the western hemisphere, while only a few are truly widespread. It also transpires that the vast majority of naturally occurring plant species arrived either inside visiting birds or stuck to their feathers. Indeed, many seeds are able to survive passage through the digestive system of a bird, while others possess barbs or an adhesive coating that promote sticking to feathers and other parts of the body. Even caked mud on the feet of a bird might harbor the seeds or fruits of an aquatic herb. Other plants, typically those inhabiting coastal regions, arrived via ocean currents. Some of Darwin's more notable 'backyard' experiments at Down House, his 7.2-hectare (17.75-acre) country home in Kent, demonstrated that many fruits and seeds are adept at floating, and remain viable after long periods in salt water. Last but not least, some fruits and seeds were carried to the islands by wind. These typically are very tiny in size (e.g., orchid seeds) or have plumes (e.g., many members of the sunflower family) that allow them to be carried aloft over quite some distance. As a result, the composition of the Galápagos flora is referred to as 'disharmonic,' which means that when compared to the mainland, there is a noticeably higher proportion of plants with adaptations for long-distance dispersal than those that either bear heavy or cumbersome seed or fruit, or that soon die if exposed to inappropriate conditions. Obviously, it takes a very special plant to successfully cross more than 1000 km (600 miles) of ocean! And we will never know how many began the journey but did not make landfall.

Once plants began arriving and establishing themselves, they occupied the various regions of the islands to which they were best suited based on their particular requirements for resources such as nutrients and light. Those that did not depend on an already mature ecosystem to survive flourished first. Ultimately, three ecological zones formed in the archipelago: the littoral or coastal zone, the

ABOVE: One of only 14 species of Galápagos orchids, the flowers of the endemic buttonhole orchid *Epidendrum spicatum* are small and green, deprived of the complex interdependent relationships with pollinators and root fungi that evolved in many continental species.
FAR RIGHT TOP: A zone of lush vegetation covers the humid highlands of the larger islands, including tree ferns *Cyathea weatherbyana* growing in the dense *Miconia robinsoniana* belt of Santa Cruz.
FAR RIGHT BOTTOM: Arid palo santo, *Bursera graveolens*, forests are typical of the coastal zone, Pinta Island.
RIGHT: The attractive morning glory *Ipomea linearifolia* may go on blooming for many weeks after the last rains.

LEFT: The succulent Galápagos purslane *Portulaca howellii*, growing around Sombrero Chino, and many other salty shores, responds to the first heavy rain by transforming quickly from deathlike stumps to delicate evening-blooming flower fields.

arid lowlands and the moist uplands. The majority of the islands possess the littoral and arid lowlands zones. However, only the larger, higher islands have moist uplands, as it is the cloud cover draping these islands that provides sufficient moisture to sustain such an ecosystem. For those plants pertaining to the moist uplands, or humid zone, an added barrier may have been the crossing of the arid lowlands, where germination may well occur during wet periods only for the plant to subsequently die off due to the seasonal droughts.

As the establishing plant species began to spread, some of them exploited ecological niches unoccupied by other species. In some cases, this resulted in one ancestor giving rise to many different species, subspecies or varieties, each occupying a different habitat or isolated on a separate island. This process, referred to as adaptive radiation — a term coined by Darwin — gave rise to some of the archipelago's more successful genera including *Scalesia*, *Opuntia* (see also chapter: Hamann) and *Alternanthera*. *Scalesia*, for example, includes 15 species, with a total of 19 members when subspecies and varieties are included. *Opuntia* comprises six species with 14 varieties, while *Alternanthera* has 14 species with 20 members (including subspecies). All of these subspecies and varieties make sense when one considers the principles of Darwin's theory of natural selection (see also chapter: Grant). If plant species are evolving in the Galápagos due to isolation and the passage of time, and if the islands are relatively young geologically speaking, then one would expect there to be many closely related groups of plants defined as either subspecies or varieties, rather than having full-fledged species status. After all, the various populations of plants have had only a relatively brief time to develop differences between themselves. However, one would expect that as time goes on, some of these will change sufficiently that future generations of botanists might decide that they merit full species rank. Clearly, it takes longer still for a group of plants to diverge sufficiently to achieve the rank of genus. Indeed, in the Galápagos, there are only seven endemic groups of flowering plants that

BELOW: *Alternanthera filifolia* (left, with lava lizard in dry season) and *A. echinocephala* (middle, rainy season) belong to a genus represented by 14 species in Galápagos; the endemic passionflower *Passiflora conlinvauxii* (right) has much smaller flowers than its native and introduced relatives.

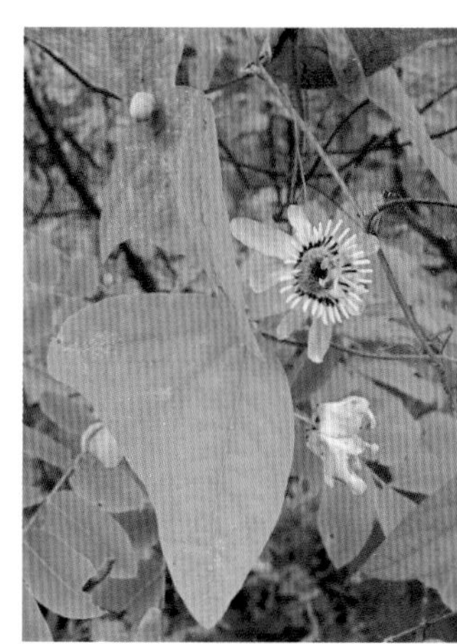

WHY DULL FLOWERS?

Scalesia villosa simultaneously visited by a carpenter bee and monarch butterfly.

The relative paucity of insect pollinators in Galápagos may explain why most endemic flowering plants — those that have evolved the longest and adapted specifically to the Galapagos environment — have relatively small, plain-looking, white or yellow flowers that produce only small amounts of pollen. Without their faithful pollinators, in the early stages there would have been little or no selection for large, attractive, brightly colored and pollen-laden flowers, as the costly production of showy floral displays would have been counterproductive. More modest flowers took less energy for growth, yet were still available to generalist pollinators. Of course, there are exceptions, as seen in the flowers of the giant prickly pear cactus and Darwin's cotton. Most native and introduced plants, on the other hand, possess showier flowers that evolved on the mainland with its more abundant and eclectic insect fauna. By the time they arrived, enough insects had become available on the archipelago to support their pollination needs in order for them to become established. The major player in this survival drama is the endemic carpenter bee, which is generally considered the most common pollinator in Galápagos. Crucially, what makes this bee so successful is the fact that it is polylectic, meaning that it visits a variety of plants for pollen and nectar, a behavior benefiting both bee and plant alike.

Conley McMullen

are thought to be so different from their mainland relatives as to represent unique genera. These include *Scalesia*, *Darwiniothamnus*, *Lecocarpus*, and *Macraea* of the Asteraceae (sunflower family); *Jasminocereus* and *Brachycereus* of the Cactaceae (cactus family); and *Sicyocaulis* of the Cucurbitaceae (gourd family). Finally, given the age of the archipelago (see also chapter: Geist), it should be no surprise that no vascular plants differ at the family level.

Reproductive biology

One area of study mentioned earlier that was still wide open for investigation at the time of my first Galápagos visit was plant reproductive biology. For this reason, as a graduate student, I decided to spend six months conducting research on Santa Cruz Island to determine how flowering plants were pollinated. Once a seed has arrived, germinated and become a mature plant on an island, reproduction must follow if a population is to become established. The prevailing theory in island biology at the time of my initial studies was that self-compatible plants would be favored in the colonization process of oceanic islands such as Galápagos. This makes a great deal of sense when one considers that if a single seed arrives at a particular point in time, the resulting plant won't have a mate with which to exchange pollen. Making things even more difficult is the fact that, compared to the mainland, the Galápagos Islands harbor few pollinating insects. So, obviously, an island plant should be favored if it can make use of its own pollen. Although this was the generally accepted theory, few studies had been conducted in Galápagos to either support or oppose this notion.

Since my beginnings as a rather 'green' field botanist, I have made many more trips to Galápagos to continue this intriguing line of investigation. The results, though not totally surprising, were gratifying. In a nutshell, of 52 species studied, the vast majority were indeed self-compatible, and not only that, most of these also had the ability to self-pollinate. In other words, not only could they make use of their own pollen, but they didn't need the services of a pollinator to move the pollen from anther to stigma. For a plant relegated to life on an island, this certainly represents making the best of a bad situation!

My studies over the years have continued to confirm these findings. But one unexpected fact that has emerged is that, while admittedly poor in comparison to the mainland, the pool of potential insect pollinators in the archipelago is larger than once thought. In fact, just because a plant may not *require* a pollinator doesn't mean that it won't be pollinated if a suitable insect is in the area. A partial list of frequent visitors to Galápagos flowers includes carpenter bees, butterflies, moths, beetles and ants. Of these, the

BELOW AND FAR RIGHT: Carpenter bee *Xylocopa darwinii* visiting *Opuntia* cactus and native passionflower *Passiflora foetida*.

carpenter bee turns out to be the most important. Birds are also known to occasionally transfer pollen from flower to flower, most graphically illustrated by some of Darwin's finches, whose facial feathers can be thickly coated in bright yellow pollen during the *Opuntia* cactus flowering season.

My interest in this subject eventually led me to investigate what goes on in the dark, since in warmer regions of the world there is often more insect activity after sunset, when the air is cooler. I well remember my summers spent studying nocturnal pollination on Santa Cruz and Pinta islands, especially the latter. There is nothing quite like having an entire island all to oneself, except for the company of an assistant, as well as that of Galápagos hawks during the day and short-eared owls at night. Being buzzed by hawk-moths at dusk never grew old and, although giant centipedes weren't my favorite nighttime companions, they were still an unforgettable component of these summers in the enchanted isles. Each night spent in the highlands, trying to keep warm (due to a jacket left back at camp, or in the case of my assistant's, shredded by a hawk), was balanced out by beautiful, starry evenings in the lowlands, where the only lights were those of ships far out at sea — I believe we saw just three in seven weeks.

Bright, warm days near the coast, bagging flowers for our pollination experiments (this is where we played the part of insects) or finding new insect visitors that had never before been reported for a particular plant species, were enthralling. And to share these experiences with not only my own student assistants, but those from Galápagos and mainland Ecuador as well, was a source of enduring pleasure.

So, what are my conclusions after more than two decades of studying pollination biology in Galápagos? First, as hypothesized, due to the possibility of having no mate and the relative dearth of pollinating insects,

ABOVE: The endemic guava *Psidium galapageium*, one of the tallest trees growing from the transition zone upward, is often festooned in mosses and liverworts.

BELOW LEFT AND RIGHT: Less than a dozen butterfly species are native to Galápagos, including the sulfur butterfly *Phoebis sennas marcellina* (left) and longtail skipper *Urbanus dorantes.*

ABOVE: Even tiny native ants seeking nectar inside the large flowers of Darwin's cotton, *Gossypium darwinii*, can act as pollinators.
RIGHT: A carpenter bee, *Xylocopa darwinii*, and cactus finch, *Geospiza scandens*, compete warily for the same *Opuntia* cactus flower.

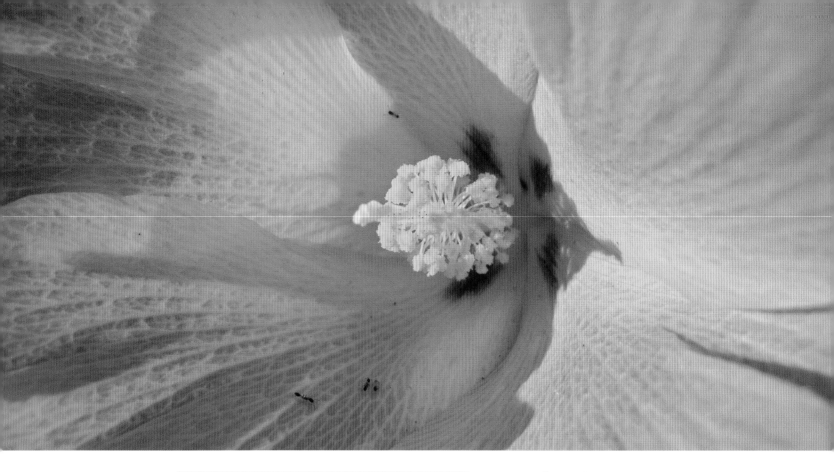

plants that are self-compatible and self-pollinating have the advantage when colonizing these islands. Second, although few by mainland standards, there are more different insects that visit Galápagos flowers than was thought years ago. Some of these make their visits only during the day (e.g., carpenter bees and butterflies), others only at night (e.g., hawk-moths, many smaller moths and most beetles), and some during both the day and night (e.g., ants). Some of these are very efficient pollinators (e.g., carpenter bees, hawk-moths), while others are quite poor at carrying pollen (e.g., ants). Third, and perhaps most interesting, it appears that the more successful Galápagos flowering plant species possess what is referred to as a mixed mating system. This means that cross-pollination will occur when pollinators are available, resulting in more variable offspring, which is always a plus in sexually reproducing organisms. But, if no pollinators are available in a particular year or at a particular location, the plants can still produce new offspring via self-pollination — the best of both worlds!

The wonderful thing about pollination studies is that each plant species presents an entirely new puzzle to solve. Plus, these studies go far beyond theoretical interests alone. Not only do they cast light on the evolution of the unique Galápagos flora, but they will help us in saving endangered species from extinction by demonstrating how they fit into the ecosystem and, when threatened by human-induced changes such as encroaching invasive species, whether the loss of reproductive output is due to a lack of appropriate pollinators or other causes. Additionally, such studies might help in determining whether seeds produced via particular modes (say outcrossing versus selfing) provide an advantage in the long-term survival and success of a particular species. Thus, reproductive studies will continue to provide a vital tool in deciding how best to increase a particular species' chances of survival in the Galápagos Islands, for these plants are indeed much more than simple wretched-looking little weeds.

Sunflower Trees and Giant Cacti

Vegetation Changes over Time

Ole Hamann

University of Copenhagen,
Botanic Garden,
Øster Farimagsgade 2 B,
DK-1353 Copenhagen K,
Denmark.

Dr. Ole Hamann is Professor of Botany and former director of the Botanic Garden, University of Copenhagen, Denmark. His research in Galápagos deals with vegetation ecology and conservation. He is a member of the Charles Darwin Foundation (CDF) and has received the CDF Merit Award for his contribution to Galápagos conservation. He has edited *Ex situ conservation in botanical gardens*, co-edited a number of books, including *Botanic Gardens and the World Conservation Strategy* and *Botanical Research and Management in the Galápagos Islands*, and authored numerous publications on Galápagos plants and conservation.

EVER SINCE MY FIRST stay in Galápagos I have been fascinated by two of the largest members of the native flora: the beautiful sunflower trees (also called giant daisy trees) of the genus *Scalesia* (sunflower family) and the magnificent giant cacti of the genus *Opuntia* (prickly pear, cactus family). With respect to their adaptive radiation and uniqueness these two plant groups are among the many botanical equivalents of the famous Darwin's finches. Indeed, the evolutionary phenomena are as evident in the flora as they are in the fauna of Galápagos (see also Chapter: McMullen).

The genus *Scalesia* is endemic, found nowhere else

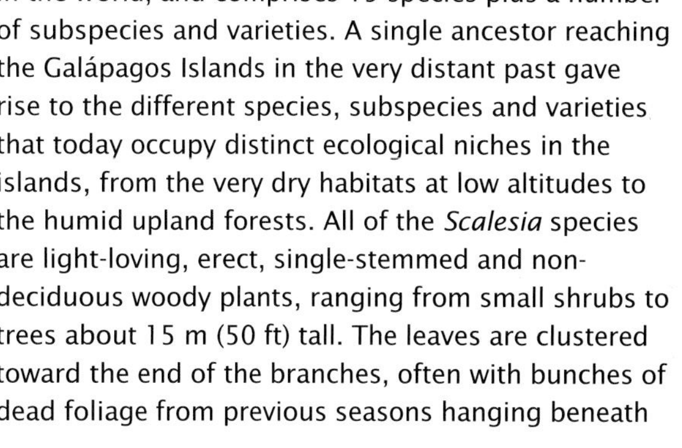

in the world, and comprises 15 species plus a number of subspecies and varieties. A single ancestor reaching the Galápagos Islands in the very distant past gave rise to the different species, subspecies and varieties that today occupy distinct ecological niches in the islands, from the very dry habitats at low altitudes to the humid upland forests. All of the *Scalesia* species are light-loving, erect, single-stemmed and non-deciduous woody plants, ranging from small shrubs to trees about 15 m (50 ft) tall. The leaves are clustered toward the end of the branches, often with bunches of dead foliage from previous seasons hanging beneath

LEFT: The dense forest of *Scalesia pedunculata* around Los Gemelos was nearly obliterated by El Niño, but its fast-growing cycle enables it to spring up again as long as invasion by introduced plants is thwarted.

the live tips, which give the plants a very characteristic look. The tiny white flowers are aggregated in small daisylike heads at the end of the branches. The fruiting bodies are small and are generally dispersed over only short distances. When compared to ordinary trees and shrubs, the life cycle of the *Scalesia* species is very short: they grow very fast, produce flowers and seeds after just a few years, and become senescent and die within a few decades. Although the evolution and phylogeny, or genetic lineage, of *Scalesia* have not yet been investigated in depth, it is reasonable to suggest that *Scalesia* have evolved from an ancestor with weedy characteristics. They are well adapted to the ever-changing volcanic environment of Galápagos.

All the *Opuntia* cacti in Galápagos are also endemic. There are six different species plus a number of distinct varieties. Their growth form ranges from prostrate to arboreal, but most of them are erect trees that, with age, develop magnificent, massive trunks with large, spineless, reddish plates of flaking bark. Some may grow to 10 m (33 ft) high or more and develop a crown with branched, greenish joints. They produce large yellow flowers in abundance at the beginning of the hot season, around December–January. In the wild, *Opuntia* cacti grow very slowly, and it may take them 50 years or more to reach maturity and start

ABOVE: Stately *Opuntia* woodland, Santa Fe.
BELOW LEFT AND RIGHT: Lush greenery during the 1983 El Niño (left) spelled doom for large numbers of cacti whose waterlogged roots rotted. Under normal conditions, the *Opuntia* habitat on Santa Fe, where the author monitors long-term changes, lays dormant during much of the year.

FLYCATCHER NICHES

Without a doubt the most eye-catching bird to be seen in the *Scalesia* forest is the brilliant male vermilion flycatcher, *Pyrocephalus rubinus* (top left), performing its territorial flight while the pale yellow female tweets from the mossy branches below. Darwin's specimens, collected in 1835, were described by ornithologist John Gould as a new species, *Pyrocephalus nanus*, due to its small size. Much more recently, an even smaller species, *P. dubius*, was proposed for the San Cristóbal form, although this bird has not been seen for many years and is presumed extinct as a result of major habitat destruction in the farming zone. Although quite distinctive in looks, both Galápagos types are currently regarded by most authors as subspecies only. Their nests consist of a small cup delicately attached to a mossy branch with spider silk, lichen and other finely woven strands.

The endemic Galápagos or large-billed flycatcher, *Myiarchus magnirostris* (left), though found from the coast to the highlands, is more typical of dense arid and transitional zone vegetation, where it stalks crawling insects by perching motionless while carefully observing its surroundings. The only cavity nester in Galápagos, in dry areas it is especially fond of hollows in dead or dying cacti, which are lined with feathers, wild cotton or other soft cushioning matter. Pairs spend a lot of time together, calling frequently and, being naturally curious, often follow people at close range.

Tui De Roy

flowering and fruiting. But once the adult state is attained, the individuals may live another 150 years or more, so some of the very large *Opuntia* are probably around 200 years old, or even older. The *Opuntia* are xerophytes (dry-loving) and well adapted to survive prolonged periods of drought. They are a primary source of food for many animals: land iguanas and giant tortoises eat the pads, and several species of Darwin's finches feed on the flowers, fruits and seeds, two of them having evolved bill shapes specifically adapted for easy penetration of these structures (see also chapter: Grant; McMullen). The animals also disperse their seeds and pollen, thus ensuring propagation. The *Opuntia* cacti are a very conspicuous element in arid areas, from sea level to high altitudes on all islands and major islets, and are the hallmark of dry landscapes in Galápagos.

Major natural events, such as volcanic eruptions, natural fires and the El Niño phenomenon, change the Galápagos vegetation over time, sometimes drastically. But since the arrival of humans a few centuries ago, the islands have also been subjected to grave disturbances caused by invasive alien species, manmade fires, and the resident human population's use of natural resources. The biological isolation of the Galápagos Islands has been broken, and many native ecosystems have been altered and modified (see also chapters: Parent & Coppois; Merlen). Consequently, the assessment of vegetation changes over time is of great conservation interest as a basis for deciding how ecosystems should be managed in order to maintain ecological processes within a natural regime, or as close to that as possible.

I first came to Galápagos in 1971 to be a staff scientist at the Charles Darwin Research Station (CDRS). One of my

ABOVE: A female small ground finch, *Geospiza fuliginosa*, incubates her eggs in a domed nest safely wedge between spiny *Opuntia* pads. LEFT: The author taking a break beneath one of his study subjects.

Photo courtesy Ole Hamann

tasks was to start investigations of the impact on the vegetation caused by introduced mammals run wild, and on the subsequent recovery when these animals were controlled or eliminated. Useful data on such changes can, over time, be obtained from the study of vegetation in permanent sample plots (quadrates). The first such plots were established by Tjitte de Vries of the CDRS in 1966, and I expanded these in 1971–72 by setting up a number of new sites in representative areas on Santa Cruz, Santa Fe and Pinta islands. Since then I have made numerous visits to the archipelago and have continued studying vegetation changes and the dynamics of plants like *Scalesia* and *Opuntia*.

ABOVE: An *Opuntia* cactus flower blooms for just one day.
BELOW LEFT AND RIGHT: Used by nesting frigatebirds on Wolf Island (left), the low, sprawling, soft-spined *Opuntia helleri* grows on Genovesa, Marchena, Darwin and Wolf islands, where there are no giant tortoises. Its growth habit is in stark contrast to the tall, flaky-barked trunks of *Opuntia echios* on Santa Cruz.

Sunflower trees on Santa Cruz

Scalesia pedunculata is the largest of the sunflower trees. It grows to 15 m (50 ft) tall and is often the dominant, or even the only, tree species in the so-called *Scalesia* forest growing in the humid zone at higher altitudes on Santa Cruz, San Cristóbal, Floreana and Santiago. This forest presents a beautiful green canopy formed by the contiguous umbrella-shaped crowns of the *Scalesia* trees, their trunks clad in epiphytic mosses, ferns and orchids, giving it a lush, fairy-tale aspect: an absolute contrast to the often stark landscapes and sparse vegetation encountered at lower elevations.

On Santa Cruz I have monitored a small section of *Scalesia* forest in the southwest part of the island regularly for over 35 years, from 1972 to 2007. The original vegetation in this area was a semi-evergreen mixed forest, with a canopy reaching about 8 to 12 m (26–39 ft) in height. *Scalesia pedunculata* was a prominent member of this assemblage, together with other native trees such as *Psidium galapageium*, the endemic guayava tree, and *Zanthoxylum fagara*, or cat's claw. The study showed that *Scalesia pedunculata* displayed a pattern of growth and survival not usually associated with forest trees: within 3.5 to 5.5 years of age the trees attained 7 to 8 m (23–26 ft), flowered and fruited. Mortality was very high initially because of the fierce competition among seedlings and saplings to reach the light and become established members of the canopy. The *Scalesia* trees were short-lived, and most individuals in the stand died when they were 15 to 20 years old, so the forest exhibited an extremely rapid turnover. All of this indicates that *Scalesia pedunculata* is a pioneer species, adept at colonizing new or disturbed ground, setting the scene early in the colonization

sequence, when the plant community is still relatively simple. However, because of a lack of competing late-succession trees, these characteristics enable it to dominate even once the ecosystem matures. In such a well-established forest the majority of the *Scalesia pedunculata* trees forming the canopy will be of the same age group. They are predisposed to a synchronous dieback when they reach senescence, such as when the forest is hit by a strong El Niño. Since there is little regeneration of *Scalesia* underneath a mature canopy, the new generation only develops when the old one has died off, and the cycle starts over again. In effect, two times during the study of this forest a so-called self-cyclic succession was observed, with a new cohort of *Scalesia* trees succeeding the old one.

The strong El Niño of 1982–83 had a pronounced impact on this forest. All the old *Scalesia* trees died during this nine-month period, as well as many of the *Psidium galapageium*. Then, after two years of drought, thousands of *Scalesia* seedlings emerged. However, very few of them survived to maturity, because at the same time an invasion of alien plant species took place, in particular introduced guayava trees (*Psidium guajava*) and Cuban cedar (*Cedrela odorata*). The result was that during the next decades these aggressive invaders outcompeted the *Scalesia*, primarily because they formed a rather dense, shading canopy, which prevented seedlings and saplings from reaching maturity. Since 2003, no *Scalesia pedunculata* specimens have been recorded in the plot, and by 2007

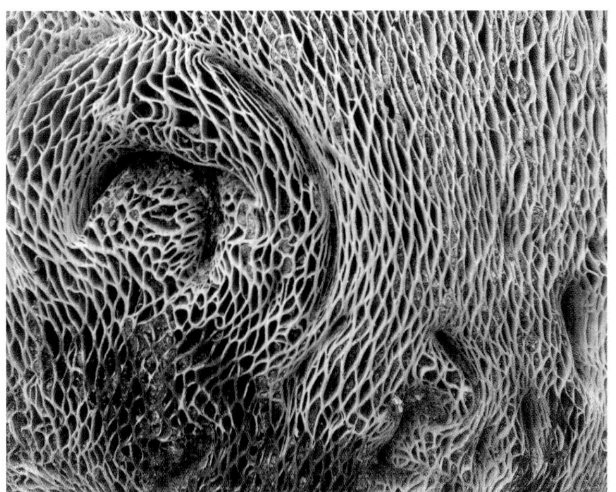

ABOVE: The survival of the Santa Fe land iguana, *Conolophus subcristatus*, rests on obtaining moisture from fallen cactus pads during long droughts.

LEFT: The remains of a dead cactus reveal a fibrous framework giving the trunk its rigidity without any real wood.

BELOW: Adapted to life under arid conditions, an Española Island saddleback tortoise, *Geochelone hoodensis*, is able to stretch high in order to reach juicy pads.

ABOVE: A mature *Scalesia* forest forms a closed canopy.
ABOVE RIGHT: *Scalesia villosa* flower detail, Floreana.
BELOW: *Scalesia pedunculata* flowers on thin stalks.

it had been transformed into a *Cedrela* forest with an understory of smaller guayava trees.

Thus, the combined effect of El Niño and invasion of aggressive alien plant species effectively eliminated the original forest from this part of Santa Cruz Island. Fortunately, there are other areas in the Santa Cruz highlands where the *Scalesia* forest has not suffered such a massive takeover, although other, more insidious invaders, such as the hill blackberry *Rubus niveus* and elephant grass, *Pennisetum purpureum*, are infiltrating its understory and gradually changing the ecosystem dynamics and so thwarting its natural ability to regenerate.

Giant cacti on Santa Fe

Opuntia echios var. *barringtonensis* only occurs on the small, uninhabited island of Santa Fe. It is the most impressive of the Galápagos *Opuntia*, reaching a height of 12 m (39 ft) and developing a massive trunk up to 1.25 m (4 ft) in diameter. The conspicuous presence and dominance of these large cacti is a very distinct feature of this island, especially on the flat plains.

Photo courtesy of Ole Hamann

Feral goats had been present on Santa Fe for perhaps 200 years before they were completely eliminated by the Galápagos National Park Service (GNPS) in 1971. The alien browsers had a strong negative impact on the vegetation, so after their removal it was expected that many severely depleted plants — and the vegetation in general — would gradually recover. In 1972 I established a series of permanent plots in order to follow this anticipated recovery over time. Since then these plots have generated data on the life and death of the *Opuntia* cacti and of some other common woody species on the island.

It turned out that strong El Niño events also had a profound impact on *Opuntia*, and at the same time caused a change in the way the vegetation recovery was developing. The incipient regeneration of *Opuntia* taking place after the removal of goats was actually stopped or even reversed: *Opuntia* do not tolerate prolonged periods with excess water, and during the major El Niños of 1982–83 and 1997–98, their populations on Santa Fe (and all over the archipelago) were decimated. The waterlogged soil killed the roots of all sizes of *Opuntia*, and many large individuals also died because they became so top-heavy from water engorgement that they toppled over, while juveniles and seedlings were smothered by heavy growth of vines and herbs. By 2007 there were actually fewer *Opuntia* individuals in the permanent plots than when the study began 35 years earlier.

In the same timeframe, common native shrubby species such as *Cordia lutea* and *Lantana peduncularis* increased in abundance. The El Niño events boosted the shrub populations, so that by 2007 they had increased to three times the level recorded just after goats had been removed. In effect, on the plains of Santa Fe, the vegetation had changed over time in a way not foreseen: there are now fewer *Opuntia* but many more bushes, so that over the 35 years of study the vegetation has altered in structure, although not in composition. Thus, vegetation recovery on Santa Fe began after goats had been eliminated, but El Niño events turned out to have enduring consequences for that recovery process.

Research and conservation

Time factors are of great importance with regard to natural and anthropogenic (human-induced) disturbances and for conservation of the natural environment. The kind of alteration taking place, and its duration, are crucial for both the vegetation overall and for individual plant species within it. Different vegetation types show distinct levels of vulnerability and resilience in relation to disturbances, and species recovery patterns may differ significantly, as exemplified by *Scalesia* and *Opuntia*.

Some of the main results of this long-term study found that *Scalesia pedunculata* and other species in this genus display high mortality, rapid turnover and relatively short life expectancy, while *Opuntia* cacti, such as those found on Santa Fe, along with other arid zone trees, display low mortality, slow turnover and long life expectancy. Consequently, conditions that favor one species may at the same time negatively affect another, while adults, juveniles and seedlings within a species may react differently to the same environmental factors.

El Niño events had a pronounced impact on the individual species as well as the vegetation generally. Our study of *Scalesia pedunculata* revealed for the first time that synchronous, stand-level diebacks occur in Galápagos and that such diebacks may be triggered by a strong climatic event like El Niño. The study on *Opuntia echios* var. *barringtonensis* documented that the frequency and strength of the El Niño may in the long run alter the typical arid vegetation on Santa Fe.

The life cycle characteristics of the *Scalesia* species make them vulnerable to persistent, long-term disturbances. For example, the presence of goats in a *Scalesia* forest over just a few decades may prevent natural regeneration when all seedlings and saplings are eaten, and eventually the seed bank in the soil is lost. During the same period, the mature *Scalesia* trees may die of old age so that no new seeds are produced and no recovery will be possible. An invasion of aggressive alien plants can result in the same impact as that caused by goats, because some invasive plant species can outcompete and outshade the *Scalesia* trees.

In contrast, the life cycle characteristics of the *Opuntia* cacti enable them to survive a long-lasting disturbance such as the presence of goats: they can persist for many years as adults, even though natural regeneration during that time may be prevented. Once the goats are removed, a number of mature cacti may still be present and be flowering and producing seeds, so recovery is then possible.

Thus, long-term study has provided new insights into the ecology of Galápagos vegetation and particular species, and at the same time it has generated valuable knowledge for the practical conservation and management of the ecosystems in the future.

ABOVE: Mosses, orchids and other epiphytes festoon mature *Scalesia* trees in the Santa Cruz highlands, helping to capture fog-drip, which in turn nurtures the forest.
FAR LEFT: Fog condenses on *Peperomia galapagensis* and mosses.
LEFT: Passionflower vines flourish in the understory.
BELOW: The forest is good daylight hunting habitat for the short-eared owl.

On the Snails' Trail

Evolution and Speciation Among a Vanishing Tribe

Christine Parent and Guy Coppois

Section of Integrative Biology,
University of Texas at Austin,
1 University Station C0930,
Austin, TX 78712,
United States.
<cparent@mail.utexas.edu>

Dr. Christine Parent is an evolutionary ecologist, now a postdoctoral fellow at the University of Texas at Austin. She is interested in the patterns and processes of adaptive radiation, particularly in island systems.

Dr. Guy Coppois is a malacologist. He started investigations on Galápagos terrestrial malacofauna in 1973, focusing on systematics, distribution, ecology and adaptation. A member of the Charles Darwin Foundation (CDF) General Assembly, he served on its Board of Directors (Belgian representative) from 1987 to 2003. He is Professor of Biology at The Brussels Free University.

Laboratoire de Systématique
et d'Ecologie, Faculté des
Sciences, CP-160/13,
Université Libre de Bruxelles,
50 av. F.D. Roosevelt, B-1050,
Bruxelles, Belgium.
<gcoppois@ulb.ac.be>

Christine's experience 1999–2005

FOR OVER EIGHT YEARS now I have been studying the endemic land snails of Galápagos, during which time these creatures have truly 'grown on me.' I first came across their trail in 1999, when I spent five months on Galápagos doing research on two introduced predatory wasps. Toward the end of my field season, in the arid zone of Santa Cruz Island, I was sitting on a lava rock observing *Polistes versicolor* wasp nests when I noticed tiny white shells littered about. It was the first time they had caught my eye, but soon I began to see them almost everywhere. The shells were quite small, at most 1 cm (0.4 in), so it was no surprise that I hadn't seen them earlier. As I would learn later, not many people had noticed them before either. Now that I had a few in my hand, I could tell that these were old shells left behind by their builders a long time ago. They were pure white, bleached through long hours lying in the sun; a light pressure of my fingers would easily crush their fragile whorls — they were simply beautiful.

Later, at the Charles Darwin Research Station (CDRS) library, I came across a revealing paper published in 1966 by Allyn Smith describing the 'Galápagos land-snail fauna [as] unique in many respects.' According to Smith, there were over 50 species in one genus alone, known as *Bulimulus*. Over 50 species?! Among the famous examples of adaptive radiation in Galápagos, Darwin's finches are the most celebrated yet consist of *only* 13 species. The largest other group I knew of were beetles, with 20 species or so. I was shocked — how was it possible that I never noticed the snails before? But I was also very excited: maybe this *Bulimulus* genus could be a great topic for the study of evolutionary diversification, my main scientific interest.

After I returned to Carleton University in Canada to finish my Masters thesis on wasps, I started thinking about pursuing a Ph.D. in evolutionary ecology, and, with my supervisor's encouragement, decided to study the Galápagos bulimulid land snails. But remembering the empty shells on Santa Cruz, I became concerned upon reading that a lot of the species were now possibly extinct. Dr. Guy Coppois had done extensive work on these snails in the 1970s and 1980s and stated 'land snails are endangered, and many species are already extinct.' But other reports were claiming that not all had so far been discovered, as every single scientific expedition that looked for snails had returned with additional undescribed forms. Even while they were going extinct, I hoped that enough snails were left to study the most species-rich radiation known on Galápagos.

RIGHT: Typically clinging to low plants, live *Bulimulus* snails are still fairly common on the rain-drenched slopes of Isabela's southern volcanoes.

In November 2000, I was back on Santa Cruz Island with a simple goal: I had 10 days to find samples of as many species of bulimulid land snails as possible to bring back to Simon Fraser University in Canada, where I had started my Ph.D. a couple of months earlier. I needed live specimens to be able to extract DNA from their fresh tissues. The same strand of DNA for each specimen would then be amplified and sequenced so that I could read a small part of the genome of every individual I collected. Comparing the DNA, I would determine the evolutionary relationship among the different species, in the same way you can determine how closely related individuals are based on morphological resemblance. Once a phylogeny (a tree representing the sequence of species formation in a group) based on the DNA sequences is obtained, it becomes possible to look back in time and infer their evolutionary history: What is the sequence of species formation in this lineage? Does it match the geological sequence of island formation? Are species often formed within an island or do they always need to colonize a new island to split definitively from an ancestor? Those were among the many evolutionary questions I was hoping to solve. In addition, my goal was to gather enough information about bulimulid populations to help protect this unique island fauna.

By this time, I had learned that there are over 80 species and subspecies described in the genus. On Santa Cruz Island, this diversification had reached such extremes that almost every hill or valley might harbor a separate type. I also knew that the snails could potentially be found at all elevations (except in the littoral zone along the shore) and on all major islands. According to early reports, the difficulty of finding these small, dull-colored species leading cryptic lifestyles should be compensated by their omnipresence. I figured that the best chances to find snails alive were in the moister environments, so with the help of the Invertebrate Department staff at the CDRS, I started my search on Santa Cruz Island, walking through the *Scalesia* forest in the humid zone. I knew that Guy Coppois had found up to 11 species at a single locality in this forest. However, after spending three entire days in the highlands, walking around and searching everywhere I could imagine, I found plenty of empty shells but not a single live snail.

Once again I worried that working on these snails might not have been such a great idea after all. With the major declines already reported by Guy, maybe I was just too late. He had noted that introduced species such as rats and little fire ants (*Wasmannia*

ABOVE: Up to seven bulimulid species once shared the lush *Scalesia* forest floor habitat in the lee of the Santa Cruz summit chain, but all have mysteriously disappeared in recent years.
BELOW: A *Bulimulus ochsneri*, once extremely common around the village of Bellavista and throughout the Santa Cruz highlands.

DEEP-SEA DISCOVERIES

Strombina deroyae

Paziella galapagana

Pteropurpura deroyana

Latiaxis santacruzensis

Fusinus allyni, named after Allyn G. Smith of land snail fame.

The marine mollusks of Galápagos, though more varied than their terrestrial counterparts, are in many ways just as fascinating and unique. In the late 1960s and 1970s, the De Roy family conducted extensive dredging explorations of the deep-sea floor between the central islands, using a lightweight rig entirely designed and handcrafted by André (fishnet dredge and hand-crank winch with 1200 m, or 3900 ft, of cable mounted on the stern of their small boat). Reaching depths of 600 m (1970 ft), numerous unknown gastropods were revealed, many of them extremely delicate in form. Jacqueline corresponded with various museum specialists who eventually described dozens of species new to science. Even though the De Roys advocated the use of descriptive or location names, which they felt would have more enduring significance in the nomenclature, several are nonetheless named in their discoverers' honor. No such dredging has ever been repeated since.

Tui De Roy

BELOW: Some of the most exquisite *Bulimulus* shapes and colors could be found on Floreana, such as *Bulimulus rugulosus*.

auropunctata) were preying on the snails and, at the same time, goats, donkeys and pigs were destroying their preferred habitats. I developed a routine of turning over lava rocks, lifting the bark of trees, rolling dead trunks, scratching in the leaf litter and even digging the soil. Usually after a few minutes I would have one or two mockingbirds join me, intent on snatching the tiniest moving creatures I might uncover. Endemic flycatchers would come after the hair in my ponytail, I suspect looking for exotic nest-building material.

One morning, I started my search at low elevation in the arid zone near town. The first rocks I turned revealed shells that were dark brown instead of the usually bleached white. My heart started pounding. I dropped my backpack, as well as my GPS and my notebook, and started frenetically lifting every lava rock I could find in the area. And there they were! Finally, I had found a few live specimens. In my head everything was going very fast: I could now return with some samples, get DNA from them, convince my supervisor that this project would work, return for several complete field seasons to Galápagos, and do my Ph.D. on a subject that I now truly loved! When I left Galápagos at the end of those 10 days, I had specimens of seven species from three different islands. This was all I needed to start developing methods towards a robust phylogeny to eventually describe the relationship among species in this spectacular group.

In January 2001, I resumed snail hunting on Galápagos, and this time stayed for three months. I went back successively in 2002, 2003 and 2005, spending overall more than two years looking for bulimulid snails. During that time, I visited most of the major islands and some smaller islets, collecting over 40 species. Along with the snails, I brought back fond memories of the unique fieldwork on Galápagos. Although the fauna is reputed to be inoffensive and tame, I managed to get bulldozed by a marine turtle while sleeping in my tent on the beach, have a Galápagos hawk, who felt that I was trespassing its territory, hit my head with his claws, and be stung and bitten by mosquitoes, wasps, scorpions and centipedes. However, these 'little' inconveniences, together with the difficulties of doing fieldwork in a place where water is scarce and the sun is unforgiving, have not spoiled my unique experience on these 'Enchanted Islands.'

Guy's experience 1973–2003

If Christine had such a hard time finding live snails, this situation was not always the case. Although scant attention was directed toward invertebrates during early expeditions, museum collections nonetheless testify that collectors picked up live specimens easily, mainly on San Cristóbal and Floreana islands. During the 19th century, human settlements and invasive introduced mammals brought deep modification to the natural habitat, and one can assume this to be the main reason for gastropod disappearance. Santa Cruz Island was colonized much later, with critical habitat destruction only beginning early in the 20th century. Populations of live bulimulids were still common when the CDRS was established and even when I initiated my work.

My field investigation on Galápagos snails started with an intensive three-year stint in September 1973, occupying the government-sponsored 'Belgian seat' at the CDRS, complemented by shorter trips up to 2003. The focus of the study, initiated by Professor Van Mol, was on species distribution, comparative ecology and systematics, with the final aim of gathering enough elements to build a clear picture of the adaptations

and evolution of this complex group.

Part of my plan was also to work with the collections of Allyn G. Smith of the California Academy of Sciences who, as Christine mentioned, had already accomplished a major taxonomic revision of Galápagos terrestrial gastropods. Unhappily, this never happened, as he died just days before I arrived in San Francisco to meet him in September 1976. But Allyn had received much prior information on collecting sites and bulimulid ecology from a long-time Galápagos resident André De Roy, with whom I was lucky enough to collaborate.

André was an accomplished naturalist and an excellent observer of tiny details. He had an artist's eye and the skills that go with practical imagination. His wife Jacqueline also made major contributions by networking with experts around the world. Together with their children, Gil and Tui, the De Roy family assembled an incredible bank of observations on Galápagos flora and fauna when exploring the islands. André was fascinated by bulimulid land snails, and noted many details I was able to confirm in my own findings later. His experiences and observations on Galápagos land snails were exceptional, and we had unforgettable hours of discussions.

I first started work in the arid zone on CDRS territory. Even in the dry season, I soon found live specimens of pale brown *Bulimulus akamatus* hiding in cavities under lava blocks, while *B. reibischi*, slender white with a wrinkled surface, was estivating (dormancy similar to hibernation, but induced by drought rather than cold temperatures) usually on, or under, tree bark. Every month, I collected in this area over a full year, using quadrate sampling techniques for comparison with other sites chosen in the various vegetation zones at different elevations. Today, this original sampling spot sits partially beneath the CDRS meeting hall and, as far as I know, living snails have vanished from the whole area. It is sad to recall, too, that in this same place I also found the well-preserved remains of another extinct species, the giant rat *Megaoryzomys curioi* [Lenglet & Coppois, 1980]. Such experiences remain in your mind forever.

In the 1970s, the variety of bulimulid land snails on Santa Cruz was extraordinary. It was so much more complex than what was observed on the other islands, though it is possible that San Cristóbal and Floreana may have been comparable before human settlement, but this we will never know. On Santa Cruz, many species were still abundant, some distributed over huge areas, others limited to tiny ranges, such as a single valley or one side of a hill characterized by a local microclimate or habitat. For this reason, I spent a lot of time crisscrossing the island, analyzing distribution patterns along altitudinal transects and taking monthly samplings to be able to compare populations. GPS was not available at the time, so

ABOVE: Many *Bulimulus* species evolved among Floreana's numerous volcanic cones.

ABOVE LEFT: Old, bleached shells reveal that even minute islets, such as Devil's Crown near Floreana, once harbored thriving land snail populations.

BELOW: From luxuriant highland habitat to arid lowlands, dozens of species adapted to tiny microhabitats on Santa Cruz.

ABOVE: A live *Bulimulus* clings to a fern frond on Cerro Azul, southern Isabela.
BELOW: As young volcanoes continue to grow, a quiltwork of lava flows such as the slope of Wolf Volcano, represent insurmountable barriers dividing populations and driving evolution.

I relied mainly on a compass. For precise transects or quadrate positioning, I used Topofil, a measuring string in 5-km (3-mile) spools (precision 10 cm, or 4 in), much to the delight of my helper who recycled the strong cotton thread once the measurements were done. But I did not forget the younger islands, and, as a final result, I was able to gather over 80 *Bulimulus* taxa from the whole archipelago.

In 1973, the road crossing Santa Cruz was still under construction as I started surveying the Cerro Maternidad area near the Gemelos craters. From this hill, I could hear dynamite blasts as a straight line was being carved through the *Scalesia* forest — the future connection to Baltra airport. This hill, or at least its southern slope, was very interesting to me, harboring a healthy refuge population of *Bulimulus cavagnaroi* at the western limit of its range, as indicated by bleached shells elsewhere in the *Scalesia* forest. Another population lived in a small central valley near Cerro Coralon, mixed with *B. gilderoyi*, a species named in 1972 by Van Mol after its discoverer (André De Roy's son). This species was only ever encountered in this tiny lush depression and has since vanished. I was often alone when working in those remote places (this was still permitted in those times), and I enjoyed the little Galápagos rails (*Laterallus spilonotus*) that often came to take a closer look at me crawling through the undergrowth. Sometimes, they were as close as a foot and a half from my face.

On the leeward side of Cerro Crocker and Puntudo (north slope), snail distribution followed fast-changing vegetation gradients, with many species restricted to horizontal bands characterized by precise local climatic conditions. For example, at its upper limit the humid *Scalesia* forest gave shelter to species like the common *Bulimulus ochsneri*, with up to seven sympatric species (species with a common or overlapping range) out of a total of 11 in the region. Further down, *B. tanneri* enjoyed a much wider distribution in the northern transition and arid zones.

But already there were early signs of human-induced destruction. In the rolling fern/sedge zone, I located only a few bleached shells along with charcoal, signs of uncontrolled fires set across the highlands in the 1960s. On the south slope where National Park land is replaced by farms, in my first year I observed one of the last populations of *Bulimulus blombergi* still alive, one of the largest species at 20 mm (0.8 in). A small strip of *Scalesia* forest survived at El Occidente along the rough road to the western farmlands. Here, the conspicuous white crinkle-shelled snails hung from trunks and low branches. All around, original native vegetation was already replaced by grassland for cattle. Less than a year later, the forest was still there but the undergrowth was drier and the ground littered with empty shells. Sadly for these endemic gastropods, what had become farmland was, for them, probably the most richly adaptive area in the whole archipelago.

Habitat destruction and the negative impact of introduced predators such as little fire ants (*Wasmannia auropunctata*), rats and possibly mice, plus competitors like the recently introduced slug pest *Vaginulus plebeius* (Veronicellidae) have all taken their lethal toll. In my view, these are the main reasons explaining most extinctions in this group, although for some species other causes are still to be found, especially where the above doesn't appear to apply, such as the small islets off Floreana. The process of speciation within each island was possible because of the complexity of available habitats and inter-island colonization events, probably also aided

by volcanic activity periodically redefining natural boundaries. But there are many challenging dilemmas that remain unsolved regarding this extraordinary level of adaptive radiation. Unhappily, present conditions have outpaced the survival ability of many *Bulimulus* species, and perhaps it will be too late to answer those questions.

How many species are extinct in this endemic group of gastropods? It is difficult to declare when a species is extinct, and one should always keep in mind that perhaps somewhere a few specimens are still alive ... for a while. I discovered this in my search for one particularly emblematic species, *Bulimulus achatellinus*, known only from a few shells. Hugh Cumming visited Galápagos on the yacht *Discoverer* in 1829-30 and collected one live specimen on San Cristóbal Island, most probably inland from Fresh Water Bay (no precise location was given). Another live specimen was collected in 1890, and five shells were found in the 1930s, all in the same area. After almost a century with no live *B. achatellinus* seen, I made two expeditions to carefully search the site, without success. But on my third attempt, in 1987, I hit gold. I had not much time, riding an old motorcycle on a day trip, which was not easy as it had rained in the highlands and the trail was slippery and muddy. Coming out of Cerro Verde village, the trail was narrower and I intended to take samples a little further along, when suddenly I saw it. No mistake was possible — I recognized its characteristic shell from a distance, attached to the leaf of a vine covering the bushes. Unbelievable! I almost fell from the motorcycle, stopped the engine and after calming down searched all surrounding vegetation and ground: only two live specimens were found ... not even old shells on the ground.

The best part of any study is often the field time, but this must be followed by analyses in the laboratory. These showed that vegetation structure and composition in microhabitats are determining factors in the local distribution of sympatric species, and that climate-driven zonation is also important on a bigger scale. Comparisons between taxa were made using biometric methods including multivariate analyses of the shells and anatomical data (factor and discriminant analyses). With no X-ray available, I had to draw each shell to scale, using a WILD binocular microscope fitted with a camera lucida, complemented by anatomical data mainly concerning the genital tracts. These analyses led to an intra-generic pattern of relationships between the species that show similarities with the results Christine found later using DNA analysis. But this is another story.

ABOVE: A tiny islet inside Beagle Crater on Isabela could become an opportunity for speciation. BELOW: Land snails among fern spores, Alcedo Volcano, Isabela.

From exhilarating discoveries to sobering conclusions — a joint view

Phylogenetic analyses were used to elucidate the relationship between species within this group, and to determine their sequence of colonization and

LEFT: André De Roy's collections show general characters unique to each of three islands, whereas species diversification reflects the islands' age and habitat opportunities: least diverse on young Isabela (opposite page), most colorful trend on Floreana (middle), and greatest variety on Santa Cruz (left).

ABOVE: *Bulimulus gilderoyi* was restricted to a small valley near Cerro Coralón (top), Santa Cruz.
BELOW: *B.* sp. nov. *tuideroyi* (top) disappeared before being formally described; *B. blombergi* (bottom) vanished in the 1970s. All shown life size.

speciation on Galápagos. Results show that bulimulids arrived first on Española or San Cristóbal (the oldest currently existing islands). The lineage then colonized younger islands successively, as shown by the younger species found there. The total number of species found on a given island is determined by a combination of biogeographical factors such as island area, elevation and habitat diversity, all of which have a positive effect on bulimulid species diversity. However, it is striking to note that the young islands of Fernandina and Isabela together form over 60% of Galápagos total land area, but only 12 (about 17%) of the 71 total described species in the genus *Bulimulus*. Although the relationship between species diversity and island age is marginally nonsignificant, this pattern suggests that at least some of the youngest islands have not reached equilibrium in species diversity.

A surprise was to discover that inter-island colonization is not such a rare event, and although some island assemblages consist exclusively of taxa formed within-island, most stem from the combination of at least two to three colonization events followed by further speciation. Interestingly, by separating island diversity into species coming from other islands and species formed in situ, a clear biogeographical pattern emerges. Species diversity resulting from colonizers is mostly determined by geographical distance between islands, along with island size. In other words, it is dictated by how far and how big the target islands are. Conversely, for species resulting from within-island speciation, the only determinant factor is plant diversity. In a sense, for bulimulid species to form within islands, it doesn't matter how large, how old, how far, how high an island is, the key is habitat diversity for new species to exploit. This finding is significant because it highlights the importance of considering the process of species diversification when trying to understand what factors promote and maintain biodiversity, with broad implications for many islandlike systems and fragmented habitats. Such systems are becoming the norm in our modern world.

Between us, we also unraveled some of the mysteries of the amazing morphological diversity amongst bulimulids. Although generally small (under 3 cm, 1.2 in) and dull (ranging from white to dark brown), their shapes often vary widely, even in the same location, though certain types seem to prevail at some elevations. After detailed morphometric analyses, and later by taking radiographs of over 2000 shells belonging to some 40 species (using an X-ray instrument to measure each shell inside and out), a pattern emerged. Species with 'slender' shells are more likely to be found at lower elevations in the arid zone, whereas those with 'fatter' shells tend to

STORYTELLING SHAPES

X-ray reveals adaptations to climate zones, with taller spires and smaller apertures prevailing in arid coastal areas, and wide-bodied shells in humid highlands. The bubble-like shapes sheltering inside the empty *B. olla* are minute *Tornatelides* land snails, a group distributed on widely dispersed Pacific islands, probably carried by birds.
Tui De Roy

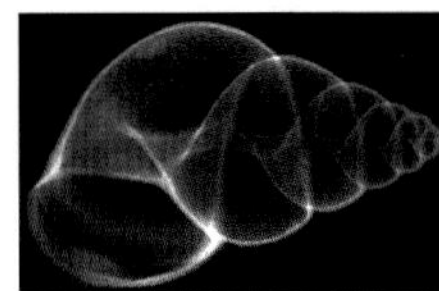

Bulimulus ustulatus, **Floreana**

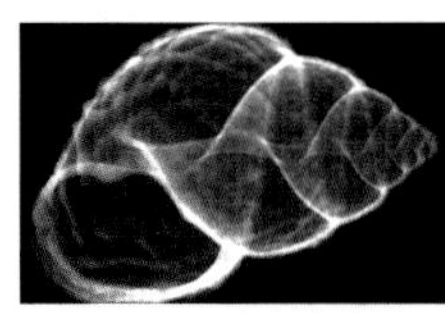

Bulimulus darwini, **Santiago**

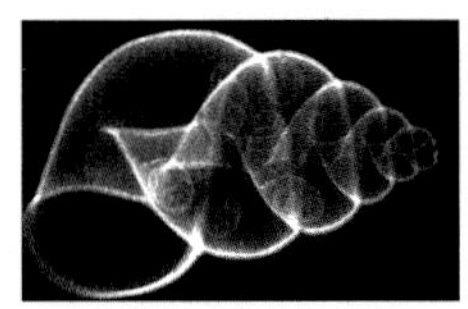

Bulimulus olla, **Santiago**

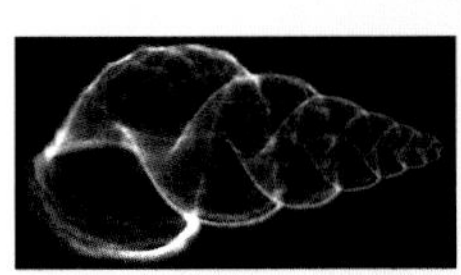

Bulimulus sculpturatus, **Santiago**

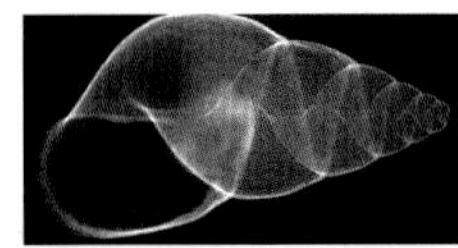

Bulimulus eos, **Santa Cruz**

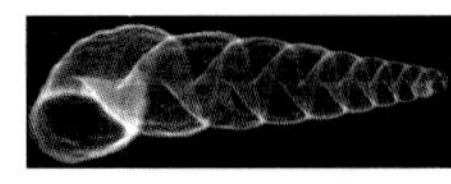

Bulimulus reibischi, **Santa Cruz**

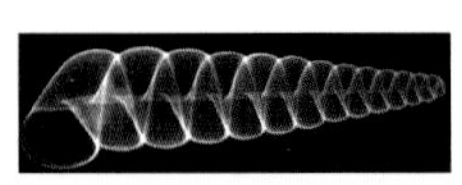

Bulimulus chemnitzioïdes, **San Cristobal**

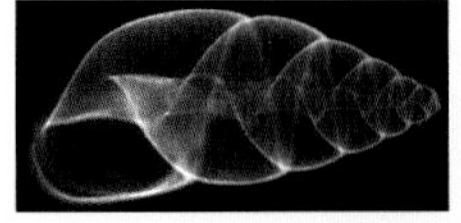

Bulimulus planospira, **San Cristobal**

Photos courtesy Christine Parent

occur in the humid highlands. One hypothesis is that even though the least 'costly' shell to build (in terms of material required versus living space within) is a blunt conical shape, in the very harsh environment of the arid zone a slender shell allows the skinless mollusk to maintain its shell opening as small as possible thus reducing moisture loss. When testing for this shape-elevation relationship further *within* individual species, these patterns paralleled the morphological trend across the entire radiated group, providing evidence for the continuity between the micro and macro evolutionary scales, a concept still sometimes debated.

Because of our jointly acquired knowledge of bulimulid snails, we are now in a better position to assess the overall health of the different populations. Our work has led to the update of their endangered status, with a total of 57 species now on the IUCN Red List of Threatened Species. On Santa Cruz Island in particular — the hub of tourism and concurrent resident human population expansion — the rich bulimulid fauna has largely disappeared during the past 30 years. Twenty-seven taxa were studied on this island in the early 1970s, and just a decade later no live populations could be found of all but a few. In highland areas in particular, both in the farmland and native habitats where numerous species were once extremely abundant, no live snails can now be found.

Another striking observation is that Santa Cruz, San Cristóbal and Floreana islands, where human settlements have been established longest, are the most seriously affected, with 32 species either categorized as endangered or critically endangered among the 43 species that have been assessed for these islands (another 17 remain to be assessed). Extensive defoliation by goats can have a severe impact too, as noted on Alcedo Volcano on Isabela, where at least four species were rapidly reduced to remnant patches. But there are hopes for their recovery in the wake of recent goat eradication. On Santiago Island, snail species composition appears to be relatively intact, despite the extensive past habitat destruction due to introduced goats and pigs, now reversed through eradication over the last decade. Similar efforts may have come too late on dry islands such as Española and Santa Fe, where few, if any, live snails are present even after goat removal and subsequent spectacular vegetation recovery.

Habitat loss is clearly a continuing threat. On Floreana and San Cristóbal, while farmed areas have become devoid of live bulimulids, remnant patches of forest still host populations of a number of species endemic to these islands, and there remains the opportunity to protect them at these sites. Land snails can be directly impacted by many invaders, especially potential predators, or indirectly due to habitat transformation by introduced plants. In one study, we found that the number of introduced plant species on a given island was a significant predictor of the proportion of endangered and critically endangered bulimulid species, though it is unclear whether this is a causal relationship or if another independent factor is responsible for the correlation. Indeed, live snails are almost extirpated from the remnant *Scalesia* forests and other areas on Santa Cruz that appear to be otherwise intact, the wave of extinction having swept across the island with little regard to apparent human disturbance. With scant direct evidence at this point to implicate specific causes, the remarkable disappearance of the bulimulid fauna remains largely a mystery.

Like many other land snails, bulimulids probably do not need large patches of habitat to survive. However, species with greater density and wider distribution have higher chances of long-term survival. We are now using a combination of population genetics and biogeographical analyses to better understand the potential impact of reduction and partitioning of suitable habitat on the genetic diversity and structure of different populations. Ultimately, we hope to be able to better define the habitat needs of the different species in this remarkable group so that more informed decisions can be made toward their protection.

On a few islands, there is still a chance that this fascinating mollusk fauna may be preserved, but continued efforts to understand the causes of their demise on islands such as Santa Cruz is necessary. Until those causes are better understood, habitat protection is all that can be accomplished, and it may not be enough.

ABOVE: We may never know how many species once inhabited the tranquil landscape of San Cristóbal, one of the oldest islands but also the most affected by human colonization. Of 57 known Galápagos *Bulimulus*, the majority have not been seen for years, with 26 species considered Critically Endangered, 22 either Vulnerable or Endangered, and the remainder awaiting assessment.

BELOW: *Bulimulus achatellinus* of San Cristóbal was only ever known from a total of nine specimens, one collected in 1829-30, one in 1890, five during the 1930s and the final two found by the author in 1987, offering a glimmer of hope for other 'lost' species.

Inshore Fishes
The Case of the Missing Damsel
Jack S. Grove

c/o Zegrahm Expeditions,
192 Nickerson St. #200,
Seattle, WA 98109,
United States.
www.JSGrove.com
<jsgIMAGES@aol.com>

Dr. Jack Grove is a marine biologist and professional naturalist, and is the senior author of *The Fishes of the Galápagos Islands*, published by Stanford University in 1997. He is also a Research Associate in the Section of Fishes at the Natural History Museum of Los Angeles County and a Founding Partner of Zegrahm Expeditions. Jack is a member of the General Assembly of the Charles Darwin Foundation (CDF) and a member of the Darwin Scientific Advisory Council of the Galápagos Conservancy. When not traveling, he manages his own photography business in the Florida Keys, where he lives with his wife Paulina.

BELOW LEFT: Impressive schools of very large and extremely curious almaco jacks, *Seriola rivoliana*, used to keep divers company.
BELOW RIGHT: Colorful king angelfish, *Holacanthus passer*, mingle with yellowtail surgeonfish, *Prionurus laticlavius*.

IT WAS MARCH OF 1983 and Galápagos was in the grips of the most powerful El Niño Southern Oscillation (ENSO) event ever recorded. The islands experienced record-breaking rainfall and the repercussions of the extraordinarily high sea surface temperatures (+4.5°C, or +8.1°F) were to be associated with droughts in Australia and Africa and storms from Polynesia to the western shores of the Americas [Robinson & del Pino, 1985]. The biological impact of the 1982-83 ENSO in Galápagos was best documented in the terrestrial realm: the carcasses of pinnipeds, seabirds and marine iguanas would bear witness to its severe negative effects (see also chapters: Romero & Wikelski; Trillmich). But the marine ecosystems of the archipelago were poorly understood, and the disappearance of a damselfish for the most part went unnoticed. Known only from the Galápagos and Cocos Island, the Galápagos damsel (also known as blackspot chromis or blackspot damsel), *Azurina eupalama*, may be the emblematic 'canary in the coal mine' signaling change in the eastern tropical Pacific.

El Niño is a warm surface flow that moves in from the northeast and usually appears off the coast of Peru and Ecuador around Christmas time (see also chapters: Sachs; McMullen). Every two to seven years, the warm flow is inherent with a large-scale climate fluctuation that may persist for up to 18 months. These quasi-periodic ENSO events have a dramatic impact on all life in the Galápagos. What triggers the ENSO remains unclear; however, a connection to global climate change is not only plausible, it is likely. The link between the evolution of marine iguanas and the ENSO has already been established [Steinfartz et al, 2007]. The ongoing search for the missing damsel will garner more interest as the interconnected nature of Galápagos fishes and global climate change becomes more evident.

There are numerous literature references to Charles Darwin being the first to collect Galápagos fishes. But it was William Beebe who was first when it came to seeing their vivid colors under the surface of the sea. His hard-hat dives from the deck of the *Norma* in 1923 represent the birth of marine ecology in Galápagos. His accounts of the life he observed beneath these enchanted seas continue to inspire visitors and marine naturalists and biologists. Indeed, until the doors of opportunity were opened by the advent of scuba, our capacity to study the behavior of fish and other marine life in their natural Galápagos habitats was limited.

Photo courtesy www.JSGrove.com

Mark Jones

ABOVE: Female streamer hogfish, *Bodianus diplotaenia*, and rainbow wrasses, *Thalassoma lucasanum*, in a typical reef scene.
LEFT: One of the only known photographs of the Galápagos damselfish, *Azurina eupalama* (taken by the author), a species not seen since the 1982–83 El Niño.
BELOW: Red-lipped batfish, *Ogcocephalus darwini*, (left) and spotted eagle ray, *Aetobatus narinari* (right).

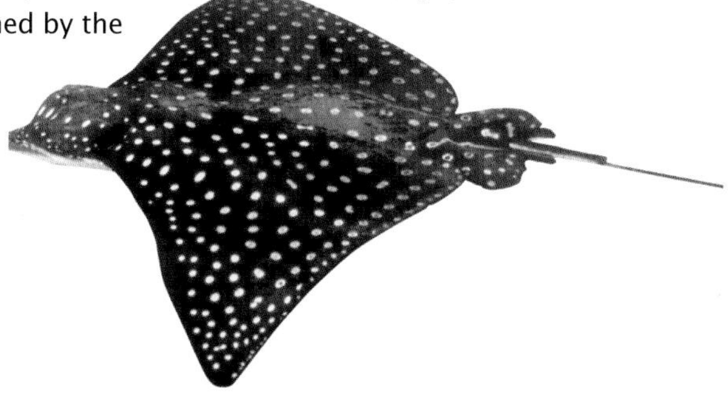

LOBSTER TALES: THE THREE GALAPAGOS SPECIES

ABOVE: Red spiny lobster in shallow rocky reef habitat.

RED SPINY LOBSTER *(Panulirus penicillatus)*

Most abundant species, found in tropical oceanic islands of the Indian and Pacific Oceans, preferring shallow rocky habitat with high wave energy. Congregating in large groups in caves by day and foraging at night, it is omnivorous, feeding on many living and dead reef organisms. It grows quite quickly, reaching sexual maturity at about 23 cm (9 in) total length in about three to four years.

GREEN SPINY LOBSTER *(Panulirus gracilis)*

Restricted to western South America, it is found in deeper, murkier waters than other species, preferring calm bays and inlets, but with otherwise similar habits as the red lobster, though less gregarious.

SLIPPER LOBSTER *(Scyllarides astori)*

Endemic to the eastern tropical Pacific region, this is a solitary, slow-moving, slow-growing species which may take up to eight years to reach sexual maturity at 22 cm (about 9 in). It seems to feed mainly on sea urchins, so may be an important regulator of these algae grazers.

REPRODUCTION: Lobsters produce vast numbers of eggs — up to 2 million in the green lobster; 750,000 for red lobsters; 350,000 for the slipper lobster — which females carry on modified feet structures for up to three months. After being released into the water column, the larval phase may last up to a year, but it is not known whether those that settle in Galápagos are actually produced there or come from elsewhere, a factor bearing important management implications.

FISHERY: Once so common lobsters could be pulled from rock pools without the fisher getting wet, small-scale catches probably date back to the earliest settlers. Industrial exploitation began in the 1960s, with hand-caught daily catches delivered by free-divers to a shore-based freezing plant on Santa Cruz Island, and industrial fishing boats plying coastal waters using surface-supply dive gear, especially around Isabela. The industry died as stocks collapsed, and the last fishing boat left in 1984. A modest population recovery followed, enough to sustain local fishers. Through the 1990s, the rapid growth of the artisanal fishing sector (spawned by sea cucumber fishing) vastly increased efforts to fish lobsters. Catches peaked in 2002 and have dropped steadily since, due to chronic overfishing. Today, there is an open season for spiny lobsters from September to December, mainly for export to the United States market, whereas slipper lobsters are taken year-round for local consumption only. Landings of this species are not inspected, but are thought to reach 13 tonnes (14 tons) annually, with declines reported. Both minimum size limits and a ban on landing females with eggs are largely ignored for all species.

Alex Hearn & Tui De Roy

TABLE: Commercial spiny lobster catch records (all weights shown refer to tails only) compiled by Alex Hearn and colleagues at the Charles Darwin Research Station show the clear correlation between rising market prices, diver days per kilogram landed and dwindling annual catches.

	Effort returns (kg/diver/day)	Landings (tonnes)	Market price (US$ per lb)
1995	7.3	98,000	
1996	7.7	57,000	
1997	6.7	65,300	3.60
1998	5.8	31,000	6.30
1999	7.0	54,400	7.80
2000	9.5	85,000	9.10
2001	7.0	66,000	10.00
2002	5.9	51,400	10.60
2003	7.0	45,800	10.40
2004	4.6	25,700	10.40
2005	4	34,300	10.80
2006	4.9	29,500	13.00
2007	not measured	32,000	no records

Although the oceanographic setting of the archipelago is eclectic, its marine environment is isolated within an already isolated biogeographical region. The eastern tropical Pacific (ETP), an ecological unit ranging from Baja California to Ecuador, has been separated from the tropical Atlantic since the closure of the Isthmus of Panama about 3.1 million years ago [Coates & Abando, 1996]. Given the isolation thus characterizing this region, it is hard to believe that, according to the IUCN Red List of Threatened Species, more species are in trouble in the ETP than any other marine region of the world. This reality is a result of over-fishing, coastal development and ENSO events.

To the west of the Galápagos Archipelago is the world's widest deep-water barrier to the dispersal of marine inshore organisms [Robertson, Grove & McCosker, 2004]. Most of the resident fishes in Galápagos have distributions overlapping with the ETP; others clearly arrived from the cooler Peru-Chile biogeographical province, and some from the distant Indo-Pacific. There is also a small percentage of fish in common with the Caribbean, while the remainder are either endemic to Galápagos or restricted in range to a triangle formed between Galápagos, Malpelo and Cocos islands (see also chapter: Hearn). The connection between ocean currents and the zoogeographic affinities of the fish fauna is apparent. The bignose and shortnose unicornfish, *Naso vlamingi* and *N. brevirostris*, bluelined wrasse, *Stethojulis bandanensis*, yellow spotted burrfish, *Cyclichthys spilostylus*, Klein's butterflyfish, *Chaetodon kleini* and Valentine's sharpnose puffer, *Canthigaster valentinii*, are all species that originated in the tropical western Pacific and all are considered vagrants here, usually observed around the northernmost islands where average sea surface temperatures (SST) are highest. An exception is the moorish idol, *Zanclus canescens*, also a tropical species from the west, which has become well established throughout the islands since the major 82/83 ENSO. From the continental coast to the south, came resident species like the Pacific knifejaw, *Oplegnathus insignis*, the Peruvian grunt, *Anisotremus peruvians*, and the rusty damselfish, *Nexilosus latifrons*, no doubt conveyed by the Humboldt Current

ABOVE: Schools of barberfish, *Johnrandallia nigrirostris*, and streamer hogfish, *Bodianus diplotaenia*, swirl around giant oysters clinging to the lava substrate at Cousin's Rock, a favored dive site.

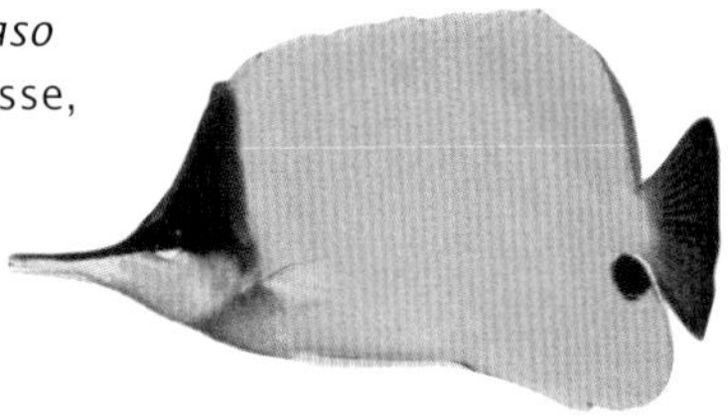

Mark Jones

LEFT AND ABOVE: Both the longnose butterflyfish, *Florepiger flavissimus*, and moorish idol, *Zanclus comutus*, typical tropical Indo-Pacific species, frequently appear in El Niño conditions, the latter having become well established in recent years.

ABOVE: The Galápagos Marine Reserve is a safe haven for green turtles, *Chelonia mydas agassizi*, seen here among Pacific creole-fish, *Paranthias colonus*, and leather bass, *Dermatolepis dermatolepis*, along the deep walls of Wolf Islands.
BELOW: Trumpetfish, *Aulostomus chinensis*, seen here in its golden phase.

system. These latter species are most common along the western shores of Isabela and Fernandina, where SST is the lowest.

Fish extinctions

All naturalists are intrinsically interested in the discovery of new species, but in this age of extinction [Quamann, 1996], the disappearance of a species is also intriguing to a natural historian. The ebb and flow of evolution and extinction among terrestrial life forms is well documented in the scientific literature — for example, in Galápagos at least two forms of giant rodents evolved here only to disappear again, before they were ever sighted by humans [Steadman & Zousmer, 1988]. Extinction in the sea, however, is far more difficult to verify. One of the first case studies of such likelihood among the fish fauna is that of the Galápagos damsel, *Azurina eupalama,* last observed in Galápagos in 1983. Other marine species were decimated as well. In 2007, IUCN added 73 species of Galápagos seaweeds to the Red List, 10 of these receiving the most threatened status of critically endangered.

Fishing pressure is undoubtedly also responsible for the decline in Galápagos food fishes. The reduction is most apparent in the abundance and average size of commercially valuable fishes such as grouper, snapper, mullet and wrasse. The drop in shark populations in the islands is attributable to the illicit trafficking in fins, and to vulnerability due to their migrating to distant unprotected waters (see also chapter: Hearn). The history of Galápagos fisheries is one of boom and bust that has provoked serious declines in sea cucumber, lobster and grouper, as well as other species [Hearn & Murillo, 2007]. Seven of the 15 fish species that Darwin collected in the islands have commercial value and all seven have been over-exploited. But how do we explain the disappearance of a plankton-feeding damselfish that has no value in the market?

Baseline is a term used in ecology to describe how things used to be. By determining what species are present, with the passage of time it becomes possible to gauge the health of an ecosystem by documenting population changes in those species. Gerard M. Wellington compiled the first baseline study of the Galápagos marine environment in 1975, but unfortunately no records exist of what the fish species densities were before human exploitation. Thirty years after Wellington's work was released, the term 'shifting baseline' was coined by fisheries biologist Daniel Pauly to describe the way that radically depleted fisheries have been evaluated by experts, using the state of the fishery at the start of their careers as the baseline, rather than the stocks in their untouched conditions. Galápagos is no exception, and we shall never know what the population dynamics of the fishes were before exploitation came into play.

In June of 1975, when I first arrived in Galápagos as a deckhand aboard a 16-m (52 ft) sailing yacht, there was only one vessel operating as a live-aboard dive boat. My first dive was at Leon Dormido, or Kicker Rock, adjacent to San Cristóbal Island. My life would be forever changed by this experience. Like most visitors, I was impressed by the stark beauty of the islands and by the fearlessness of the wildlife, but it was the awe-inspiring abundance of life in the seas surrounding the archipelago that became a driving force in all my activities. In 1997, my obsession with the ichthyofauna led to the publication of its first comprehensive treatment; *The Fishes of the Galápagos*, coauthored with Dr. Robert Lavenberg. Since that time, much of my career has revolved around the Galápagos World Heritage Site and its fishes.

According to FishBase, which is updated monthly, there are 30,700 known fish species in the world, comprising the most diversified group of vertebrates on our planet (www.FishBase.org). Not surprisingly, they also represent the most diverse vertebrates in Galápagos. When I began my research in the mid-1970s, 289 kinds of fish, from 88 families, had been recorded [Walker, 1966]. Credit is due to the collaborators of FishBase (October, 2008) who have increased that number to 475 species — 33 of which are endemic — from 126 families.

During my first seven years in the archipelago as a naturalist guide aboard the M/V *Bucanero,* my employment enabled me to make consistent observations of fish around the islands, resulting in the documentation of their population dynamics and fluctuations. The *Bucanero* sank in the Guayas River of Ecuador before the term 'shifting baseline' had become accepted [Pauly, 1995], but in those days one did not need to be an accredited fisheries biologist to recognize the impact that fishing, both commercial and recreational, was having on resident fish stocks. I would like to thank Dr. Daniel Pauly for giving me the inspiration and advice to place my early observations into perspective in more recent times.

In the 1970s and early 1980s crewmembers were allowed to fish from tour boats and their tenders (locally called pangas) whenever they had free time. All of the tour boats provided their guests with fresh fish, usually caught at the same locations set aside by the National Park as visitor sites. The crew also took fish home, and the *Bucanero* and several of the other larger vessels periodically returned for resupplying in Guayaquil, Ecuador, with their spacious walk-in freezers generally loaded to capacity. This recreational fishing was concentrated in near-shore habitats close to the anchorages, and nowhere was fishing prohibited. It was not until quite recently that no-take zones were established [Banks, 2008].

Even today, there is still no minimum size restriction for any of the commercialized

SEA CUCUMBER FISHERY

Of at least 30 Galápagos species, *Isostichopus fuscus* (left) is the only sea cucumber ever to be legally harvested. Ranging from Baja California to Peru, it lives on shallow rocky bottoms, feeding on organic detritus and possibly performing an important ecological function by cleaning surfaces that may assist the establishment of juvenile corals, sponges and other sessile invertebrates.

The fishery began informally in the early 1990s, initially driven by migrant fishers who had stripped the resource along the Ecuadorian coast. Not until 1999 was it opened officially, and it has been characterized by overexploitation, conflict and mismanagement ever since. Catches peaked in 2002, with 8 million sea cucumbers taken in two months, but that figure dropped to 5 million, 3 million and 1.5 million in subsequent years. Plummeting densities are reflected in sea floor counts, for example at Fernandina: 161 individuals per 100 m^2 (just over 1000 square ft) in 2001, down to 100 in 2002 and 45 in 2003. Currently, between 7 to 11 sea cucumbers survive in the same surface areas. The fishery was closed in 2006, but reopened under pressure in 2007 (1.2 million caught) and again in 2008 (0.8 million caught) despite the lack of resource recovery.

A broadcast spawner, releasing eggs and sperm into the water column, the species requires high population densities for successful reproduction. This added to generally slow growth rates and sporadic recruitment makes recovery extremely difficult. Galápagos fishers are now illegally targeting less valuable species, while management efforts are suffering from shifting baselines, in which the original high densities are being overlooked, setting ever lower threshold limits to appease industry demands.

Alex Hearn

BELOW: Red-tailed triggerfish, *Xanthichthys mento*, Wolf Island.

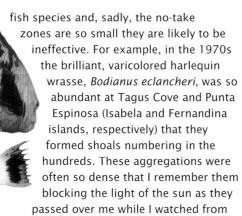

RIGHT: Each harlequin wrasse, *Bodianus eclancheri*, typical of cold western waters, displays a distinct pattern of orange, black and white.
FAR RIGHT: Adult and juvenile white-tail damselfish, *Stegastes leucorus beebei*, with green sea urchins, *Letchinus semituberculatus*, and red Panamic cushion stars, *Pentaceraster cumingi*, Tagus Cove, Isabela.
BELOW: Large groupers, or bacalao, *Mycteroperca olfax*, have become very rare.

fish species and, sadly, the no-take zones are so small they are likely to be ineffective. For example, in the 1970s the brilliant, varicolored harlequin wrasse, *Bodianus eclancheri*, was so abundant at Tagus Cove and Punta Espinosa (Isabela and Fernandina islands, respectively) that they formed shoals numbering in the hundreds. These aggregations were often so dense that I remember them blocking the light of the sun as they passed over me while I watched from the rocky bottom. During the most recent decade, a snorkeler is considered lucky to see more than one or two of this spectacular species typical of the cool western Galápagos waters.

There is a paucity of data to provide detailed evidence of the shifted baseline in Galápagos fisheries, but the reality is evident at the fish markets of Puerto Ayora on Santa Cruz or Puerto Baquerizo Moreno on San Cristóbal. Many species being sold, such as almaco jack, *Seriola rivoliana*, and several sea basses and groupers, are not large enough to be sexually mature, while much-favored species like bacalao, *Mycteroperca olfax*, and norteño, *Epinephelus cifuentesi*, are increasingly rare in any size. It is especially poignant that the latter species was not even described until 1993, and was named after Miguel Cifuentes, one of the most renowned conservationists to serve as Director of the Galápagos National Park and President of the Charles Darwin Foundation. What must be done to ensure that the norteño and other Galápagos species of commercial value do not

BELOW: The Puerto Ayora fish market shows ever deeper and smaller fish species for sale.

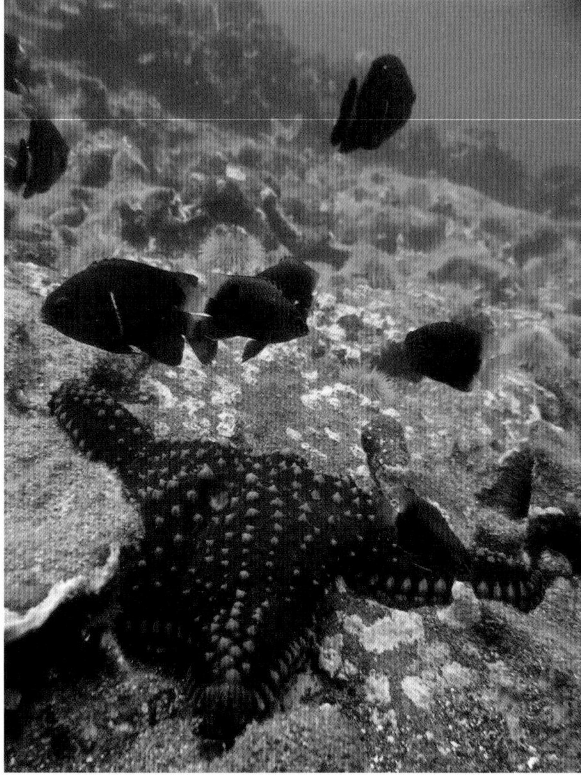

go the way of the Steller's sea cow of the Bering Sea, the last individual slaughtered for food only 17 years after the animal was first observed?

In the 1970s and 1980s, the fishermen in Puerto Ayora called me 'Pez Loco' or 'Crazy Fish' because I'd walk through the village with a briefcase in one hand and a plastic bag to gather fish in the other. My normal routine was to make a stop at the Charles Darwin Research Station library to do some research and then drop in at restaurants that sold fish in the village. In exchange for supplies from the mainland, fish that were unidentified were set aside for me.

In 1982, Dr. Richard Barber of Duke University installed an expendable bathythermograph (XBT) onboard the *Bucanero*. At the time, this was the state of the art device used by oceanographers to record temperature at depth. Increasing SSTs had long been known as a characteristic of an El Niño event, but on 14 August 1982, I couldn't believe my eyes when the SST reading at Cape Berkeley was 30°C (86°F) at the surface and it did not drop until the probe hit the thermocline at 70 m (230 ft). The combined effects of abnormally high SSTs and nutrient depletion caused by the absence of localized upwelling made the water crystal-clear, probably devoid of the plankton that sustained *Azurina eupalama*. This species had survived countless ENSO events throughout its evolution, yet some unknown factor prompted its demise in 82/83, while the population size of the

LEFT: Large leather bass, *Dermatolepis dermatolepis*, mingle with young streamer hogfish, *Bodianus diplotaenia*, in a luxuriant garden of sea fans and endemic black coral along the western shores of Isabela and Fernandina islands, where cold Cromwell Current upwellings nourish frequent plankton blooms.

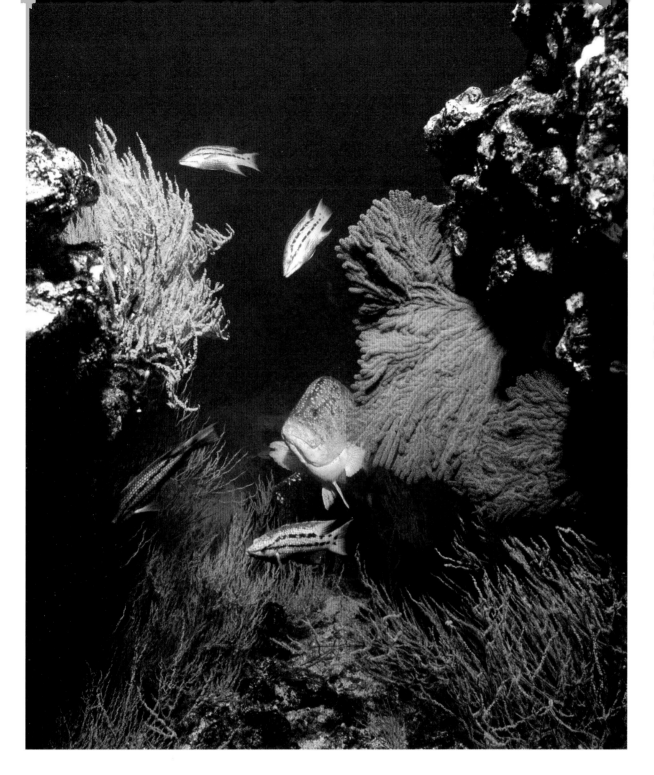

scissortail damsel, *Chromis atrilobata*, the closely related planktonivorous species with which it usually fed, was only slightly reduced by the sea change.

Predicting similar changes in the future is only conjecture, but it is worth pondering the implications of the extinction of a damselfish in Galápagos. If this species is rediscovered, it is all the better, but if not, the extinction will be further evidence that marine life here is at far greater risk than we have assumed. There is no silver bullet to ensure the preservation of the amazing life in the Galápagos seas, but there is one unequivocal reality accepted by all marine conservationists: if we are not successful in our efforts to set aside more areas of the oceans as marine preserves and maximize the no-take zones within them, the outlook will be dim. We have many reasons to protect the Galápagos Islands, among them to secure their reputation as an icon of conservation, marine and terrestrial. The task ahead is enormous, but if it cannot be accomplished in Galápagos, what hope is there for life anywhere?

It is provocative to consider what Charles Darwin would have encountered had scuba been available aboard the *Beagle*, but it is essential that we postulate what sea change still lies ahead. It is not simply in the best interest of Ecuador to preserve the riches of these enchanted seas, it is inherent to the survival of life on our planet. In the words of Aldo Leopold (1968), 'we must construct a narrative that places us in the community of the earth.'

BELOW: Stone scorpionfish, *Scorpaena plumieri mystes*, Wolf Island.

Mark Jones

Shark Migrations
Discovering the Golden Triangle
Alex Hearn

Biotelemetry Laboratory,
Department of Wildlife,
Fish & Conservation Biology,
University of California,
Davis, One Shields Ave,
CA 956161, United States.
<arhearn@ucdavis.edu>

Dr. Alex Hearn began work in Galápagos in 2002 in the fisheries monitoring program, researching growth and population dynamics of exploited marine species such as sea cucumbers and lobsters, and the application of science for fisheries management. He developed the shark research program in 2006 and is currently a postdoctoral scholar at the University of California, Davis (UC-Davis), but maintains his affiliation with the Charles Darwin Foundation (CDF) as an adjunct scientist.

SHARKS HAVE ALWAYS BEEN considered a top priority in the management of the Galápagos Marine Reserve (GMR). Their condition as apex predators positions them as key species in the functioning of the marine ecosystem. But their slow growth, late onset of sexual maturity and low reproductive output also make them vulnerable to overfishing. Although sharks are fully protected in Galápagos, local dive guides have reported declines in their populations over the last decade, which they attribute to fishing pressure. Shark mortality from fishing occurs in three ways: illegal shark fishing by local fishers targeting fins for the lucrative Asian market; bycatch from industrial vessels (from mainland Ecuador or other nations) entering the GMR illegally; and migration of the sharks to unprotected waters where they are vulnerable to a range of fishing fleets.

There are 30 species of shark described for the GMR, occupying open water (e.g., blue sharks and thresher sharks), deep sea (e.g., lanternsharks and dogfish) and coastal areas (e.g., Galápagos sharks and whitetip reef sharks), ranging in size from the diminutive, and as yet unnamed, deep sea catshark, *Bythaelurus* species B, to the gargantuan whale shark — at over 15 m (50 ft), the world's

BELOW AND RIGHT: For reasons still largely unknown, scalloped hammerhead sharks, *Sphyrna lewini*, gather in large schools along the deep walls of Darwin and Wolf islands, but their numbers in recent years have decreased as fisheries target sharks for the lucrative sale of their fins.

largest fish. Certain species, such as the scalloped hammerhead, have acquired iconic status, fueling a burgeoning dive industry that is already worth up to an order of magnitude more than the socially conflictive, poorly managed, moribund export fisheries for sea cucumber and lobster. Indeed, along with the oceanic Marine Protected Areas (MPAs) of Cocos (Costa Rica) and Malpelo (Colombia), Galápagos forms a 'golden triangle' for hammerheads, where schools numbering in their hundreds may regularly be observed by divers who visit from all over the world.

No one knows for sure why hammerheads gather in such large numbers, but the schools are generally made up of females, with the largest ones closest to the center. Males tend to swim around the fringe of the school, and may dart in to attempt mating with the biggest females. So these schools may serve a social or reproductive function, although they also occur close to 'cleaning stations' where individuals will glide into the reef displaying their white undersides and allow smaller reef fish to groom them.

The seeds for the shark research program at CDF were sown in July 2003, when I had the incredible good fortune of spending almost an hour swimming with a young 4-m (13 ft) whale shark at the anchorage site in Wolf Island while on a research cruise studying lobster growth patterns. Not only was this an amazing

Photo courtesy Pete Oxford — www.peteoxford.com

ABOVE: Always shy, a curious hammerhead shark makes a brief pass for a closer look, as creole fish scatter.

LEFT: Using a pole-mounted dart while free diving, one of the author's team deftly affixes an ultrasonic tag to the back of a whale shark, *Rhincodon typus*, in an effort to learn about its mysterious movements.

ABOVE: A hammerhead monitored by the Migramar team turned up at Cocos Island, off Costa Rica, 14 days after last being detected near Darwin Island, nearly 700 km (435 miles) away.
BELOW: A free-diver tags a shy hammerhead using no scuba gear, as sharks are afraid of bubbles.

experience, but it sparked a series of questions in my mind: where had the shark come from? How long would it stay in the area? Where would it go next? The same questions apply to the schools of hammerheads and Galápagos sharks regularly seen at Wolf. It would be another two and a half years before we could begin to answer these questions.

In early 2006, an informal partnership between CDF, Fundación Malpelo (Columbia) and PRETOMA (Costa Rica) was born with the support of Conservation International (CI), and the technical advice of Dr. Pete Klimley, from UC-Davis, who pioneered hammerhead behavioral studies in the 1970s and 1980s. I participated in an expedition to Malpelo to fit ultrasonic tags to hammerheads. These are small cylindrical devices, about 10 cm (4 in) in length, which are tethered to a dart that is inserted behind the dorsal fin of the shark, using a spear gun or pole spear. As sharks tend to be wary of bubbles produced by scuba divers, all tagging was carried out by free diving. The ultrasonic tags emit a coded signal every two minutes, which is detected by receiver devices placed at strategic locations on the seabed, whenever a tagged shark comes within a 200-m (650 ft) radius of these sites. Upon retrieving the receivers (every four to six months), a record of the date and time that each tagged shark was present is obtained. This information is used to study differences in daytime and nighttime presence, site fidelity, movements between sites, and can also serve to understand the group dynamics of sharks — whether certain individuals tend to remain together.

Photo courtesy German Soler — fundacionmalpelo.org

A collaborative approach

The success of the pilot study in Malpelo provided the springboard for a similar initiative in the Galápagos Islands, and other initiatives in Cocos and Coiba (Panama). It was clear from the outset that in order to fully understand the migratory patterns of sharks research must break national boundaries. Ecuador, Costa Rica, Panama and Colombia were each developing National Plans of Action for their shark resources, and yet, if the shark populations of each nation were in fact connected, then coordinated policies would be essential for their management and protection.

A regional network was created, whereby institutions at each study location use the same technology and share data. In this way, a shark tagged in Galápagos, were it to migrate to Cocos, would be detected by the receivers there. The network, with its web page (www.migramar.org), is committed to working together to understand the importance of MPAs and specific sites within them for sharks and other migratory species.

In Galápagos, a tri-institutional research team involving CDF, the Galápagos National Park Service and UC-Davis, was provided with seed funding from CI, with the simple aim of evaluating whether shark tagging was feasible here. In July 2006, 18 hammerheads, three Galápagos sharks and one whale shark were fitted with ultrasonic tags at Darwin and Wolf islands. Two receivers were installed at the southeast corner of Wolf, covering an area of approximately 1000 m (3300 ft), and a further two receivers were placed at Darwin Arch, covering a similar area. Additionally, 12 Galápagos sharks were fitted with satellite tags in order to track their

Map created by Cesar Peñeherrera, Darwin Foundation/background photo courtesy Sterling Zumbrunn

migratory pathways, in partnership with George Shillinger of Stanford-TOPP.

One year later, the results showed that almost every hammerhead tagged remained at the site for at least several days, but after that there appeared to be a constant movement between the study areas at both islands. The sharks also displayed a clear preference for these locations during daytime hours — consistent with studies carried out by Dr. Klimley at a seamount near Baja California, Mexico, where hammerheads swam up to 32 km (20 miles) out to sea each night to feed, and returned along the same route in time for dawn. Hammerheads are known to feed on open-water species of squid.

However, the most surprising results came in via telephone from Randall Arauz, president of PRETOMA in Costa Rica: two of the hammerheads had appeared at Cocos Island, on the same day in March 2007, only 14 days after one of them was last detected at Darwin — a straight line distance of nearly 700 km (435 miles). The sharks remained at Cocos for approximately one month, and one was subsequently detected once more at Darwin, the first solid evidence of a migration between these MPAs. This spawned a new set of questions: did the two sharks travel together? Did they travel directly or spend time in the open ocean? Why was only one shark detected back at Darwin?

The Galápagos sharks displayed a shorter residency at their tagging site, but also moved between the two islands. One of the three tagged individuals

ABOVE: Using ultrasonic tags that are detected by fixed underwater acoustic receivers placed at key locations, the Migramar project links work from four countries, shedding new light on the distant movements of sharks between some favored haunts.

ABOVE: Female scalloped hammerhead, *Sphyrna lewini*.

A SEA TURTLE SAFE HAVEN

The green turtle of the Galápagos (above) represents a quandary for taxonomists. Due to its unusually dark coloration and other features, it has been tentatively named the black turtle, *Chelonia agassizii*, yet geneticists find it identical to the widespread Pacific green turtle, *C. mydas.* Whichever is correct, the large population nesting in the archipelago is a conservation treasure-trove compared to turtles worldwide. Most sea turtles are in precipitous decline due to a plethora of manmade causes, ranging from direct hunting and nest poaching to plastic bag ingestion, fisheries bycatch, pollution-induced diseases, coastal development, plus predation, trampling and light pollution on nesting beaches. With the banning of longlines from the Galápagos Marine Reserve and the eradication of nest-raiding pigs in critical areas, these islands represent a true safe haven. But sea turtles have survived since before the age of dinosaurs by ranging far and wide across the oceans. Flipper tagging and more recent satellite tracking show that while some appear to be quite sedentary, Galápagos green turtles, like the sharks found in the islands, regularly travel thousands of miles into danger zones along the coasts of Central and South America. This tragedy was graphically illustrated by DC235, a female tagged by CDF researchers while nesting on Las Bachas Beach, Santa Cruz, in February 2004. She was subsequently drowned on a tuna longline hook near Panama in December 2005. Other satellite tracks show turtles moving far into the Pacific, south and west of the islands. How or why they travel so far remains a mystery. Galápagos offshore waters are also on the migration routes of leatherbacks and olive ridleys (below left) from the north, while hawksbills (below right) are infrequent inshore visitors.

BELOW: Olive ridley (*Lepidochelys olivacea*)

ABOVE: Hawksbill (*Eretmochelys imbricata*)

Photo courtesy Mark Jones

TOP and ABOVE: The Darwin Arch, where hammerheads and whale sharks congregate; the best way to observe them is to remain still and breathe as little as possible to avoid blowing bubbles.

was detected at Gordon Rocks, off Santa Cruz Island, several months later. The results of the satellite tracks showed that several individuals migrated down into the heart of the GMR, while one swam out into the open ocean, covering a distance of more than 1100 km (685 miles) in little over three months. The whale shark was detected only once more, at the same site where it was tagged, eight weeks later.

The results of this first tagging experience in Galápagos far exceeded expectations, and paved the way for a much larger project. Currently, receiver sites have been established around the entire island of Wolf in order to study fine-scale movements, but the array has also been expanded to include 12 sites throughout the archipelago. Almost 100 hammerheads have been fitted with ultrasonic tags, and we are now working on increasing the numbers of whale sharks and Galápagos sharks. A local project studying foraging patterns and group dynamics of whitetip reef sharks in Santa Cruz Island has just begun. The results from these studies will provide us with valuable information on site fidelity and seasonal movements both within the GMR and between the regional MPAs.

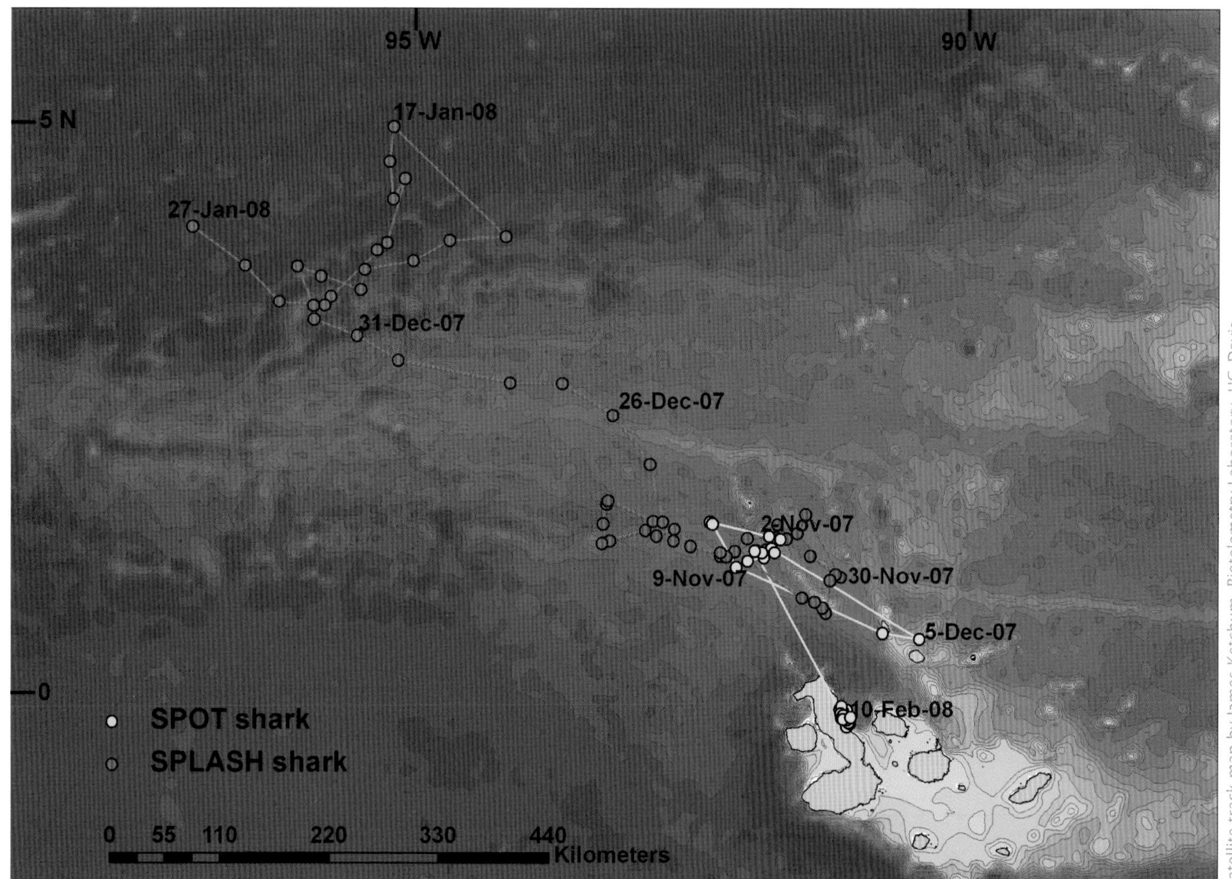

Satellite track map by James Ketchum, Biotelemetry Laboratory, UC-Davis

LEFT: Electronic tags were attached to two hammerhead sharks nicknamed 'Spot' and 'Splash,' revealing their fascinating travels during several months. The devices store crucial data while also transmitting information at intervals to a passing Argos satellite, such as depth, temperature and location, which is downloaded by the research team.

BELOW: New work has begun to map the habits of the inshore white-tipped reef shark, *Triaenodon obesus*, vulnerable to gillnets (left).

Migratory corridors

One question that is not answered by ultrasonic tags is that of migratory corridors. Ultrasonic tags tell us when an individual is at a particular site, but does not reveal the route it took from one place to the next. For long-distance migrations in open water this is a key question, as the debate between industrial fishers, authorities and conservationists on how to reduce shark mortality rages on, based sometimes more on emotion than hard fact.

In order to track a shark over long distances and extended periods of time, satellite tags are used.

ABOVE: Hammerheads may cruise at the surface or dive deep during their travels. They are known to feed on open-water species of squid, hunting mainly at night.

RIGHT: Rising from the depths well north of the main Galápagos Platform, the sheer walls of Wolf Island are swept by strong oceanic currents, an attractive spot for pelagic sharks.

BELOW RIGHT: The plankton-feeding whale shark, seen here with a green turtle near Darwin, is the largest fish in the world's oceans, yet almost nothing is known about its life history.

However, tags cannot send information to satellites while under the surface. One solution is to use a pop-up tag, which can be attached to the shark in the same way as the ultrasonic tags. This uses a light detector, depth gauge and clock to estimate the position of the shark by relating day length with latitude; and by taking the time of highest light intensity as midday and relating this to Greenwich Mean Time in order to approximate longitude. After a predetermined period, the tag pops off and floats to the surface, from where it sends the stored information to a satellite. The data must be corrected for currents and sea surface temperature, but this gives a good general pattern over long spans of time and distance. However, hammerhead sharks seem to have the ability to shed these tags very quickly, resulting in only a few weeks of data in return for a large investment.

A more suitable option is to use tags that send positional information directly to satellites. However, in order for this to work, the shark must spend some time at the surface, and the tag must be attached to the dorsal fin of the shark, usually with screws. This involves taking the shark out of the water in a sling, and ensuring that it is constantly ventilated with seawater pumped through its gills. Hammerheads are ram ventilators — this means that they require movement to breathe, making them particularly vulnerable to fishing gear. Many of the hammerheads caught accidentally during an experimental longline fishery by local fishers in Galápagos in 2004 died on the lines because their restricted movement prevented them from oxygenating their gills effectively.

No research team had yet been able to successfully place these tags on hammerheads, but, in November 2007, two males were brought aboard, fitted with satellite tags and returned to the water unharmed in under five minutes. These provided daily positions for over three months. Initially, they spent most of their time in open waters close to Wolf Island, then one individual swam out into the open ocean in a northwesterly direction, covering a distance of 1500 km (930 miles), whereas the other migrated south into the main archipelago, and spent over a month around the eastern coast of Isabela. Four more male hammerheads were fitted with these tags in July 2008. Their movements can be followed on the Migramar website.

Despite the promising results, our work is only just beginning. In July 2008, a hammerhead from Malpelo, 1180 km (733 miles) distant, was detected at both Darwin and Wolf islands, after having first swum to Cocos, thus closing the connectivity loop between the three points of the golden triangle. We are now working closely with a group of trained dive guides in order to create a database of shark distribution and abundance throughout the region and over time. This will provide a baseline from which changes can be measured. It may also be a starting point from which to work backward and attempt to model the shark populations before human impact began.

Locally, residency behavior and movement patterns of sharks can be used to design efficient patrolling strategies within each MPA, suggest changes to the coastal usage zonation schemes, and perhaps provide a backdrop for the design of oceanic no-take zones — a theme currently being developed by my colleague James Ketchum for his Ph.D. thesis at UC-Davis.

Regionally, the collaboration and coordination between the research institutions will strengthen the advice provided to authorities and stakeholders in the development of Shark Action Plans and cross-boundary management measures to protect these remaining hotspots of marine biodiversity. For me, after seven years of applied marine research in the GMR, much of it related directly to the thorny management issues of heavily exploited fisheries such as sea cucumbers, spiny lobsters and grouper (bacalao), lifting the lid on the mysteries of shark migratory behavior has cast a whole new light on my Galápagos experience.

Photo courtesy Mark Jones

LEFT AND BELOW: Impressive feeding frenzies in the deep offshore waters around Fernandina, Wolf and Darwin may involve many different predators. Large gatherings of pelagic silky sharks work together with rainbow runners, wahoo and tuna, and sometimes dolphins and sea lions as well, plus boobies raining down from above.

Giant Tortoises

Mapping their Genetic Past and Future

Adalgisa (Gisella) Caccone & Jeffrey Powell

Department of Ecology and Evolutionary Biology,
Yale University, New Haven,
CT 06520-8105,
United States.
<adalgisa.caccone@yale.edu>
<jeffrey.powell@yale.edu>

Dr. Adalgisa Caccone is an evolutionary geneticist at Yale University where, since 1998, she has headed the Molecular Systematics and Conservation Genetics Laboratory. She received her B.S. from 'La Sapienza' University of Rome and her Ph.D. from Yale University.

Dr. Jeffrey Powell is a Professor in the Department of Ecology and Evolutionary Biology at Yale University. He received his B.S. from the University of Notre Dame and his Ph.D. from the University of California, being the last student of Theodosius Dobzhansky.

GIANT TORTOISES ARE THE largest terrestrial vertebrates on the Galápagos Islands. They furthermore represent the largest exothermic (cold-blooded) terrestrial herbivores alive on earth. These animals played a key role in Darwin's thinking, and also occupy a crucial position in the insular ecosystem. Similar giant tortoises once roamed all of the world's continents, except Antarctica, yet all went extinct largely during the Pleistocene era, from 1.8 million to 10,000 years ago — except those on two remote oceanic archipelagos nearly on opposite sides of the globe: Galápagos and the Seychelles (Indian Ocean). Only Galápagos have a number of giant tortoises that differentiated morphologically and genetically survived to the present day, allowing for close evolutionary study.

Our own involvement with these extraordinary animals began almost accidentally. In the spring of 1991, Jeff received a call from the Association of Yale Alumni asking if he would accompany a tour group to Galápagos. 'Why me?' was his incredulous response, having no previous experience in these islands. But for an evolutionary biologist — even an insect geneticist specializing in fruitflies and mosquitoes — Galápagos is Mecca, so he quickly agreed. This decision marked the beginning of a lengthy Galápagos involvement for both Jeff and Gisella, his wife and long-time collaborator at Yale University.

What first piqued our curiosity was learning from Jeff's guide on the last day of his tour that a number of giant tortoises held at the Fausto Llerena Tortoise Center were old confiscated pets that could neither be used for breeding nor returned to the wild because of their unknown ancestry. Even though their variable shell shapes indicated different islands of origin, their exact lineage could not be ascertained because of taxonomic similarities between populations. On the spot, Jeff introduced himself to Linda Cayot, then resident herpetologist at the CDRS, suggesting we might be able to help using modern genetic analyses. 'I've been looking for a geneticist to work on these tortoises for years!' was her instant reply.

Back at Yale, our enthusiastic discussions soon took shape. By chance, James Gibbs was in our lab, having just finished his Ph.D. work on salamanders. With prior Galápagos field experience as an undergraduate with Peter and Rosemary Grant's landmark Darwin's finch study, he agreed to lead our first collecting expedition in 1994. Thus, the three of us became the core of the 'Yale team' that embarked on a 15-year research program that continues today.

Galápagos and their tortoises

The taxonomic history of the Galápagos giant tortoises is complex and remains unsettled and contentious. Long considered a single species with

RIGHT: Large males on public display at the Fausto Llerena Tortoise Center in Puerto Ayora cannot be used for breeding because of their unknown mixed genetic origin.

multiple races, *Geochelone elephantopus,* or more recently *Chelonoidis nigra*, various numbers of species and subspecies have in fact been proposed. Of up to 15 formally named forms that once lived on Galápagos, only 11 remain today, four having gone extinct since humans first arrived on the islands in 1535. Although the subject remains controversial, our genetic work has led us to believe that species status is warranted for almost all remaining taxa.

The most obvious differences are in shell shape. Some populations have a 'domed' carapace with a low, gently curved front and rounded profile. These concentrate mostly on the upper slopes of higher islands where conditions are generally cool and moist, with relatively abundant food supply. This compact shape is also thought to provide thermal protection against the cold. Tortoises characterized as 'saddlebacks' (due to their resemblance to ancient Spanish-style saddles) have an elevated anterior carapace opening, and a straighter lateral appearance, with extra long forelegs and necks. They are found on low, arid islands where resources are scarce and food is the primary limiting factor. This morphology allows longer upward extension of the neck, which permits browsing on higher perennial vegetation. It may also serve in antagonistic displays during inter-individual competition for scarce food or mates, since dominance is established by the individual

ABOVE: Giant tortoises awaken after spending the night in a mud puddle on Alcedo Volcano.
BELOW LEFT AND RIGHT: Once considered variable island races, distinct species emerged through genetic analysis. Domed forms such as Alcedo Volcano's *Geochelone vandenburghi* (left) evolved in humid habitat, where they graze low-growing plants, whereas saddlebacks, such as Española's *G. hoodensis* (right), are typical of low islands, adapted to browse high in the arid vegetation. Facial color also varies from jet black to yellow or pink.

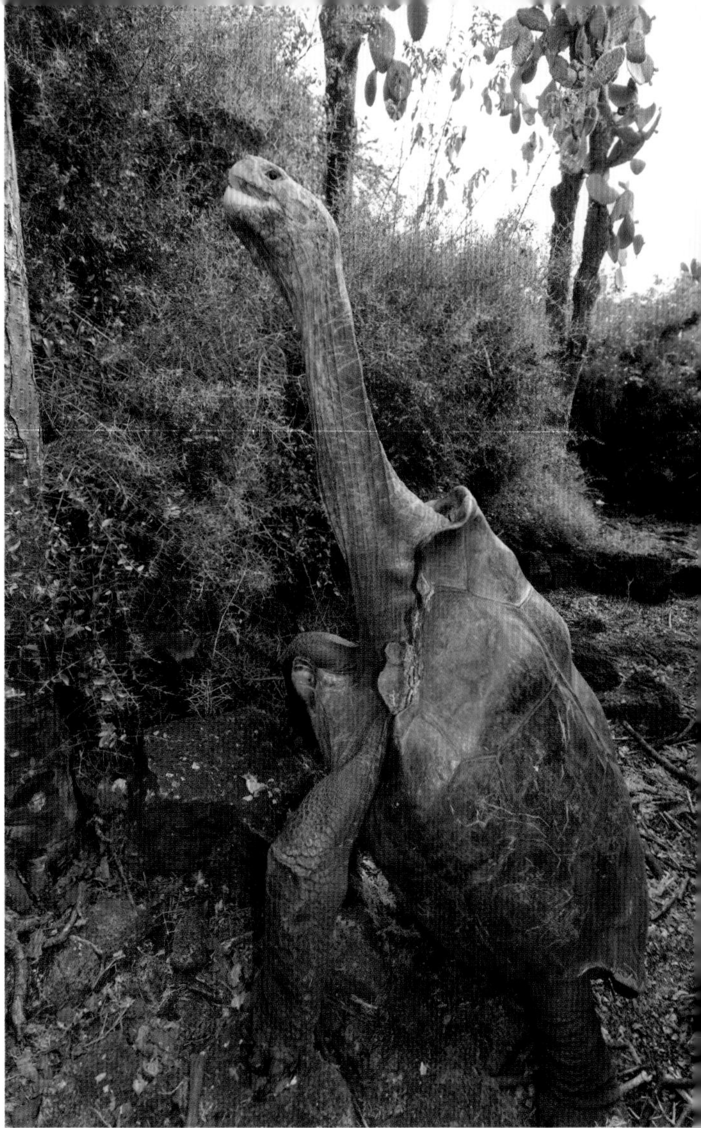

RIGHT: With its flared shell, slender legs and remarkably long neck, a male Española tortoise at the breeding center reaches high into the dry scrub to feed on scant foliage.

who can raise its head highest while gaping widely. Several researchers have argued that these distinct morphologies develop adaptively during growth, but the hereditary basis for these differences is evident in both types of tortoises sharing the same environment when reared in captivity.

Of the 11 extant types, those on the islands of Española, San Cristóbal, Pinzón and Pinta are the most conspicuous examples of saddleback animals. The clearest dome types occur on Santa Cruz Island and Alcedo Volcano on Isabela Island, with less distinctive, intermediate forms on Sierra Negra, Cerro Azul, and Darwin volcanoes, all on Isabela. The population on the northernmost Isabela volcano, Wolf, is of mixed morphology, a fact that is important to our genetic work discussed below.

Surprisingly, Santa Cruz Island, characterized by its gentle contours and no major natural habitat barriers, harbors the highest intra-island genetic diversity of giant tortoises. On this old, moderate-sized island (around 100,000 ha, or 250,000 acres) central to the archipelago, we have identified three distinct lineages among populations formerly considered a single species. These are at least as genetically distinct from each other as are tortoises from different islands assigned separate species status. Paradoxically, Santa

Giant Tortoise Species and Status

LOCATION	SPECIES	SITUATION
Española	*Geochelone hoodensis*	**Down to 15 individuals in 1960s, but recovering through breeding program; repatriated population.**
Fernandina	*G. phantastica*	**Extinct — only one individual ever recorded.**
Floreana	*G. galapagoensis*	**Extinct since mid-1800s.**
Wolf Volcano, Isabela	*G. becki*	**Large population (at least 2000), with recently discovered genetic input from other islands.**
Darwin Volcano, Isabela	*G. microphyes*	**Medium-sized population (estimate not available).**
Alcedo Volcano, Isabela	*G. vandenburghi*	**Large population (5000–10,000); habitat restoration following goat and donkey eradication.**
Sierra Negra Volcano, Isabela	*G. guntheri*	**Fragmented population with unclear genetics; limited captive breeding — some poaching.**
Cerro Azul Volcano, Isabela	*G. vicina*	**Several separate populations with unclear genetics; limited captive breeding — some poaching.**
Pinta	*G. abingdoni*	**One surviving individual.**
Pinzón	*G. ephippium*	**Unable to reproduce in the wild due to black rats, population increasing with repatriation program.**
San Cristóbal	*G. chathamensis*	**Population restricted to one end of the island; limited captive breeding.**
Santa Cruz – La Caseta	*G. porteri*	**Large population (3,000–5,000); continued problems with pigs and fire ants.**
Santa Cruz – Cerro Fatal	*G.* sp. nov.	**Recently differentiated by genetic analysis. Less than 100 individuals — some historic poaching.**
Santa Fe	*G.* sp.	**Extinct.**
Santiago	*G. darwini*	**Population increasing with repatriation program — habitat improving following eradication of pigs, goats and donkeys; continued problems with black rats.**

Source: Linda J. Cayot, Washington Tapia, Adalgisa Caccone

Cruz also supports the largest human population in Galápagos (currently more than 20,000 and anticipated to double by 2013), thus its unique biota is under inordinate pressure from human activities, including the conversion of substantial natural habitat for farming and ranching.

One of the largest remaining populations of domed tortoises is represented by the named species *porteri*, with an estimated 3000–5000 individuals around the La Caseta region. A second domed population is much smaller, comprising less than 100 individuals, and is found in an isolated location on the eastern slope around Cerro Fatal, recently separated from La Caseta by the development of private agricultural lands. This beleaguered cluster, the subject of historic poaching, is genetically quite distinctive and thus warrants naming as a new, previously unrecognized, species with its closest relatives on San Cristóbal Island. This calls for priority conservation action, considering its precarious status.

The third population, at Cerro Montura, near the northern Santa Cruz coast, has a saddleback carapace, and only three individuals have ever been seen. They are very closely related to those on nearby Pinzón Island, but whether this connection is ancestral or human-assisted is impossible to know.

ABOVE: The rainy season on Alcedo Volcano is a time of great activity for giant tortoises; feeding, drinking, mating and occasionally showing aggressive displays between males.

FAR LEFT: Polished smooth by generations of tortoise traffic, a boulder provides sparse drinking opportunities for a saddleback, *Geochelone ephippium*, on arid Pinzón Island, where fine garúa drizzle is often the only precipitation.

LEFT: Dome tortoises on high islands, such as *G. vandenburghi* on Alcedo, can drink copious amounts of water from temporary rain pools.

ABOVE: Major tortoise pathways run along the rim of Alcedo Volcano, where the largest population survives. BELOW: Little is known about the habits and survival status of *G. microphyes* on Darwin Volcano.

Dispersal and colonization

Before humans arrived less than 500 years ago, the only means of tortoise dispersal was drifting in the ocean, presumably after being washed to sea by flash floods. Galápagos tortoises are not good swimmers, but they do float with their heads above water and, since they can stay alive without food or water for six months or more, currents could potentially carry them long distances. Capable of storing viable sperm for long periods, at least one gravid female must have arrived from South America, some 1000 km (600 miles) distant. Major currents run north along the mainland coast, then veer westward at the equator toward Galápagos. Likewise within the archipelago, strong currents run northwesterly from older to younger islands. Our genetic work suggests the arrival of tortoises in Galápagos around 2–3 millon years ago, which coincides with the age of the oldest extant islands (San Cristóbal and Española). Yet this pre-dates the origin of most of the other islands occupied by tortoises today, so there must have been subsequent dispersals within the archipelago.

Through our genetic work, we have determined that the closest living relative to the Galápagos tortoise is the Chaco tortoise (*Geochelone chilensis*). This is the smallest of the three South American tortoises, but also the most drought-adapted, making it more suitable for Galápagos. It is unlikely to be the direct ancestor of the Galápagos tortoise, only the closest living relative, whose lineage has several extinct giants that were perhaps more likely candidates.

This raises the question of whether gigantism of tortoises on remote islands (Seychelles and Galápagos) is a result of in situ evolution of larger sizes or a prerequisite for colonizing such remote places. Good evidence on the Seychelles indicates the latter. The Seychelles are very low islands that have been submerged several times as ocean levels have risen and fallen. The fossil record indicates that each time these islands became submerged, a giant tortoise went extinct; as the sea level dropped, a new giant tortoise appears with no small intermediates. This strongly suggests that gigantism is required to survive the long journey across extensive stretches of ocean. Given the extinct giants in the Chaco tortoise lineage, there is no reason to believe that it was not a giant that initially colonized the Galápagos as well.

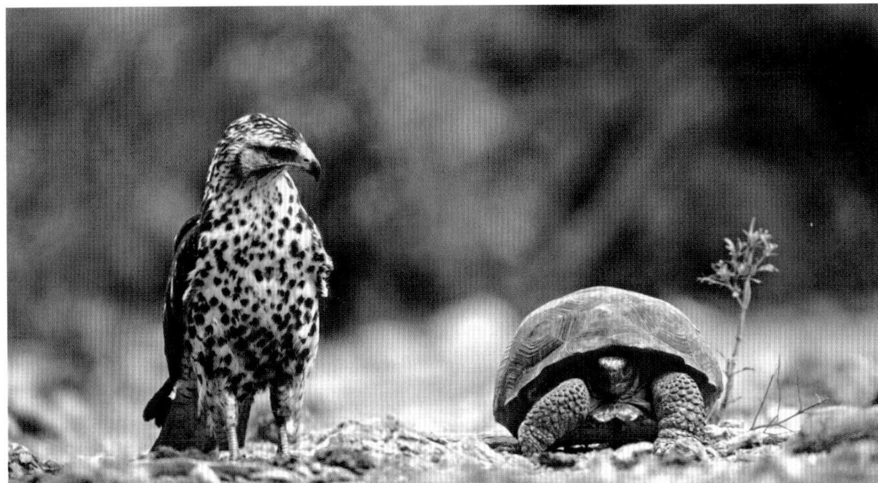

Genetics of a captive breeding program

There is no evidence of human presence in Galápagos before their discovery by Europeans in the 16th century. However, from the 17th to the 19th century, whalers and buccaneers collected tortoises as a source of fresh meat; records exist for the taking of some 40,000 tortoises, and it has been suggested that as many as 200,000 may have been removed. With only an estimated 20,000 tortoises at most remaining today, obviously this exploitation had a disastrous effect. As noted by Cayot and Tapia later in this volume, by the late 1960s the population of tortoises on Española had dwindled to 12 females and two males, due to past hunting pressure and food competition from introduced goats. These 14 tortoises, plus another male donated by the San Diego Zoo, became the nucleus of a captive breeding program that, 40 years later, can be hailed as one of the most successful species restoration efforts ever undertaken.

As geneticists, we were interested in testing how much genetic diversity these 15 captive breeders had imparted in the repatriated population. Our analysis revealed very uneven breeder contributions to the current gene pool on the island, with an *effective* population size of about one-fifth of the breeding population, which could eventually lead to inbreeding and other genetic problems. By examining the young before versus after release on the island, we showed that unevenness of parental contribution is due not to differential survival of offspring of different parents, but rather to nonrandom breeding in the captive population, a problem that should be rectified by controlled mating.

In the late 1960s, it was noted that the tortoises on Pinta (species *abingdoni*) had succumbed to human-induced pressures (hunting and goats) and were for all practical purposes extinct. Despite considerable search efforts, only a single male was found. In desperation, he was brought into captivity at the CDRS in 1972, where he resides to this day. As the last of his kind, subsequently nicknamed Lonesome George (LG), he has become a major tourist attraction as well as a conservation icon. In an effort to save at least some of the Pinta genes, he was placed with two females from Wolf Volcano (Isabela), in the hopes that they might, in due course, produce hybrid offspring.

When we first carried out preliminary genetic analyses on LG we were surprised to discover that his closest relatives were in fact not from the geographically closest tortoise population on Wolf Volcano, but from Española, the most distant island. Later, we extracted DNA from bones and skins of museum specimens collected on Pinta in 1905–06 and kept at the California Academy of Sciences in

ABOVE: No larger than flattened tennis balls, giant tortoise hatchlings are vulnerable to natural predation by Galápagos hawks when they first emerge from the nest (left), but are only the object of passing curiosity when two or three years old, Alcedo Volcano. BELOW: The grunting sound of mating males echoes around the Alcedo caldera during the rainy season.

ABOVE: Broken shards of hand-blown bottles attest to tortoise gathering forays by sailors of past centuries in the far interior of many islands.
BELOW LEFT AND RIGHT: After baffling researchers for many years, DNA analysis by the authors' team (extracting blood sample at right) is casting new light on the highly variable *G. becki* tortoises found on the slopes of Wolf Volcano of northern Isabela, an area once heavily frequented by whalers, where the genetic signature from other islands implies hybridization with tortoises brought here by people.

San Francisco. The results provided a useful genetic profile of Pinta tortoises. When we later investigated the genetic makeup of the wild tortoise population of Wolf Volcano on northern Isabela, to everyone's utter surprise this information allowed us to identify an individual that is a first generation Wolf–Pinta hybrid.

The only possible explanation for this amazing discovery can be sketched from scant historic records left by tortoise gatherers of past centuries. The current-ripped waters immediately adjacent to the Wolf coastline are often still rife with whales today and were the whalers' prime hunting grounds, not to mention their last port of call before sailing west into the Pacific. According to logbooks, tortoises from one island were sometimes placed on another for safekeeping, and they were also occasionally tossed overboard to lighten cargo loads. These may have later reached shore, as would be the case when such cumbersome ships may have run aground, possibly explaining why this area may have received a far greater influx of human-transported tortoises than any other in the archipelago. Even in more recent times, it is possible that Galápagos residents have transported tortoises between islands; they are prized for their oil and meat and, while illegal, poaching still occurs. Thus, it is probable that the heterogeneous nature of the Wolf tortoise population is the result of human-mediated introductions.

Since the estimated tortoise population on this volcano is at least 2000, and given that our initial genetic sample was very small (27) relative to this total number, the chance of finding additional Pinta genes is quite high. In collaboration with the Galápagos National Park (GNP), a large expedition comprising 42 men deployed in nine teams, was mounted in December 2008 to provide a much more thorough sampling in an effort to detect more hybrids, and even to determine the remote possibility of 'pure' Pinta tortoises still surviving there. Should genetic analyses identify such individuals, they will be brought to the GNP headquarters in hopes of beginning a selective breeding program (including LG, if he is willing) toward rescuing a majority of the original *abingdoni* species genome.

Extinction may not be forever: Bringing back the Floreana tortoise

When Charles Darwin landed on the island of Floreana in 1835, tortoises were already rare, being the staple food of recent settlers. Large numbers of shells were found lying around the settlement, causing Robert FitzRoy, captain of the HMS *Beagle*, to remark in his notes 'what havoc has been made among these helpless animals.' By the mid 1800s, Floreana had lost all its tortoises.

As with Pinta, DNA-based work from our group provides new insights on the history of this species and raises hopes of someday giving Floreana back its tortoise too. Using bones from museum specimens, we first clarified the evolutionary history of this taxon, confirming its genetic uniqueness at species level. On the slopes of Wolf Volcano we again found recent Floreana ancestry, with 15 tortoises out of a sample of less than 100 clearly carrying Floreana genes. This genetic data now offers us the rare opportunity to go back to Wolf, identify as many Floreana-related individuals as possible, and hopefully begin a program to resurrect this extinct species through selective breeding after an absence of more than 150 years!

LEFT: DNA studies have demonstrated that Lonesome George, the last surviving Pinta tortoise, *G. abingdoni*, is more closely related to the Española tortoise, *G. hoodensis*, than *G. becki* of Wolf Volcano, as previously thought. BELOW: Alcedo tortoises graze the grassy meadows on the caldera floor in a scene that could pre-date the rise of mammals.

Over the past 15 years, genetic work in our laboratory has provided DNA markers to distinguish taxa, and has supplied insights into tortoise evolution. Crucially, it has also aided in the development of conservation strategies. We could not have done all of this without the help and collaboration of scientists from CDRS and GNP. Especially Linda Cayot, Cruz Marquez, Howard and Heidi Snell, and Wacho Tapia, who are among the many who were always ready to listen and give us advice. Tom Fritts and Peter Pritchard also have been an invaluable source of help and information. Many postdocs and graduate students joined our group and never left it, even though they now have their own research agendas and laboratories scattered all over the world (Australia, Belgium, Canada, Greece and Italy). They always come to help when we need them, by either participating in field trips or by analyzing data and providing insights on possible interpretations of recent findings. Luciano Beheregaray, Ylenia Chiari, Claudio Ciofi, James Gibbs, Scott Glaberman, Michel Milinkovitch, Nikos Poulakakis and Michael Russello are the ones that produced and analyzed the data — they are our strength and our future. This research program has been a testament to the importance of long-term studies. Paramount to our success has been the technological innovations that allowed us to retrieve genetic data from museum specimens, a profound credit to the importance of these collections. As biodiversity continues to be lost globally, we need to use the most modern technological advances in efforts to protect our heritage.

Marine Iguanas

Life on the Edge

L. Michael Romero and Martin Wikelski

Department of Biology,
Tufts University, Medford,
MA 02155, United States.
<michael.romero@tufts.edu>
ase.tufts.edu/BIOLOGY/
faculty/romero/

Department of Biology,
Building Z, Room Z818,
University of Konstanz,
D-78457 Konstanz,
Germany.
<martin.wikelski@
unikonstanz.de>
www.princeton.edu/~wikelski

Max Planck Institute for
Ornithology,
'Vogelwarte Radolfzell,'
Schlossallee 2,
D-78315 Radolfzell, Germany.
<martin@orn.mpg.de>
http://www.orn.mpg.de/
migration

Dr. L. Michael Romero is a Professor in the Biology Department at Tufts University. He did his graduate work at Stanford University, followed by postdoctoral work at the University of Washington in Seattle. For the past 20 years, he has focused on the physiology, behaviour and ecology of vertebrate stress responses.

Dr. Martin Wikelski is Director of the Department of Migration and Immuno-ecology at the Max Planck Institute for Ornithology. He conducted his graduate studies on the evolution of body size in Galápagos marine iguanas in Bielefeld, Germany, and went to the University of Washington for postdoctoral work. Wikelski held professor positions in the United States at the University of Illinois and at Princeton University before returning to Germany.

MARINE IGUANAS, *Amblyrhynchus cristatus*, are unique in many ways. Their oddest behavior is that they graze on algae beds, rather like cows in a grassy pasture. Although there are reports of marine iguanas consuming terrestrial plants, some of our recent work suggests that they may not be able to gain much nutritive value from this source, since another analogy with cows is that they can only digest their food with the help of endosymbiotic bacteria in their digestive tract. In iguanas these microorganisms appear to specialize in breaking down marine plants, so their gut bacteria may not be able to digest terrestrial vegetation all that well.

Marine iguanas are also famed for their diving abilities, although only the largest animals of each island (thus only around 5% of the population) descend as much as 30 m (98 ft) below the rough surface waters to reach better pastures. Depending on location and food conditions, however, most dives range within 1–5 m (3–16 ft) only, lasting anywhere from 2 to 45 minutes. To determine if diving is stressful to them, we took blood samples while they were underwater, but we could not detect any increase in stress hormones, suggesting that diving is just like regular grazing to marine iguanas. Previous literature showing a 'dive response,' in which they slow down their heartbeat, appears to take place only when animals are forcibly submerged, though we

RIGHT: The Galápagos marine iguana is the only lizard in the world that routinely swims and dives for its food.

LEFT: Using small lobate teeth set very close to the edge of its jaw, a marine iguana grazes short green and brown algae off the seabed while rainbow wrasses dart in for dislodged invertebrates. BELOW: Prior to the major El Niño of 1982–83 the iguana colonies of Fernandina Island were truly enormous, but their numbers have not recovered to the scale seen in this photograph from the 1970s.

intend to investigate further whether or not these old studies reflect the majestic free-diving prowess of our beloved critters.

Meanwhile, we have been studying why so few animals actually dive, and this appears to be dependent upon foraging efficiency. The waters in Galápagos are usually quite cold, down to 9°C (48°F) in some iguana feeding areas. Being cold-blooded lizards, this limits how long they can spend actively foraging because, as their body temperature drops, so does their ability to sustain their bite rate. Only the largest individuals can maintain a sufficiently high activity rate to make foraging at depth a winning proposition: most iguanas simply forage intertidally, taking advantage of the low ebb. Also, because they swallow seawater liberally with their food, they need to eliminate enormous amounts of salt, and for this marine iguanas have developed the most efficient salt-excreting glands in the terrestrial realm. It is one of our distinct Galápagos pleasures to observe the strange little spluttering geysers of sneezing marine iguanas against the sunset!

Marine iguanas need to spend as much time as

RIGHT: A huge Isabela male in prime breeding condition can weigh up to 12 kg (26 pounds) and relies on his size to attract the most females.
BELOW LEFT: At the height of the mating season, males do not feed or leave their small temporary breeding territories, which may cover just the surface of a large boulder, displaying by licking the substrate and bobbing their heads vigorously.
BELOW RIGHT: Termed behavioral thermoregulation, an iguana avoids overheating by adopting a 'sky pointing' posture that reduces the amount of sunshine striking its body.

possible at a sufficiently high body temperature to enable their gut bacteria to process food optimally. Thus, whenever they emerge from the cold wave-wash, they bask on the lava by exposing their entire body perpendicular to the warming rays of the sun. Their black color helps absorb heat rapidly (but when they turn reddish during the mating season their rewarming times, are vastly slowed down). Once their desired temperature has been reached (36–38°C or 96.8–100.4°F), to avoid overheating they simply reorient themselves facing directly into the sun in a 'sky pointing' posture, thus shading their flanks while raising their bodies to reduce contact with the hot substrate — this efficient trick is often referred to as 'behavioral thermoregulation.'

Another striking, though often overlooked, feature of marine iguanas is that on some islands they remain tiny throughout their lives, while on others they can grow to gigantic proportions: the largest Genovesa Island males weigh less than 1 kg (2.2 lb), while on Isabela Island, prize males tip the scales at a whopping 11 kg (24 lb) or more.

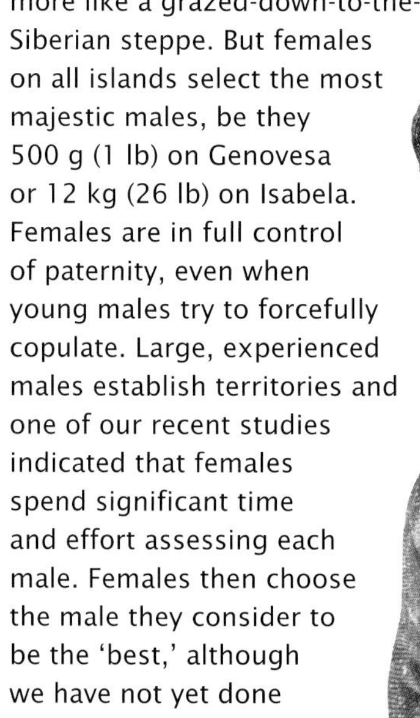

This greater than tenfold difference is among the largest in any species of vertebrates.

Why such extraordinary range? Our recent data suggest that body size is a tradeoff between successful reproduction and survival. In the nutrient-rich upwelling waters of southwestern Isabela, food is superabundant — this is like a tall-grass prairie for iguana-buffalos. Genovesa, on the other hand, looks more like a grazed-down-to-the-ground Kansas or Siberian steppe. But females on all islands select the most majestic males, be they 500 g (1 lb) on Genovesa or 12 kg (26 lb) on Isabela. Females are in full control of paternity, even when young males try to forcefully copulate. Large, experienced males establish territories and one of our recent studies indicated that females spend significant time and effort assessing each male. Females then choose the male they consider to be the 'best,' although we have not yet done

the paternity studies to verify that these males do, in actual fact, sire more young. What is clear is that females usually select the largest of the territorial males. Thus males have to grow, grow, grow to gain the reproductive advantage. This sexual selection pressure appears to be the mechanism explaining large male body sizes.

Females, too, have to become large to lay the biggest eggs possible. Large eggs mean large hatchlings, and large hatchlings are more difficult for lava herons or snakes to eat. On Genovesa, all females lay just one egg, comprising up to 25% of their body mass. On Isabela, females allocate the same proportional amount of material to their eggs, but here this is enough to produce six eggs per female.

The El Niño effect

Given the selective pressures described above, one would think that all marine iguanas would become giants, but then comes El Niño. With sea surface temperatures reaching up to 31°C (88°F) across the archipelago, sometimes for months on end, most algae die off, with only low-nutrient, poorly digestible types remaining available to the iguanas. Under these conditions of extreme food shortage, most individuals become severely stressed, beginning with visible muscle wastage.

In 1997, we visited a number of islands throughout the archipelago at the height of a major El Niño that lasted about five months. Although iguanas everywhere were suffering, Punta Espinosa on Fernandina Island in the west seemed especially hard hit. The exposed nature of the point, where great concentrations of iguanas can normally be seen, made it easy to observe the numerous dead and dying lizards. When we returned in 1998, many months after the El Niño had ended, the carcasses littering the ground had turned into skeletons. Punta Espinosa had become an open-air marine iguana graveyard.

ABOVE: Nesting females fight over space to lay their eggs in the sand. BELOW: A brightly colored Española male nods his open-mouthed display among a group of smaller, drab females.

ABOVE LEFT: Severely stressed by El Niño conditions, an emaciated male, though barely able to walk, may still recover if cold conditions return in time. Such large individuals are at greatest risk of starvation, as more food is required to sustain them. ABOVE RIGHT: Carcasses littered the colony at Punta Espinosa, Fernandina, after the 1997–98 El Niño. BELOW: The exceptionally bright colors of breeding Española males comes as a temporary tradeoff to efficient heat-absorption, best achieved through black skin, a metabolic requirement to offset cool sea temperatures.

Stress response to famine

As horrible as the 1997 El Niño was for the marine iguanas, it was an outstanding opportunity for us physiological ecologists. Because iguanas need to bask in the sun after eating, we can walk down the coastline and readily count them to obtain good population estimates. Comparing these counts before and after severe El Niños tells us that as many as 90% of iguanas may die, although survival differs greatly between islands. What determines which iguanas will live and which will die became an intriguing question, considering that all are faced with identical famine conditions, yet some cope adequately even as their neighbors starve to death.

Although research is ongoing, our current work suggests that physiological responses to stress play a major role, with survival appearing to depend upon an individual's ability to release stress hormones. One of the critical elements in this regard is corticosterone (the equivalent to cortisol in humans). Hormonal responses to stress are nearly identical in all vertebrate species, and corticosterone has a bewildering repertoire of effects in the body. Most researchers believe its release is critical for surviving unpredictable, stressful events, yet the precise mechanisms for this currently remain unknown. However, corticosterone release can be a diagnostic tool to determine how much stress an individual is experiencing, measurable in blood samples. Catching iguanas is not particularly difficult, but we had to do it quickly, as corticosterone release is the culmination of a cascade of hormonal events that begins in the brain, a process generally completed in two to three minutes. So if we could collect a blood sample within three minutes of initiating capture, we were reasonably confident that the corticosterone concentrations reflected the original levels when the animal was still at rest.

Our data indicate that corticosterone levels are generally low under La Niña feasting conditions, but during El Niño famines an individual's body condition worsens and corticosterone rises. One critical function of this hormone is that it stimulates the breakdown of protein, converting it to glucose. Glucose, in

turn, is the primary energy molecule for the body, fueling most metabolic processes. Without glucose from food, an animal will first convert stored fat and, when this runs out, resort to protein in muscle tissues. We believe that we were observing iguanas in the process of metabolizing proteins as a last ditch effort to survive.

This makes sense. Because the termination of an El Niño is unpredictable, holding on for just a few more days, even if it means metabolizing essential proteins, might make the difference between survival and death. Just as in humans, however, individual iguanas release different amounts of corticosterone in response to different stressors, including famine. We discovered that these levels predict their probability of survival, with animals releasing the largest amounts actually being the most likely to die. In other words, the higher the corticosterone, the earlier the individual starts to metabolize protein, and the less likely it can extend its survival for those last critical weeks. So who lives and who dies appears to depend upon the strength of the corticosterone response to famine.

Mortality is also related to body size, with the largest animals most at risk of starvation. This starts to explain why iguanas exhibit different sizes on different islands. Females want to be bigger to produce more eggs and males likewise to attract females, but El Niño prunes the largest animals from the population. In an attempt to compensate for being big, large animals shrink in length during severe El Niños. We have found that marked iguanas are shorter after an El Niño than they were before it, even after they have otherwise recovered. How shrinkage occurs is not known, and it is a focus of our current research, but we believe that iguanas are reabsorbing calcium from their vertebral bones. Shrinking may serve to protect an animal's body condition (they need less food to support a smaller body), delaying the worsening starvation process and its associated rise in corticosterone.

Other stress factors

Marine iguanas are not only stressed by food shortage, but also by Galápagos hawks. These ferocious predators can develop habits of hunting in groups, which we observed on Santa Fe Island at one of our long-term field sites. Here, over a period of six years, a family of hawks grew from two to five. Hunting together, they confused the marine iguanas, thus gaining exceptional success compared to the early years when there was only the original

LAVA LIZARDS — THE SMALL COUSINS

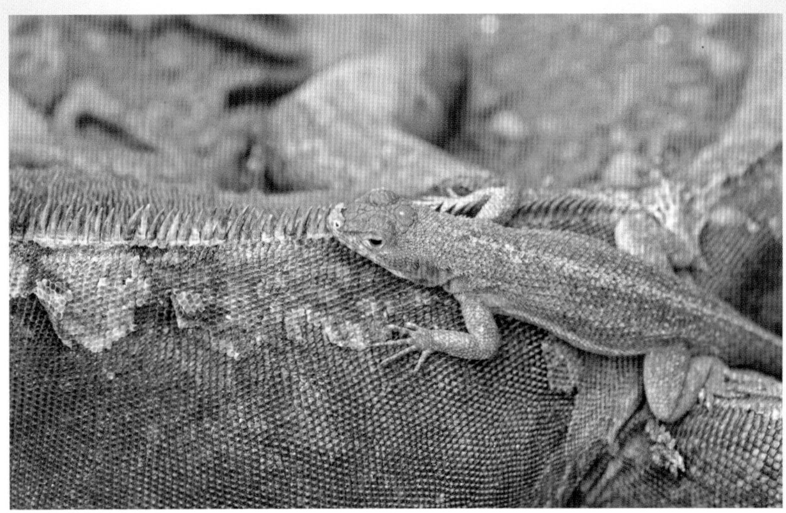

A female Galápagos lava lizard, *Microlophus albemarlensis*, basks on the warm back of a marine iguana.

Much smaller and strictly terrestrial, lava lizards have diversified into seven different species, one of them widespread on a number of islands (Santa Fe, Santa Cruz, Baltra, Santiago, Rabida, Isabela, Fernandina), and the others each isolated on a single island of their own (Española, San Cristóbal, Floreana, Pinzón, Marchena, Pinta). Hardly a square yard of arid lava terrain is left unclaimed by one of these fiercely territorial, robust and busy little characters.

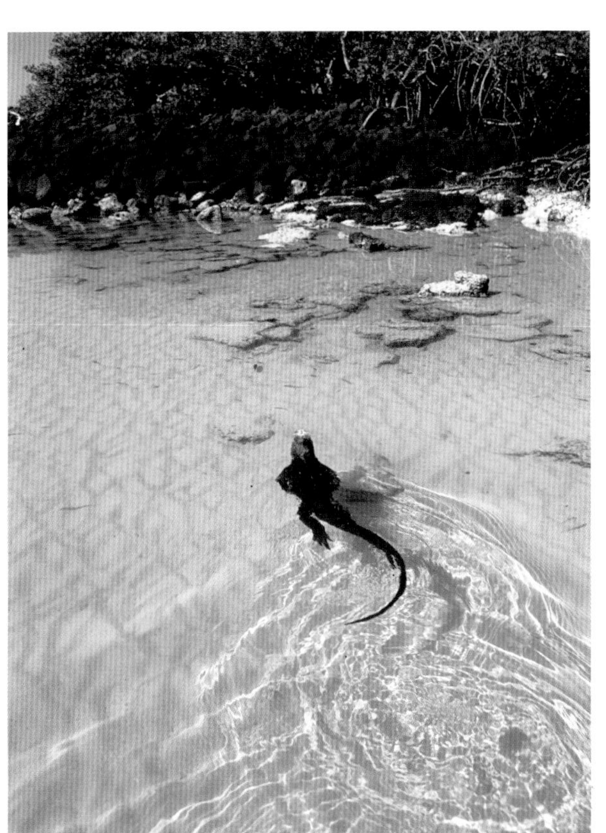

LEFT AND OPPOSITE: At the onset of the hot season in December, when the breeding season begins, large males can be seen traveling long distances, both by walking and swimming, along the coasts of Santa Cruz and other islands in search of preferred mating areas. Such energy expenditure can have dire consequences if an El Niño sets in.

TOP AND ABOVE: During severe El Niño conditions, some iguanas eke out an existence eating terrestrial plants, although their digestive system is adapted to process lush green alga that thrives when upwelling is strong.
ABOVE RIGHT: Great blue herons are major predators of small hatchlings.
RIGHT: Females are especially vulnerable to predation by Galápagos hawks when nesting on open sand away from lava crevices where they can escape.

pair. But the Santa Fe marine iguanas have partially compensated for the situation: we have also observed them listening to sentinels, Galápagos mockingbirds who utter a loud, distinctive alarm call for hawks. As soon as the mockingbirds cried 'hawk,' marine iguanas disappeared among the boulders. This response came as a complete surprise to us. There are few accounts in the literature of one species using the alarm calls of another, and, although lizards hear reasonably well, none for a reptile. We became aware of this while making detailed observations of marine iguana reproductive behavior. When a hawk flew over, the iguanas would dive for cover. But when a hawk landed out of sight of the iguanas, they still disappeared. The only possible cue was the mockingbird alarm calls. Detailed observations and experiments eventually proved this hypothesis correct.

El Niño devastation and hawk predation are natural phenomena and, not surprisingly, marine iguanas have adapted to deal with them in remarkable ways. Adult iguanas can expect to survive several El Niño famines during their lifespan of up to 30 years (short for a reptile this size and half that of land iguanas) because these events, historically at least, occur approximately every seven years. Humans, however, have introduced dangers that the iguanas are not adapted to cope with. For example, on 17 January 2001, the oil tanker *Jessica* ran aground on San Cristóbal Island. Within a week, roughly 2.8 million liters (750,000 million gallons)

Photo courtesy Mark Jones

of diesel and heavy bunker oil spilled into the sea and spread throughout the southwestern islands of the archipelago.

Even though oil was visible in tide pools, and 70% of individual iguanas at our Santa Fe study site (which was directly downwind and down current of the spill) showed residue on their skin, the oil killed few animals immediately. The absence of carcasses shortly after the accident led many observers, and the media, to conclude that its effects had not been drastic. We were fortunate, however, to have been collecting corticosterone samples from these iguanas just three days before the spill and could thus compare concentrations in blood taken just prior and immediately after it happened. We weren't surprised to find that marine iguanas responded to the oil with large increases in corticosterone. They were stressed. Because levels of this magnitude were associated with approximately 40% mortality during El Niños, we predicted substantial delayed effects.

This proved correct, and in the end we found the oil spill had a devastating impact. Over 60% of the marine iguanas on Santa Fe Island did not survive the year, even though normal survival in a non-El Niño year is over 95%. Furthermore, the oil spill killed different animals than does an El Niño, when the largest and oldest iguanas are the most likely to die, leaving the younger animals to survive and reproduce when conditions improve. In contrast, both young and old animals succumbed to the oil. Eight years later, the iguana populations on affected islands have still not completely recovered.

Predation

Dogs and cats provide a second example of the dangers introduced by humans. Island species that evolved without the presence of mammalian carnivores often lose their ability to recognize large mammals as potential predators. Marine iguanas are no exception. They do not initially identify dogs and cats as threats. Furthermore, unlike their stress response during famines and oil spills, they do not release corticosterone when stalked, even though this hormone is normally very important in helping animals survive predator attacks — if a mainland iguana were stalked by a dog or cat, its corticosterone would increase dramatically.

We explored potential reasons for this lack of a corticosterone response by simulating a predation attempt by a human. Marine iguanas appear tame (they don't run away when approached) because they also don't recognize humans as potential predators. Similar to the lack of fear response to dogs and cats, the iguanas do not release corticosterone during experimental chasing by humans. Only after handling the animal does corticosterone increase to expected levels. The same happens when iguanas survive attacks by real feral predators, yet they never seem to learn, and the distance at which they start running from such potential threats remains insufficient to effectively avoid predation. Although successfully eradicated by the Galápagos National Park in recent times, feral dogs once roamed Floreana, Santa Cruz and parts of Isabela islands, while cats are still rampant on all inhabited islands except Baltra. Both prey heavily on marine iguanas.

In conclusion, the Galápagos marine iguanas are a terrific example of animals highly adapted to their habitat. They have evolved to successfully cope with the major constraints of island living, have adapted their body sizes to utilize the food available to them, have developed mechanisms to choose appropriate mates, and have learned to use mockingbirds to warn them of lurking hawks. They have also evolved to cope with the major stress in their lives: El Niño famines. However, as with many tame island species, they may not have the time to adequately evolve and adapt to human-induced changes, including pollution and introduced predators. In the end, much more information will be needed if we are to conserve the world's only seafaring lizard.

ABOVE: Feral dogs, once common on Isabela, Santa Cruz and Floreana, can have a major impact on marine iguanas, who demonstrate no fear response when stalked. BELOW LEFT: Sociable by nature, iguanas pile together in a classic scene along the shores of Fernandina Island. After returning from feeding, whether on wave-washed rocks or by diving, they spend hours basking in the sun to reach a body temperature of 36–38°C (96.8–100.4°F) that enables them to efficiently digest their food. BELOW RIGHT: The stranded tanker *Jessica* spills its oil cargo in 2001.

Heidi Snell/Visual-Escapes.smugmug.com

Land Iguanas

Emergence of a New Species

Gabriele Gentile

Biology Department, Tor Vergata University of Rome, Via della Ricerca Scientifica 1, 00133 Rome, Italy. <gabriele.gentile@uniroma2.it>

Dr. Gabriele Gentile is a researcher at Tor Vergata University of Rome, where he teaches courses in Conservation Genetics and Nature Conservation. He is a zoologist with a broad interest in evolutionary biology. Since 2003, he has been coordinating a project aimed at the conservation of the land iguanas from Galápagos, in close collaboration with the Galápagos National Park (GNP) and the Charles Darwin Research Station (CDRS).

OCEANIC ISLANDS FREQUENTLY ACT as natural laboratories where evolution's experiments are most apparent. In fact, a combination of geographic isolation, restricted area and limited resources may prove crucial for the development and diversification of particular forms and their adaptations. It is for these reasons that, over the past two centuries, the Galápagos Islands have been the subject of innumerable studies, all contributing to unveil the uniqueness of these islands.

Iguanas are certainly among the most emblematic species of Galápagos. Two strikingly different forms exist: marine and terrestrial. Marine iguanas (*Amblyrhynchus cristatus*) occur in most islands of the archipelago, whereas land iguanas once inhabited many areas of the archipelago: *Conolophus pallidus* on Santa Fe, and *C. subcristatus* on Santa Cruz, Baltra, Plaza Sur, Santiago, Rábida, Isabela and Fernandina. Unfortunately, these are now restricted to just a few localities on some islands, and gone from others, such as Santiago and Rábida, their current distribution having been much reduced due largely to the direct and indirect impact of human activities across the archipelago.

BELOW: The Santa Fe land iguana, *Colonophus pallidus*, is found on just that one, very dry island. RIGHT: The Galápagos land iguana is the most widespread, found on Santa Cruz, South Plazas, Baltra/Seymour (seen here), Santiago (extinct), Isabela and Fernandina, though its coloration varies slightly from island to island.

Conservation and genetics

I first met Howard and Heidi Snell in the year 2000. That year I also saw my first land iguana. I was impressed by Howard and Heidi's story: two young herpetologists who, in 1977, came to the CDRS to offer their contribution in saving some populations of land iguanas from the brink of extinction. That is when I decided I would start studying these animals too. Taking advantage of a 'Brain Gain' program financially supported by the Italian government, in 2003 I set up a team of people who have been collaborating together ever since. Howard and Heidi, and also Michel Milinkovitch, Cruz Marquez and Wacho Tapia, were in this original group from the beginning. At that time, the prime goal of the project, in cooperation with the GNP and the CDRS, was to use molecular genetics to improve conservation of land iguanas. In fact, land iguanas are classified as VU-D2 in the IUCN Red List (2007) meaning they will become 'critically endangered or extinct in a very short period.' The project rapidly became a multidisciplinary

program of investigation that took advantage of collaborations with many other institutions all over the world, including the Laboratory of Epidemiology, Parasitology and Genetics (LEPG) of the GNP, the California Academy of Science, the University of Leeds in the United Kingdom, and others.

One of the issues we wanted to address was the origin and diversification of the land iguanas in Galápagos. We did not start from scratch. Other scientists had previously showed that marine and land iguanas share a common ancestor that lived some time at least 11 million years ago when the two lineages started diverging. In particular, Kornelia Rasmann and collaborators had previously suggested that this ancestor had already colonized the islands when the split between the marine and terrestrial lineages occurred. They also noted that the ancestral species must have lived in an archipelago that did not possess its current shape, as none of the existing islands were formed before 5 million years ago. Since that original split, both types shared a parallel evolutionary history, with no clue to further diversification prior to the Pleistocene to be documented along either of the two lineages. In fact, during that epoch repeated rises and falls in sea levels determined the conditions for land iguana dispersal and diversification throughout the major western and central islands of the archipelago.

The pink iguana

I first heard of this 'myth' in 2000, when I was just about to hike up Wolf Volcano, the highest and one of the youngest shield volcanoes on Isabela Island. At that time, I was collaborating on a program based at Yale University aimed at conservation genetics of giant tortoises. Howard alerted me to the possibility of running into a peculiar type of land iguana with a strange phenotype: pink all over with black stripes on the rump. In fact, this form had been seen on just a few occasions since its

ABOVE: Normally solitary, females on Fernandina associate briefly with a large male (second from left) during the mating season, sharing his territory and spacious lava burrow for a few days.
BELOW: The pink iguana of Wolf Volcano is a new species whose formal description is in press.

A VOLCANIC NURSERY

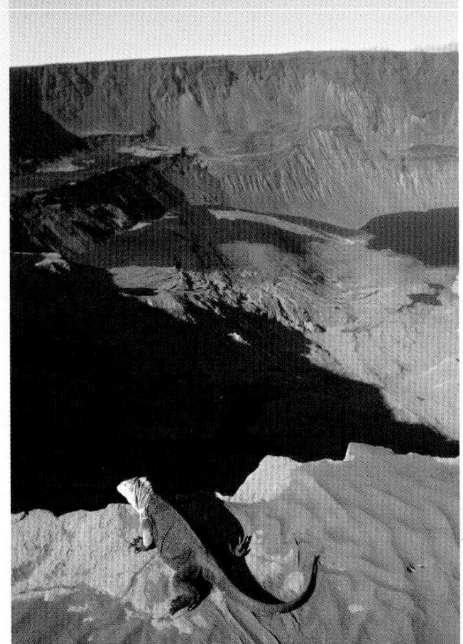

Fernandina Island, free of introduced predators and still virtually pristine, harbors by far the largest remaining population of Galápagos land iguanas. Amazingly, when the July nesting season arrives females from all over the island migrate to the summit area at nearly 1500 m (4900 ft), a trek that may take them three days of almost nonstop walking when coming from the coastal region. Many females choose open areas of sun-warmed volcanic ash near the edge of the caldera, but a substantial number continue on down into this gigantic, 800 m to 900 m (2600–2950 ft) deep pit. They scramble down vertiginous crumbling rock walls, across gaping crevasse fields and over bare lava flows to reach their favorite nesting sites on the plantless caldera floor, where temperatures are higher than on the rim above. In a matter of a week or two, the females compete for the best spots to bury their eggs before reversing their grueling journey. Three months later, the tiny hatchlings emerge into a barren world, from where they must escape quickly if they are to find cover from heavy hawk predation. Living on their yolk sac reserves, to survive they must reach the vegetated areas on the outer slopes of the volcano.

Tui De Roy

first sighting in 1986, when park rangers mentioned having accidentally encountered it during a field trip to the remote northern slope of Wolf Volcano.

I did not see any pink iguanas during the 2000 expedition, but the idea of this strange animal had become stuck in my mind and could not leave it. So, in 2005 and 2006 I organized two field trips aimed at searching for this intriguing animal and collecting data on it. Regrettably, I could not participate in the first trip, during which time four 'Rosadas' ('pink' in Spanish) were located and sampled — a great start.

Much encouraged, I would not miss the mission in 2006 for anything in the world. We had decided to start the hike from Piedra Blanca on the north shore. One of us had to remain at a beach camp, to carry out the haematological analysis of the samples that would be transported down 14 km (9 miles) every day by an assistant — from the upper camp at the edge of the gaping caldera, 1700 m (5500 ft) above sea level. With so few pink iguanas sampled the previous year, we all were worried about the success of this latest trip.

I remember the emotion I felt the first time I saw a pink iguana. It was a mixture of excitement and worry: What if it runs away before we collect the data? What if this is going to be the only one we can get? We had started collecting data immediately, although we had not reached the crater and the camp was not yet established. A second pink iguana was seen and captured right after the first. I remember that, despite being exhausted from the arduous hike, we only settled into camp after we made sure we hadn't missed any other iguanas nearby.

TOP: Tiny hatchlings are well camouflaged, but many fall prey to Galápagos hawks and snakes, their natural predators.
RIGHT: Males fight fiercely over prime territories, sometimes battling for hours or even days.

The pink iguana is certainly a strange animal! As soon as I saw the first individual, I realized that this was very different from the syntopic (living in the same place), well-known yellow form (*Conolophus subcristatus*) in many respects. It was not just a matter of color pattern, as there are other idiosyncrasies in the morphology that distinguish the two. The dorsal head scales are flat or almost flat in the pink form. Additionally, a prominent adipose dorsal crest is evident in males (present but less conspicuous in females), with small conical scales that never reach the size of those present in the spiny crests of yellow iguanas.

The color pattern is very distinctive, too. The head is pink; the body presents a typical black-striped pattern in the lower dorsal region. These irregular stripes, which are less evident on the ventral surface, run along the dorsal-ventral axis and their number is variable. In some individuals, the black marking may join to form a more complex pattern. The tail is dark.

Before hiking up Wolf Volcano, we had already genotyped the four pink iguanas sampled from the previous year. Results were striking. More than 7% of a large portion of mitochondrial DNA was different from the other two taxonomically recognized species (*Conolophus pallidus* and *C. subcristatus*). For comparison, the same genomic region shows differences of less than 2% between these latter two species. This result was highly unexpected.

The 2006 expedition turned out to be a great success. During 15 days spent on top of the volcano, we were able to observe and sample a total of 32

ABOVE AND BELOW: Nesting females inside the Fernandina caldera, an area later covered by fresh lava in the 1978 eruption.

RIGHT: The only known home of the pink iguana, Wolf is the highest and also one of the youngest shield volcanoes in Galápagos, last erupting in 1982. BELOW: Restricted to the northern sector of the volcano between about 900 m and 1700 m (2950–5600 ft), a pink iguana basks near the rim of the caldera. Although often caught in light clouds, the habitat here is arid, as most rain falls along the slopes at lower elevations.

individual pink iguanas, 21 males and 11 females. We also observed something that had not been noted before: the particular pattern of nodding, or 'head bobbing' behavior, demonstrated by pink iguanas. This display is important in iguanas as it is used in territoriality and courtship. Land and marine iguanas exhibit a very different pattern of such head movements. Even the two known land iguana species, *Conolophus pallidus* and *C. subcristatus,* each show their specific pattern. We were able to film nine pink iguanas displaying their head-bob, and all showed the same, highly distinguishable pattern, some elements of which are more reminiscent of marine iguanas than their terrestrial cousins.

Once back in the lab at Tor Vergata University in Rome, we immediately genotyped the new samples. All shared the same type of mitchondrial DNA, very divergent from those of the other two species of land iguanas. The lack of polymorphism, or varibility, was also reflected in an analysis of genetic variation at several microsatellite loci of the nuclear DNA. In fact, the pink iguana showed only about half of the genetic variation observed in the other two land iguana species. These results were evidence of the importance of genetic drift as the primary evolutionary factor of the pink iguana.

Of course, this led to many more questions. For example, is the pink iguana genetically isolated from the syntopic populations of yellow iguanas? Could this form possibly be the result of hybridization between marine and land iguanas? When did the pink iguana originate? What are the evolutionary relationships with the other Galápagos land iguana species? We could answer those questions taking advantage of the

recent advances in population genetics and molecular phylogenetics.

Although the pink and the syntopic yellow iguanas share 26% of alleles, none of the pink iguanas that we investigated was a first-generation hybrid between the two forms. Furthermore, none of the pink individuals incorporated genes from the other form at least in the last two generations, and only one yellow individual, which we also sampled in the area, showed possible mixed ancestry with a pink grandparent.

According to our genetic tests the pink and yellow iguanas of Wolf Volcano are genetically isolated, and hybridization between the two forms is indeed rare and has not impeded independent evolution and differentiation. That the yellow and the pink forms may occasionally exchange genes is not surprising, if considering that, on Plaza Sur Island near Santa Cruz, it has been observed that marine and land iguanas, two genera morphologically, ecologically and genetically very different, can still hybridize, generating a viable hybrid. Anyway, our data clearly reject the hypothesis that the pink iguana originated by hybridization between marine and land iguanas.

When we reconstructed the genealogical relationships between the pink iguanas and the rest of Galápagos land iguanas we were surprised again. In fact, in a treelike representation of the genealogical (phylogenetic) relationships, the pink iguana lineage resulted basal to the rest of all remnant populations of land iguanas. This means that the pink iguana represents the only remnant of a lineage which we estimated to have differentiated more than 5 million years ago. The pink iguana is the only evidence of ancient diversification along the Galápagos land iguana lineage. No analogous evidence has been found in marine iguanas, so far. In fact, the pink iguana documents one of the oldest events of divergence ever recorded in Galápagos species.

The combination of a long evolutionary

ABOVE AND LEFT: Some of the most notable characteristics of the pink iguana are its striking, often barred coloration and its fleshy, nearly spineless crest. Tracing its lineage through genetic analysis, the author concluded that this species shows one of the oldest cases of diversification among Galápagos vertebrates, having split from the other two land iguanas some 5 million years ago, when the island where it now lives did not exist.

RETURN OF THE BALTRA LAND IGUANA

The story of the Baltra land iguana population is marked by three contrasting twists of human history. First, in 1932 and 1933, an odd little scientific experiment by an American expedition saw several dozen iguanas transferred from their native island to nearby Seymour in an attempt to discover why they were common on one and absent on the other, in spite of the two islands' similarities and close proximity. Second, a combination of habitat destruction, heavy traffic and introduced mammals (dogs, cats, rats, goats) as a consequence of the World War II American airbase on Baltra, caused the population's sudden demise between 1948, when they were still common, and 1954, when they were gone — one of the fastest known cases of an island extinction. Third came the successful eradication of cats, together with a thriving captive breeding program, using Seymour Island survivors to repopulate their island of origin. Today, the repatriates are reproducing successfully, but some problems remain, especially escalating roadkills as the Baltra airport grows into an ever-busier hub for both tourists and residents.
Tui De Roy

ABOVE: A Baltra land iguana feeds on seaside heliotrope, *Heliotropium curassavicum*, growing along the shores of Seymour after the rains. BELOW: Cactus fruits and even pads, consumed spines and all, are an important source of food and moisture, especially on small, dry islands such as South Plazas.

history since the split of the marine and land lineages, the phylogeographic pattern of genetic diversification, and the restricted number of islands currently inhabited by land iguanas would indicate that the present differentiation and distribution of land iguanas throughout the archipelago is largely due to founder effect, geographic isolation and extinction. These three factors could well explain the observed lack of ancient mitochondrial polymorphism. According to this view, despite the long evolutionary time, the present-day genetic variation would have occurred within the time frame of the age of the presently emerged islands. If we do not consider the pink form, our data support this scenario. However, given the present distribution and pattern of genetic differentiation of land iguanas throughout the archipelago, the pink iguana clearly represents a conundrum. As far as we know, it occurs only on Wolf Volcano, which is considered younger than the oldest Isabela volcano, Sierra Negra, and almost as old as Cerro Azul (350,000 years). This would imply that the pink form, or a direct ancestor, colonized Isabela very recently (within the last 500,000 to 300,000 years) from a presently existing island where it has since gone extinct. Since the pink iguana is not known from anywhere else in the western islands, it would be most conservative to speculate a direct arrival to Wolf Volcano from the central islands. Such dispersal is not supported by the pattern of genetic variation of marine iguanas, for example, or by the prevailing ocean currents capable of dispersing such animals during torrential flood events, but it has been considered one of the routes of colonization of Isabela by giant tortoises. In historic times, land iguanas inhabited Santiago and Rábida, two central islands, whose radiometrically determined age are 780,000 and 920,000 years, respectively. Land iguanas became extinct at Santiago during the last half century, whereas fossil records from Rábida indicate that their extinction on the island occurred well before the discovery of the archipelago by humans. However, although Charles Darwin documented the occurrence of land iguanas at Santiago, and several older residents of Galápagos have reported seeing land iguanas on this island as recently as the 1950s and 1960s, none of those accounts mention pink iguanas.

Although Darwin, when he visited the Galápagos in 1835, collected a number of plant and animal species that have since vanished from those locations, he did not encounter the pink iguana. However, he remained

FAR LEFT AND LEFT: Almost nothing is known about the life habits of the pink iguana, seen here feeding on small herbs (left) and *Solanum erianthum* berries (right). Neither breeding behavior nor young have been observed, and nesting sites remain unknown. Despite several expeditions, only 32 individuals have been identified, suggesting the species may be endangered.

in the archipelago only five weeks, during which time he did not explore many sites, including Wolf Volcano. More surprising is that several scientists visited this area in recent decades yet missed the pink iguana too. This is certainly another conundrum.

It should now be clear to the reader that the pink iguana of Galápagos is a very special organism. It is much differentiated and distinguishable in many respects from the other two known land iguana species. For these reasons, we believed that it deserves appropriate taxonomic recognition as a new species. So, we elevated the pink iguana to new species rank, with the description currently in press.

The discovery of a new species of megafauna in one of the most intensely studied archipelagos in the world is truly astounding. Besides the taxonomic implications, the pink iguana is extremely important as it carries a substantial evolutionary legacy and opens a diachronic window — a kind of genetic chronicle over time — on the evolution of Galápagos iguanids, which was primarily a history of extinctions and recolonizations, most of which left few traces in the fossil records and in the genes of remnant populations.

The pink iguana is rare. Its restricted distribution and low population density are alarming. Both introduced cats and black rats, known predators of reptilian eggs and young, are numerous on Wolf Volcano, and so far we have found no juveniles or other evidence of successful reproduction in this enigmatic species. Our findings call for a conservation program aimed at evaluating the risk of extinction of this newly recognized taxon which, based on currently available data, would be assignable to the 'critically endangered' category by meeting criteria B and C of the IUCN Red List. Therefore, we will continue to investigate this unique species. There are many other questions that await answers. In agreement with the GNP, a management program for this species could also be envisioned, to prevent such a rarity from going extinct just as we make its acquaintance. To achieve this, we desperately need more funds, and are actively looking for supporters.

LEFT TOP: The ritualized head-bobbing threat display of the pink iguana is unique, and may serve as a behavioral barrier to interbreeding. LEFT BOTTOM AND BELOW: Though attacked by the resident male land iguana, a marine iguana on South Plazas sired several hybrids (left foreground) by mating with land iguana females.

Darwin's Finches

Studying Evolution in Action

B. Rosemary Grant and Peter R. Grant

Department of Ecology and Evolutionary Biology,
Princeton University,
Princeton, N.J. 08544,
United States.
<rgrant@Princeton.edu>

Dr. Rosemary Grant was initially trained at the University of Edinburgh, received a Ph.D. from Uppsala University, and was a research scholar and lecturer with the rank of professor in the Department of Ecology and Evolutionary Biology at Princeton University until she retired from teaching in 2008.

Dr. Peter Grant is the Class of 1877 Professor Emeritus in the Department of Ecology and Evolutionary Biology at Princeton University, having trained at Cambridge University and the University of British Columbia. Before joining Princeton in 1986, he taught at McGill University and the University of Michigan.

THE YEAR 2009 MARKED the 150th anniversary of Darwin's *On the Origin of Species by Means of Natural Selection*, a book that established the scientific basis for understanding how evolution occurs. Darwin did not completely solve the problem of explaining the book's title, the origin of a new species, because, among other reasons, he lacked knowledge of genes, chromosomes and DNA. Nevertheless, he developed the outline of an enduring theory of species formation as a three-step process. Speciation, according to his reasoning, begins with the colonization of a new environment, the population then diverges from the parent population as it becomes adapted to its environment through natural selection, and finally some degree of inviability or sterility evolves between them. He went further and suggested that investigations of closely related species might help us to understand the processes when he wrote, 'Those forms which possess in some considerable degree the character of species, but which are so closely similar to some other forms, or are so closely linked to them by intermediate gradations, that naturalists do not like to rank them as distinct species, are in several respects the most important to us.'

The finches of the Galápagos are one such important group. They appeared to make little impression on him while he was in the archipelago, but he did not forget them. Only two years later, back in England, he wrote: 'seeing this gradation and diversity of [beak] structure in one small, intimately related group of birds, one might fancy that, from an original paucity of birds in the archipelago, one species has been taken and modified for different ends' [Darwin, 1842].

Since Darwin's time, new techniques and insights from the fields of genetics, behavior and ecology have shed light on the general question of how species of animals, plants and microorganisms are formed.

BELOW: Closely related ground finches have developed specialized beaks to deal with different size seeds. From left to right: large ground finch, *Geospiza magnirostris*, Genovesa; medium ground finch, *G. fortis*, Santa Cruz; small ground finch, *G. fuliginosa*, Santa Cruz.

The finches that were named after him much later, in recognition of his scientific specimens and his insights about their evolution, continue to play an important role in these discoveries. They constitute an extraordinary radiation, uniquely suitable for the scientific study of Darwin's fancy, how 'one species has been ... modified for different ends' [Lack, 1947; Bowman, 1961, 1983; Grant, 1999; Grant & Grant, 1989, 2008]. Fourteen species have been derived from one over a relatively short time, and each is adapted to a different ecological way of life. Today, these closely related species occur both separately and together on different islands, which allows detailed study of each of Darwin's three stages of species formation. Some of the islands are in near-pristine condition, therefore any changes that occur on them are the result of natural causes rather than those induced through human activities. And changes do occur, both in the finches and in their food supply, because climatic conditions fluctuate strongly as a result of oscillations between droughts (La Niña), sometimes lasting as long as two years, and severe and prolonged rainfall (El Niño) (see also chapter: Sachs). A final reason that makes them so suitable is that none of the species has become extinct as a result of human interference. This cannot be said for many other radiations elsewhere in the world.

Thirteen of the Darwin's finch species live in the Galápagos Archipelago and one more lives on Cocos, an isolated island 700 km (435 miles) northeast of Galápagos. Who were the ancestors? Genetic evidence points to a group of seed-eaters (*Tiaris* and their relatives), allied to tanagers, as the closest mainland relatives. The same evidence also suggests the ancestors arrived in the islands about 2–3 million years ago. At that time, the archipelago was a very different place from what it is now. Three million years ago it consisted of perhaps only five islands. According to evidence from the minute 'shells' of foraminifera taken from cores at the bottom of the ocean, from east and west equatorial Pacific, the climate was hotter and probably wetter at that time. New islands were progressively formed from a hot-spot underlying present day Fernandina, as well as from a spreading center to the northeast (see also chapter: Geist). The climate became gradually drier, and the vegetation changed to a mixture of dry deciduous forest in the lowlands and evergreen cloud forest in the highlands. As the number of islands increased, so did the number of finch species. Thus, environmental change has been a prevalent feature throughout the history of the finch radiation.

Which is the oldest species? Again genetic comparisons among the finches offer an answer: the warbler finch. Also relatively old are the vegetarian and sharp-beaked ground finches, and the Cocos Island finch, which appears to have arisen in the Galápagos Archipelago. The five species of tree finches and the other five species of ground finches evolved later, during just the last million years.

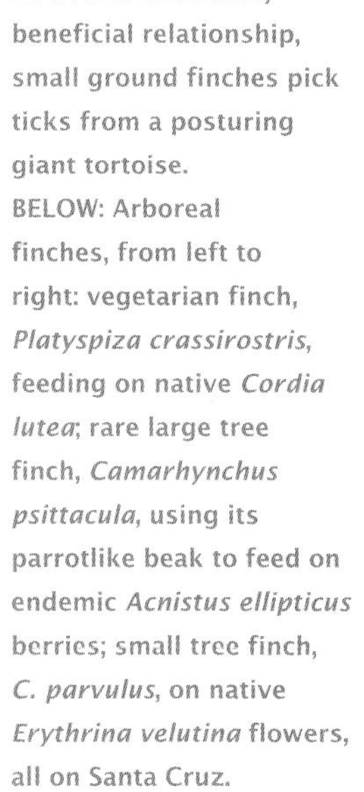

ABOVE: In a mutually beneficial relationship, small ground finches pick ticks from a posturing giant tortoise.
BELOW: Arboreal finches, from left to right: vegetarian finch, *Platyspiza crassirostris*, feeding on native *Cordia lutea*; rare large tree finch, *Camarhynchus psittacula*, using its parrotlike beak to feed on endemic *Acnistus ellipticus* berries; small tree finch, *C. parvulus*, on native *Erythrina velutina* flowers, all on Santa Cruz.

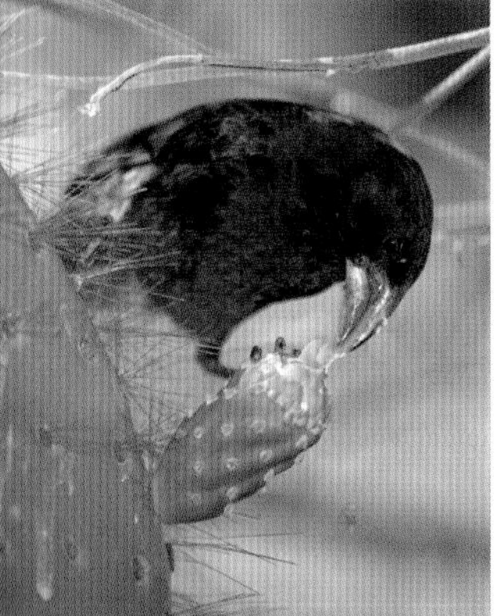

ABOVE: Ground finches with longer beaks have also adapted to differing ends. Left to right: large cactus finch, *Geospiza conirostris*, Genovesa; cactus finch, *G. scandens*, Santa Cruz; sharp-billed ground finch, *G. difficilis*, Genovesa. BELOW LEFT AND RIGHT: On Wolf and Darwin islands a form of *G. difficilis* dubbed the vampire finch has developed the habit of drawing blood from the wing feathers of nesting boobies, and breaking unattended eggs by rolling them against rocks.

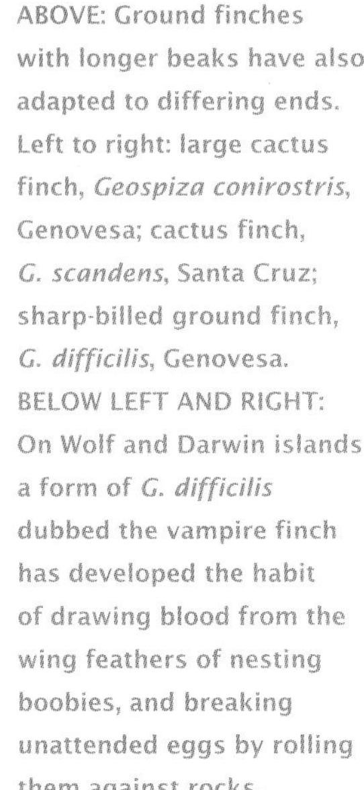

In plumage, Darwin's finches are rather dull and uniform. They build similar nests, and have similar courtship displays, yet the species differ conspicuously from each other in body size, and in beak size and shape. Beaks act as tools, and finches use their different beaks to exploit different foods in the dry season. For example, with their long beaks, cactus finches probe *Opuntia* cactus flowers and fruit to feed on nectar, pollen and seeds. The large ground finch with its massive, deep beak can crush larger and harder seeds than any other species — such as those produced by *Tribulus* and *Cordia* plants. At the opposite extreme, the fine needle-like bill of the warbler finch is a specialized tool for feeding on small insects, picking them up like forceps.

Soon after the finches first colonized the islands, they began to diverge from their mainland relatives. We can only infer what happened at that time. A factor that must have contributed to their success is their flexibility in developing new feeding skills under very different conditions from their original habitat. We see signs of remarkable behavioral adaptability, especially in the dry season when food becomes scarce. For example, the woodpecker finch uses a twig or cactus spine to extract beetle larvae from holes in branches, and even has the ability to modify the tool to the appropriate size. On the far northern islands of Wolf and Darwin, the sharp-beaked ground finch pecks at the base of wing and tail feathers of nesting boobies, draws blood and drinks it. And if they cannot crack open an egg of the unfortunate booby with blows of the beak they brace their heads against the ground and kick it against a sharp rock to crack it open, whereupon they gain a rich reward of calories and moisture. Nowhere else in the world are these feeding behaviors to be seen in any bird.

Many theoretical papers have been written about the profound genetic changes that might be expected

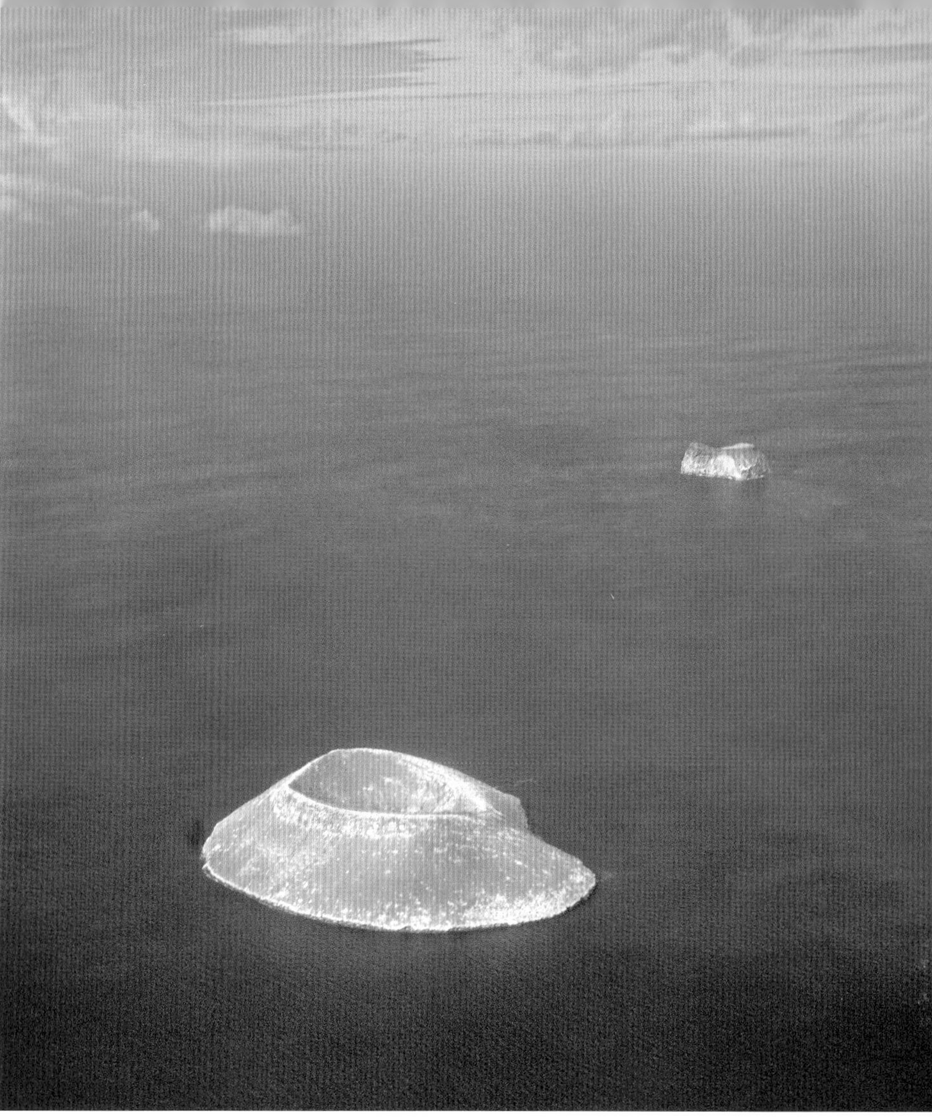

during and after the founding of a new population. Actual colonization of a new location has rarely been witnessed because it occurs exceedingly rarely on a human time scale. We were lucky to be in the right place (Daphne Major Island) at the right time (1983), when three males and two females of the large ground finch, *Geospiza magnirostris,* colonized the island. These birds and 14 of their offspring died in the next two years, leaving a single sister and two brothers. Later they bred with each other, and so did their offspring. As expected from such intense inbreeding their offspring did not survive well. But immigration continued, at a trickle. Genetic analyses tell us that the immigrants came from the neighboring islands of Santa Cruz and Santiago. One immigrant, a male, was exceptionally fit. He bred repeatedly with the residents and introduced not only his genes but also a novel song type to the island. Gradually the population increased in size, and today it comprises approximately 30 breeding pairs. However, even 25 years after the population was founded the large ground finches have not changed much in body size or beak characteristics. Their normal foods were present on the island when they arrived, and so they have not had to adapt to a new food supply, or emigrate. Evolution might be much more rapid in other situations where immigrant finches encounter novel feeding conditions on a new island.

Divergence

Evolution by natural selection occurs when the following three conditions are met. First, there must be variation in a character, for example the size or shape of the beak. Second, as a result of environmental change some individuals survive better than others owing to the particular traits they possess, for example conditions whereby birds with large beaks survive better than those with small beaks. Third, in the following generation the survivors pass on this trait, for example large beaks, to their offspring because the trait is genetically heritable.

Darwin understood these factors, and even anticipated our findings remarkably closely, when he wrote to clarify what he meant by natural selection: '... I would now say that of all birds annually born, some will have a beak a shade longer, and some a shade shorter, and that under conditions or habitats

ABOVE LEFT: Medium ground finches, *G. fortis*, show highly variable beak sizes. All ground finches share colorations ranging from mottled grey in juveniles and females to black in males, getting darker with age. Bill color reflects age and reproductive state; all-black indicating full breeding condition.
ABOVE: Daphne Major seen from the air, the site of the authors' ground-breaking research.
LEFT: Using its long beak, a cactus finch, *G. scandens*, one of the species studied on Daphne, can feed on *Opuntia* cactus nectar, pollen, undeveloped ovules and seeds, enabling it to breed even in dry conditions when others fail.

ABOVE: A breeding male cactus finch patrols his territory in prime coastal *Opuntia* cactus forest on Santa Cruz, a habitat it shares with at least six other species (small, medium and large ground finch, small tree finch, vegetarian and woodpecker finch).
BELOW: The typically domed nest of a small ground finch is balanced among the thin branches of a *Croton scouleri* shrub, holding a maximum clutch size of four eggs.

of life favoring longer beak [sic], all the individuals, with beaks a little longer would be more apt to survive than those with beaks shorter than average.' [Darwin 1867, in Burkhardt et al, 2005].

Daphne Major is an ideal island for the study of divergence by natural selection, owing to its small size and large populations of two species: the medium ground finch (*Geospiza fortis*) and the cactus finch (*G. scandens*). Finches do not migrate, many can be captured, measured, uniquely banded and observed over many years because they are conspicuous, and individuals vary markedly in size and beak shape. By measuring parents and their adult offspring we determined that beak size and shape, and body size are all strongly inherited in both species. We then followed the survival and reproduction of banded individuals across 36 years. During severe droughts, the majority of finches may die, and the interesting question is why some survive and most do not. In 1977, for example, when more than 80% of medium ground finches died, the answer was that survivors were large birds with large beaks capable of cracking large and hard *Tribulus* seeds. These were the only seeds remaining in relatively high abundance after most of the small and soft seeds had been consumed, and were too large and hard to be cracked by birds with small beaks. In the following year, the surviving large adults mated with each other and produced offspring that were, like their parents, large.

An unprecedented amount of rain had fallen over an eight-month period during the exceptional El Niño event of 1982–83, resulting in prolific growth of grasses, vines and many more species of plants, all producers of small and soft seeds. Vines and herbs smothered *Opuntia* cacti and the low-lying *Tribulus* plants, which provide seeds for the cactus finch and the medium ground finch respectively. When the next drought occurred in 1985, the resulting seed bank consisted mainly of small seeds. Thus the island had been converted from a producer of predominantly large and hard seeds to a producer of small and soft seeds. This time, birds with small and relatively pointed beaks had an advantage in being able to feed rapidly and efficiently on small seeds, and as a consequence of this and their small body sizes and relatively low energy needs, they survived better than large birds.

Over the next 30 years, droughts occurred at intervals of four to six years, selection oscillated in direction, and whether large-beaked birds or small-beaked birds survived best depended on two things: the types of seeds available to them, which in turn depended on the preceding conditions of rainfall and vegetation, and the abundance of large ground finches, which compete for *Tribulus* seeds with medium ground finches during droughts. Average beak size and shape characteristics of both medium ground finches and cactus finches changed repeatedly due to natural selection. By 2008, the beaks of medium ground finches had become smaller and more pointed, on average, than in 1973 when we began our study, and the beaks of cactus finches had become generally smaller and blunter.

Following Daphne populations over 36 years has shown us that evolution by natural selection of both species (cactus and medium ground finches) can occur over short periods of time when the environment changes. Furthermore, it can be measured, and it can be interpreted. Evolution means change in the genetic composition of a population.

What are the genes involved when finch beaks evolve? A beginning has been made in answering this question. Our colleagues, Cliff Tabin and Arkhat

LEFT: A small ground finch, *G. fuliginosa*, plucks the minute fruiting bodies of *Cryptocarpus pyriformis* salt bush, Santa Cruz.

BELOW: A ground finch with a very large vicelike beak cracks a hard *Maytenus octogona* fruit on Santa Cruz. Showing intermediate characteristics between medium and large ground finches, *G. fortis* and *G. magnirostris*, this individual may be a hybrid as they are known to interbreed, though rarely. Ongoing studies on Santa Cruz are delving deeper into the genetic exchange between these two species.

Abzhanov, have discovered two genetic pathways that affect early development of Darwin's finch beaks. Interestingly, the same genes are important in the development of facial features of other vertebrate animals, including humans. The two genes produce a bone morphogenetic protein, Bmp4, and calmodulin, CaM, respectively. These function as signaling molecules during cellular differentiation. They affect growth along different beak axes and, importantly, they do so differently in different species. Bmp4 influences the depth and width of the beak, whereas the calmodulin-signaling pathway is involved in beak length development. The two genes are expressed independently in the developing embryo, and so a change in just one produces a beak of a different shape.

A barrier to interbreeding

Speciation is completed when two populations that have diverged in geographical isolation (for example, on different islands) are so different that when they

ABOVE LEFT: The vegetarian finch, *Platyspiza crassirostris*, has the largest body of all Darwin's finches, but its cutting beak is adapted to a diet of juicy leaves, buds, flowers, berries and even the bark of alkaline-rich manzanillo or poison apple tree, *Hippomane mancinella*, seen here.
TOP RIGHT AND RIGHT: The woodpecker finch, *Camarhynchus pallidus*, has developed woodpeckerlike habits, hammering and peeling dead bark and extracting wood-boring insects with the use of a spine or twig in lieu of a long tongue.
BELOW RIGHT: The Cocos finch, *Pinaroloxias inornata*, is the 14th Darwin's finch, restricted to Cocos Island some 700 km (435 miles) northeast of Galápagos.
BELOW: Genetic analysis suggests the small, insectivorous warbler finch, *Certhidea olivacea*, is the oldest of the 13 species of Darwin's finches found in Galápagos today.

later encounter each other on the same island they can co-exist with little or no interbreeding because they do not recognize each other as potential mates, or they fail. Alternatively, they might interbreed but their offspring survive poorly or are infertile. We found that a barrier is formed by differences in beaks and song; a finch does not court and pair with another if it is very different in beak size or shape, or in some features of the song. Unlike beak size and shape, which are largely genetically determined, song is learned during a short period of time between 10 and 40 days after hatching. The time coincides with the last few days in the nest and the fledgling period when young are fed by their parents and the father is singing. Song is thus learned early in life in association with parental beak characteristics and, in typical Lorenzian imprinting fashion, influences the choice of a mate later in life.

The barrier to interbreeding is not watertight; under some circumstances it leaks. This happens rarely when a young finch learns the song of another species, for example as a result of another species taking over the nest, 'inheriting' an egg that is left behind and raising the offspring to independence; or when a dominant close neighbor of another species sings loudly and thus renders the paternal song ineffective. In these cases, a young bird learns the song of its foster father belonging to another species, and if it survives it mates with a member of that species and produces hybrid offspring. At the start of our research we expected hybrids to be relatively unfit, but to our surprise we discovered that they not only survived, they were capable of breeding, and successfully too. Under favorable food conditions they did as well as the parental species, and they reproduced with one or the other parental species depending on paternal song type. This is an important finding because it shows that a genetic barrier to interbreeding is not fully formed during the period when the species are geographically isolated from each

other. In fact, a genetic barrier may take millions of years to evolve. Until this barrier is fully formed the two species may exchange genes, which is biologically significant because it enhances the genetic, body size and beak shape variation of a population on which selection can act.

This critical time in speciation, when previously separated populations come together, can be thought of in terms of fission through divergent natural selection, and fusion caused by hybridization. Fission completes speciation, while fusion causes

it to collapse. The environment — principally feeding conditions that determine survival — plays an important role in the outcome, and since these conditions oscillate back and forth, so do the fission and fusion tendencies of the species. Darwin's finches are enlightening because they give us a glimpse of how the long-drawn-out process of speciation works.

There are also strong implications for conservation. As we mentioned at the beginning of this article, none of the Darwin's finches has become extinct as a result of human activities. This fortunate circumstance, combined with their diversity, makes them unique. However, some *populations* have become extinct, and the species persists only because it survives on other islands. The sharp-beaked ground finches (*Geospiza difficilis*) on Santa Cruz and Floreana have vanished, whereas those on six other islands have survived. The largest form of the large ground finch (*G. magnirostris*) on Floreana and San Cristóbal became extinct some time after Darwin's visit in 1835, and, more recently, the warbler finch (*Certhidea olivaca*) on Floreana has either followed suit or is close to doing so. The mangrove finch on Isabela is so endangered that a captive breeding program is being considered to pull it back from the brink of extinction. All these populations are, or were, on islands inhabited by humans who have directly and indirectly altered their own and the finches' environment.

In the face of these known extinctions, our long-term studies of finches on Daphne and Genovesa deliver a strong message to conservationists. Both species and environments change over short intervals of time, and to conserve them in their natural state we must keep them capable of further change. An anonymous quote, frequently but erroneously attributed to Darwin himself, sums up this process: 'It's not the strongest of the species that survive, nor the most intelligent, but the one most responsive to change.'

FAR LEFT: Young finches such as this fledgling medium *G. fortis* begging to be fed by its male parent on Santa Cruz, learn the song characteristic of each species from hearing their fathers sing during a critical few weeks of their early life. This will enable them to recognize their own species as adults and avoid interbreeding.
ABOVE LEFT: The song of the warbler finch in the Santa Cruz highlands is distinctive, with variations heard island to island, and even at different altitudes.
BELOW: The mangrove finch, *Camarhynchus heliobates*, shares similar tool-using habits with the woodpecker finch, but is now restricted to just two mangrove forests along the western coast of Isabela. Heavily impacted by introduced-rat predation and parasitic botflies in the nest, this critically endangered species is down to less than 100 individuals, although pest control is showing promising results.

A Most Unusual Hawk
One Mother and Several Fathers
Patricia Parker

Department of Biology,
University of Missouri–
St. Louis,
8001 Natural Bridge Road,
St. Louis, MO 63121,
United States.
www.umsl.edu/~parkerp

WildCare Institute,
Saint Louis Zoo,
1 Government Drive,
St. Louis, MO 63110,
United States.
<pparker@umsl.edu>

Dr. Patty Parker is the Des Lee Professor of Zoological Studies at the University of Missouri-St. Louis, and Senior Scientist at the Saint Louis Zoo. In 1990, she began working on the long-term study of the Galápagos hawk and its population genetics, phylogeography and social behavior, and added research into disease threats to Galápagos birds since 2001. She has published numerous papers on these topics with her students and collaborators, and received the Order of Scientific Merit for her work from the people of Santa Cruz in 2008. She is a fellow of the Animal Behavior Society, the American Ornithologists' Union, and the American Association for the Advancement of Science.

THERE ARE FEW PLACES on earth where a hawk will land inches above your head to see what you are doing. Charles Darwin remarked that shooting Galápagos hawks (*Buteo galapagoensis*) for his collection was complicated by the fact that they would fly in and perch on his rifle. This striking fearlessness helps to make this bird one of the world's best-understood raptors. It is the only resident hawk on the Galápagos Islands, occurring nowhere else, and is the top predator in the terrestrial ecosystem. Like all *Buteo* hawks, it is a soaring hawk with broad tastes in prey. Today, the species nests on eight islands: Pinta, Marchena, Santiago, Santa Fe, Española, Pinzón, Isabela and Fernandina. In the past, it also inhabited Santa Cruz, San Cristóbal and Floreana, but has now completely disappeared from those islands, almost certainly as a result of human colonization.

We have seen hawks feed their nestlings the flesh of every vertebrate known to occur on the islands, including marine iguanas, young giant tortoises,

RIGHT: The dominant land predator in the insular ecosystem, a Galápagos hawk, *Buteo galapagoensis*, feeds on a fresh marine iguana kill.
FAR RIGHT: Completely unafraid by nature, juvenile hawks check out a rare visitor to the summit of Fernandina, landing on André De Roy's head to investigate.

sea lion placenta and even meat scavenged from the goat eradication programs. Hawks also prey upon invertebrates, such as the giant centipede and the harlequin grasshopper. These eclectic feeding tastes and fearless ways, along with a predilection for easily caught chickens, made hawks the bane of a Galápagos farmer's existence, who soon eliminated them from the human-inhabited islands.

I began studying this species in 1990 in collaboration with John Faaborg and Tjitte de Vries. At that time I was an offsite geneticist helping answer questions about these hawks' unusual cooperative nesting behavior. In particular, John and Tjitte and their students wanted to know which males within the hawk's multimale breeding groups were actually reproducing.

On some islands, Galápagos hawks establish and defend year-round territories in groups consisting of one adult female and multiple adult males, a social system called cooperative polyandry (many males to one female) (see also chapter: Valle). We have seen up to eight males in one of these groups, an arrangement that remains very stable from year to year. As with all hawks and owls, the females are larger and more aggressive than males. In the Galápagos hawk, the female truly runs the show, controlling the males' behavior through a variety of vocalizations, ordering them to attack intruders (including humans) that have wandered too close to the nest, or to bring more food to her and the chicks. The noticeably smaller males — only about two-thirds her weight — interact quite peacefully within the group. Every year, each

ABOVE: Hawks flock to the caldera of Fernandina in time for the October hatching of land iguanas that nest in sun-warmed volcanic ash.

BELOW: Always opportunistic, an adult hawk carries away a fresh sea lion placenta.

ABOVE: One of several males caring for a small chick on Santa Fe, all taking turns to bring food.
BELOW AND RIGHT: Being generalist predators, young Galápagos hawks are bold and inquisitive, always on the lookout for new opportunities.

group typically has a single nesting attempt in which the female lays two eggs, which she alone incubates while being fed by the males. All adult members bring food to the chicks, even though in a larger group it is impossible that all males are fathers, since the number of potential fathers is greater than the number of chicks. Most successful nesting attempts will eventually produce just one fledgling.

Before working with John and Tjitte, I had already spent years studying other cooperatively breeding birds elsewhere, but in these species the social groups expanded because offspring never dispersed from their natal territories. Being close relatives meant that by becoming 'helpers' to their parents, as they matured they could still benefit genetically by assisting with raising their own siblings. We suspected this was not the case with the Galápagos hawk, knowing that young birds leave their natal territories to wander before becoming sexually mature in their fourth year, so we set out to estimate the kinship of males within each group, and to determine who among them was the father of the chicks produced. As we expected, it turned out that cooperating males were no more closely related than a random draw from males at large. Paternity was also randomly distributed among them, showing that over the years each male has some reasonable probability of being the sire of a fledgling. Since the female copulates with all of them during any breeding cycle, they probably have no way of ascertaining which chicks are theirs or not, so they bring food just in case.

Mating systems on different islands

Since those early discoveries, my students and I began our own field research on the hawks, and have been studying them ever since. What interested me as much as

the perplexing behavior of individuals within these groups was the fact that the mating system varies dramatically from island to island. On Española, for example, the Galápagos hawks are strictly monogamous, breeding in male-female pairs just like most hawks do elsewhere in the world. In contrast, all hawks on Pinta and Marchena breed in groups with as many as five or six males, whereas the larger central islands of Santiago and Isabela contain some breeding groups and some monogamous pairs. Our studies took both a historical genetic approach as well as an ecological slant to seeking an explanation for this variation.

We now know that the Galápagos hawk is a close relative of the Swainson's hawk, a smaller *Buteo* hawk that breeds throughout a large North American range, then undertakes an enormous latitudinal migration each year to the pampas of Argentina. Our genetic studies with Jennifer Bollmer, Rebecca Kimball and others estimate that there was a single colonization of Galápagos roughly 126,000 years ago, making it the youngest endemic Galápagos lineage for which these estimates have been made (in contrast, the original colonization by finches, mockingbirds and penguins are each more than one million years ago, or multiple millions). It seems that a group of migrating Swainson's hawks was blown off course and landed in Galápagos to eventually become what we now recognize as the Galápagos hawk.

Since that fairly recent colonization, the species has established itself on 11 islands (since reduced to eight) and become more and more sedentary, so that movements between islands are now rare. In fact, the hawks on some islands are highly distinct, such as those on Española, which is the most geographically isolated population, with significantly larger-bodied

ABOVE: A monogamous pair overlooking a seabird colony guards their territory together, the female (left) being over 30% heavier than the male. LEFT: An immature hawk hovers over the sulfur fumaroles of Alcedo Volcano, Isabela.

ABOVE: A nonterritorial adult female joins a wandering group of immatures feeding on a goat during the Alcedo eradication campaign. RIGHT: A lone juvenile is viciously attacked. BELOW: A large female (center) presides aggressively over a group of males.

birds. Those restricted to small islands, such as Santa Fe and Pinzón, are not only different genetically, but also the most homogeneous. On Santa Fe, for example, many of the males are so genetically similar that we have been unable to obtain paternity results within polyandrous groups even using very rapidly evolving genetic markers for DNA fingerprinting, by which even close relatives can usually be distinguished. In contrast, the large populations on Santiago and Isabela have sufficient genetic variability that paternity studies are easily conducted. Jennifer Bollmer's work shows that, across the eight islands occupied by hawks, the genetic diversity within a population is directly related to its numbers, which is in turn dictated by the size of the island.

One of my students, Noah Whiteman, also studied the lice that live on those same hawks, themselves descendants of the lice that arrived with the original Swainson's hawk ancestors. He discovered that the hawks with low

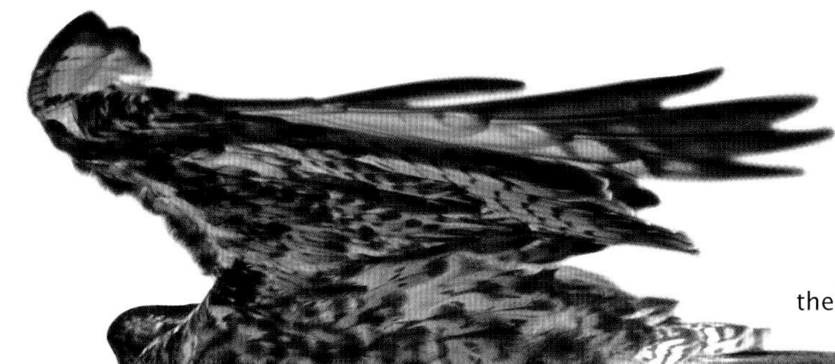

their natal territories to join a large wandering band of non-territorial birds, a mechanism that apparently allows them to overwhelm the defensive behavior of territorial adults. Lone juveniles invariably are mercilessly attacked — being driven away, severely battered, is the best outcome a hopeful young bird could experience from such an intrusion. On the other hand, when a group of several dozen

LEFT AND BELOW LEFT: Hawks may hunt from the air or on foot, taking birds, reptiles, carrion and many types of insects, from giant centipedes and caterpillars to the painted grasshopper, *Schistocercus melanocera.*

genetic diversity living on small islands have significantly more lice than those on large islands with high genetic diversity. Working with Kevin Matson, we added the final piece of the puzzle to explain this relationship. Kevin showed that the low-genetic-diversity hawks on small islands had poorer immune responses than those on larger islands with higher genetic diversity, and that this difference in immune function explained the relative abundance of parasites. This was a dramatic example of one of the presumed reasons that island populations are more susceptible to extinction than mainland populations, and the smaller the island the more so. In summary: by virtue of their restricted population size, they have lower genetic diversity, weaker immune systems and therefore are more vulnerable to parasites and other pathogens.

As logical as this finding may seem, there have been very few such real-life examples with all of the elements in place. When we reported this to the Galápagos National Park Service (GNPS), we suggested that it meant hawks on small islands may be at greater risk from potential introduction of novel pathogens than those on larger islands, and that this may be true for other bird species as well. We inferred that a safety precaution would be to consider further restricting numbers of visits by tourists and scientists to smaller islands, as this is one of the routes by which new pathogens may inadvertently arrive. This concept was implemented within a few weeks, when the Park began enforcing restrictions on tourist visits and size of research teams working on uninhabited islands as a function of the island's size.

We have studied the hawks on the island of Santiago most thoroughly, following the histories of 46 territorial groups for more than 10 years. Every hawk in each of these groups is banded with an individually coded leg band legible from a distance using good binoculars. In this manner, year after year we can monitor which adult is in which group, where young birds first breed, how long each individual lives, and many other aspects of their life histories. We know, for example, that the young birds leave

Day and Night Owls

The only native raptors on Galápagos besides the Galápagos hawk are two species of owl, the short-eared owl (above) and barn owl (below). Although neither is endemic, both have developed characteristics that differ from their continental ancestors. The barn owl subspecies *Tyto alba punctatissima* is smaller, with more contrasting colors (dark back with pronounced spotting), and is strictly nocturnal, feeding primarily on rodents, both native and introduced. The short-eared owl, *Asio flameus galapagoensis*, is possibly a distinct, very dark island species, and is mainly a hunter of birds. On islands where the Galápagos hawk is absent, the short-eared owl exhibits semidiurnal habits, such as on Genovesa where some individuals seem to specialize in catching band-rumped storm petrels as they return to their lava nesting cavities. Like hawks, both species suffer mortality on inhabited islands from a combination of factors, including traffic accidents and persecution by farmers fearing for their chickens, as well as burning of nests due to superstitions about owls bringing bad luck.

Tui De Roy

RIGHT: Before acquiring territories of their own, mottled immature hawks wander around their home island as a roaming band, overwhelming the defences of the all-dark, aggressive, sedentary adults, who can do nothing but wait for the flock to move on.
BELOW AND OPPOSITE: Eyries are built on rocky promontories or in large trees with good surrounding views and are often reused year after year, with new layers of branches added.

juveniles enters a territory, the much smaller number of resident adults can do nothing but wait for them to leave. This gang of mostly beige and brown-spotted immatures roams the island freely, until individuals reach sexual maturity at four years of age and attain their dark brown adult plumage. At this stage, using the group dynamics described above, they begin to insinuate themselves into the breeding groups, where they usually remain for the rest of their lives, which may reach 16 to 19 years. Occasionally, breeding females take a year off and are replaced for a season, but the original one may return later to reclaim her spot in the group. Females without their own breeding group may remain with a juvenile flock even as adults. Likewise, males usually spend their entire lives in the group where they first find a breeding spot, but sometimes one may leave a group and join another. We have been watching this complex social system for long enough now that we have observed many examples involving both the usual patterns as well rare events that are the exceptions to the rule, like when one group of males split, with some members taking up residence with the neighbors.

Santiago population

For many years, the group of nonterritorial birds on Santiago numbered more than 100, and they would often visit us in flocks of several dozen, curious about these two-legged creatures on their island. In 2004, Felipe Cruz asked us to work with him to monitor the effect of goat eradication on this well-known population of hawks. Under a bi-institutional project of the Charles Darwin Foundation (CDF) and the GNPS, the last goats were removed from Santiago between 2005 and 2006 (see also chapter: Campbell), so we began the formal annual hawk census in 2005. During the first four years of this census, the average size of breeding groups has dropped slightly, although this reduction is perhaps within normal fluctuations. It also seems that the residency rate of individual adults in breeding groups has declined over the last three years, suggesting that breeding adults may not be living as long. However, most noticeable was the disappearance of the nonterritorial wandering juvenile age class. At the Espumilla study area on the west coast of Santiago, our estimates of this component of the population fell from 135 in 2005 to 65 in 2006 to zero in 2007. In February 2008, there were approximately 65 young birds, almost all of which we think hatched in the breeding season of 2007. By July 2008 they too were gone.

These findings raise a number of unexplained questions that may relate to adaptations to changing or altered environments possibly having a direct effect on social behavior on the different islands. This appears to involve complex equations based on elements of habitat richness and prey diversity/accessibility, which probably affects juvenile survival and general population density, all of which in turn are likely to contribute to pressure on breeding group size and the need for more members to defend the territory from increasing intrusion by itinerant nonbreeders. It seems that each female is able to somehow assess what is the ideal number of males — from one to eight — required to ensure successful reproduction, but being hyperaggressive, she will not bend to permit another female helper.

This fascinating hawk still holds many conundrums. Through Hernan Vargas of the Peregrine Fund, we have received funding to study these relationships further through detailed comparisons of the ecology of the hawks on Santiago, where polyandry is common and goats were very recently removed, and Española, where hawks nest in pairs and goats were removed several decades ago. This funding will support the involvement of two Ecuadorian students who are committed to Galápagos conservation and who, we hope, will continue the studies of Galápagos hawks for many years to come, sustaining the local efforts of Ecuadorian partners in the conservation of this and other Galápagos endemic species.

ABOVE: Immatures ride the wind over Alcedo Volcano, Isabela.
BELOW: A large female orders the males around using loud vocalizations.

Nazca Booby Behavior

Some Evolutionary Surprises

Dave Anderson

Department of Biology,
Wake Forest University,
Winston-Salem,
NC 27109, United States.
<da@wfu.edu>
www.wfu.edu/~djanders/anderson.htm

Dave Anderson is Professor of Biology at Wake Forest University in North Carolina. His continuing studies of seabirds in Galápagos began in 1982, and the Nazca booby project is the most complete study anywhere of a tropical seabird. In addition to his work in basic science, Dave is a member of the Charles Darwin Foundation (CDF) for the Galápagos Islands.

GALÁPAGOS IS HOME TO three booby species. The red-footed booby lives and breeds throughout the tropical oceans, wherever it can find a predator-free island on which to nest. This vulnerability to predators restricts its distribution in Galápagos, because it cannot live with Galápagos hawks. The hawks are naturally absent from Genovesa, Darwin and Wolf, and each of these islands supports a large red-foot population. Hawks have been extirpated in historical times on San Cristóbal and Floreana, and red-foots colonized those islands (the satellite island Gardner in the case of Floreana) shortly after the hawk's disappearance. Blue-footed boobies are restricted to the eastern tropical Pacific (ETP), with population centers in Galápagos, the Peruvian guano islands, Mexico and Isla de La Plata near the Ecuadorian coast. Blue-foots also have problems with hawks, but, unlike the smaller red-foot, they have a vulnerable body size only during part of the nestling period. Hawks attack and eat many mid-size nestlings, but apparently not enough to repel blue-foots from hawk-rich islands like Española. While red-foots lay a single egg and so have at most a single chick per nesting attempt, blue-foots will lay two and even three eggs, and they can raise more than one chick if feeding conditions and hawks permit. Blue-foot breeding failure falls most heavily on the chicks that hatch second and third, because their larger and dominant nestmate bullies them regarding food, leading to starvation during food shortage, and because hawks take the smallest when attacking a multichick brood.

Nazca boobies are the subject of this chapter. Nazcas have been recognized as a species, *Sula granti*, only since 2002, when they were distinguished from masked boobies based on genetic, color and distributional differences, and on the fact that

BELOW LEFT: The red-footed booby, *Sula sula*, nests on isolated islands in all tropical oceans. It feeds primarily on flying fish, which it often catches on the wing.

BELOW RIGHT: The blue-footed booby, *S. nebouxii*, is a plunge-diver feeding close to shore; it nests mainly in Galápagos and islands off the coasts of Mexico, Ecuador and Peru.

they strongly tend not to interbreed where their ranges overlap. Like blue-foots, they have a limited distribution, breeding and feeding over the Nazca tectonic plate in the ETP. Half of the world population breeds on Colombia's Malpelo Island, and most of the other half in Galápagos. Nazcas share most breeding sites with one of the other two Galápagos boobies, but rarely with both.

My work on boobies began in a preliminary way in 1982, as a field assistant for Peter and Rosemary Grant and their graduate students in Darwin's finch research. In 1984, I began a formal study that centred on Nazca boobies on Española Island, and produced a doctoral dissertation in 1989. The scale of the study expanded in 1992 to achieve comprehensive monitoring of that same population every year, with a group of field technicians and graduate students present during the entirety of each breeding season. This approach, inspired largely by my formative experience in the Grant Group emphasizing study of marked individuals, has allowed us to compile complete life histories of several thousand identifiable individuals, providing a valuable resource to use in answering questions about environmental factors affecting reproductive success, natural selection of behavior, individual health, and other areas of scientific and conservation interest. We are now in the 26th year of the Nazca booby study, and along the way we have also worked on other booby species, waved albatrosses, swallow-tailed gulls, and other seabirds. To date, the project has produced 60 peer-reviewed publications.

'I wish you were dead!'

Ornithological wisdom has it that breaking out of the constricting confines of the eggshell is the hardest single thing that a bird does in its lifetime. But few birds face the buzzsaw that hatchling Nazca

ABOVE: The Nazca booby, *Sula granti*, became a species in 2002 when it was separated from the masked booby, *S. dactylatra*, a worldwide tropical seabird, seen below on Palmyra Atoll in the central Pacific. Both are powerful plunge-divers, feeding well offshore.

ABOVE: The author's study area at Punta Cevallos. BELOW, LEFT TO RIGHT: Parental shade is a matter of survival; a chick hatches with murderous intent; the first to hatch drives its sibling from the nest.

boobies do: a fight to the death right after hatching. Nazca boobies exhibit obligate siblicide, laying two-egg clutches, but with only one chick ever raised: nestmates are absolutely intolerant of one another, and one of the chicks is evicted from the nest and family within days, or even hours, of hatching.

The obligate siblicide syndrome provides a fascinating opportunity to examine the interacting behavioral adaptations of family members and the consequences of those adaptations for others. We start with the parents' situation, since they are the ones that create the two-egg clutch, that drives the syndrome. A one-chick brood is favored by natural selection. We know this because experimental suppression of siblicide reduces the fitness of parents. Although parents raise more chicks without siblicide, they produce fewer grandchicks because the experimental chicks (perhaps less fit after sharing

Photo courtesy Mark Jones

Photo courtesy Dave Anderson

food during growth) have few offspring themselves. So, parents do best with a single-chick brood. Are they maladapted with a two-egg clutch?

No. For reasons still unknown, Nazca boobies hatch only around 60% of their eggs, compared to success rates above 90% for many other bird species, including some other boobies. So if Nazcas lay a second egg, they receive a benefit in the form of insurance against the disaster of total clutch failure. From a single-egg clutch, the probability of hatching zero eggs is 0.40 or so. From a two-egg clutch, the probability falls to 0.40 x 0.40 = 0.16 (given some assumptions). That is, they increase the chance of getting at least one chick from the clutch from 0.60 to 0.84, a substantial increment. Parents receive at least one other benefit from laying second eggs, including the occasional (apparently maladaptive) adoption of an evicted offspring by neighbors. But the insurance value of second eggs far exceeds any other benefit.

Allowing that embryos cannot do much to prevent Mom from laying that insurance egg, and that the embryo's prospects for reproduction fall dramatically if that sibling is not dealt with severely, we can expect selection to favor the evolution of lethal behavior by chicks, to ensure that the food, nest, parental shade, etc., is mine and all mine! Of course, the same selection favors the same attitude in one's sibling, providing another reason, sufficient in its own right, to clobber that nestmate before it clobbers you. And so we observe the expected outcome: whichever is larger pushes the smaller from the nest, in a fit of muted squeaks that pass for war-cries, relentlessly shoving it away each time it lurches back toward safety, a fight to the death staged by Mom among her babies.

Can we identify a physiological basis for this extremely aggressive behavior? Up-regulation of androgen hormones like testosterone generally figures prominently in the expression of aggression, and Nazca boobies provide a textbook case of this action. During a fight, blood testosterone concentration increases dramatically and probably triggers or facilitates the sudden transition from cohabitation to death match. But Nazca booby babies gird for battle even before the rumble begins. Blood samples from Nazca nestlings taken immediately after hatching, before any fighting and indeed while only a single chick is present, have androgen hormone levels three times as high as closely related blue-footed boobies do, which lack the ultimate challenge that obligate siblicide represents for hatchlings.

Exposure to steroid hormones can have long-term effects on behavior in addition to the immediate 'activational' effect during siblicide, but these deferred 'organizational' effects can be difficult to detect, for several reasons. In the case of Nazca boobies, we can follow individuals throughout their lives, using numbered steel leg bands, because they return for their entire adult lives to the colony of their birth and the siblicide event is easy to detect, giving us opportunities to note these androgens' effect both in early life and subsequent adulthood. An obvious candidate for an adult behavioral trait demonstrating residual effects of early androgen behavior is the enigmatic 'NAV' phenomenon seen uniquely in Nazcas.

'NAV' is an acronym for Non-parental Adult Visitor, a purely descriptive term connoting an adult that comes to a nestling at another family's nest site and spends time with it. The complete story is exceedingly bizarre. These adults are birds that, for one reason or another, are not breeding in the current season. Such 'unemployed' members of both sexes wander their own corner of the breeding colony, actively seeking a medium-sized chick whose parents have left it unattended while they both forage at sea. The

ABOVE: A red-foot keeps a wary eye on a nest-building Nazca booby below, Darwin Island.

Jacquelyn K. Grace

ABOVE: A severely injured Nazca chick may die after a violent NAV (Non-parental Adult Visitor) attack.

Photo courtesy Jacquelyn K. Grace

Photo courtesy Jacquelyn K. Grace

LEFT: Strange NAV behaviors vary from violent aggression or attempted copulation (far left) to friendly activity, such as delivery of nesting material (left), while the chick responds by tucking its beak appeasingly under its chest.

ANOTHER NEW SPECIES

The ubiquitous small shearwaters frequently seen feeding in large rafts between the islands became one of the most recent additions to the expanding list of endemic seabirds. Known for as long as birds have been studied in Galápagos, it was always thought to be a mere variation of the Audubon shearwater, *Puffinus lherminieri*, found in the tropical Atlantic. But in 2004, just like the Nazca booby two years before, it acquired its own status as a new species, becoming the Galápagos shearwater, *Puffinus subalaris*, with genetic studies tracing its nearest relative to the Christmas shearwater of the central Pacific. It breeds in small colonies throughout the archipelago, using fissures in cliffs and lava flows, and is unusual among shearwaters by being diurnal around its nest.
Tui De Roy

NAV joins the chick at its nest and exhibits any of three behaviors. Affiliative behavior can be mistaken for a parent's attention, and includes preening the chick and offering playthings like feathers. Aggressive behavior is what it sounds like: biting, jabbing, sometimes a frenzied attack. Sexual behavior is perplexing, with both sexes acting like males, including attempted copulation. We have been unable to identify any effect, either positive or negative, of NAV behavior on NAV fitness, but aggressive NAV interactions often have a strong deleterious effect on chicks, typically leaving lacerations on their necks. In a drought year, endemic mockingbirds flock to the injured chick as a rare food source, taking macabre blood meals that shortly kill it. NAVs show the same interest in nestling blue-footed and red-footed boobies, probably accounting for the blue-foots' aversion to nesting near Nazcas and perhaps also the red-foots' insistence on nesting in trees. We wondered whether NAV behavior is an epiphenomenon of early androgen exposure, expressed despite being apparently selectively neutral.

Like the androgen exposure at hatching, NAV behavior by adults is ubiquitous. Most unemployed adults show the behavior, and almost all adults are unemployed at some point. This widespread and intense social interest in unrelated young sets Nazca boobies apart from other animals, and the species' high across-the-board exposure to androgens at hatching is unusual also. The coincidence of these two singularities is noteworthy, and we can conduct a more robust test using the histories of individual birds that have provided data on both siblicide and NAV behavior. Not every chick shows siblicidal behavior, because some two-egg clutches hatch only one chick and some food-limited mothers lay only one egg. Likewise, some adults show more NAV behavior than do others. It turns out that participation in siblicide as a hatchling (remember that siblicidal birds had an extra testosterone dose) predicts very well the amount of NAV behavior that a bird will show years later as an adult. While correlational in nature, these data certainly suggest an organizational effect of the siblicide experience on adult social behavior.

This story becomes stranger still. Another variable predicts propensity to behave as a NAV even better than siblicide, and that variable is a bird's own victimization by a NAV. Adults who themselves were visited frequently as chicks by NAVs show more NAV behavior as adults than do adults that experienced little victimization as chicks. This remarkable result connects to the 'cycle of violence' idea in human biology, positing that abusive upbringing begets abusive victimization as an adult. We are currently

RIGHT: Even after some experimental flights and diving attempts, a fledgling returns to the nest site for extra provisioning by its parents.

ABOVE: The Nazca booby female is 15% heavier than the male, and often chooses a new mate in consecutive breeding seasons, while evicting her previous partner from the territory.

evaluating this hypothesis experimentally in Nazca boobies, protecting some chicks from NAVs with portable fencing that we remove when the parents are present. How remarkable it is that studies of the evolution of social behavior in this seabird provide the only research model in a natural (noncaptive) setting to investigate this important topic in human social pathology.

Vive la différence

Should a young Nazca booby survive the siblicide experience and then the NAV/mockingbird combined attacks, the next challenge comes at independence. Fledgling Nazca boobies leave the breeding colony at around 150 days of age and are absent for some years. We know little about this part of their life. The metal leg bands worn by the fledglings have my email address on them (da@wfu.edu, supershort to fit on the band), allowing people who see our birds to let us know. In the case of fledglings, we get email reports from coastal areas of Central America and Mexico, so evidently these newly independent birds move 1000–2000 km (620–1240 miles) north of Galápagos. They apparently consider their colony of birth to be their permanent home, and after two to eight years they reappear as adults at the colony and move toward breeding status. At this point, we begin to see interesting differences between males and females that lead to a newly discovered mating system.

The time between fledging and return to the colony must be challenging for a naive Nazca booby, because about half of them die during this period, compared to 90% annual survival rate of established adults. They must learn to locate and catch food, avoid bad weather, discover how dangerous big predatory fish are, and get along with other birds without injury. Probably they spend most or all of this period at sea. Males survive this challenge better than females do: female nestlings and subadults die more often than males do, and by the time the young adults return this has shifted to about 60% male. We are currently studying the causes of this disproportionate female mortality, and suspect that the larger body size of females (around 15% heavier than males) is involved. For example, female nestlings may require more food than males to attain adult weight, thus being at greater risk of malnourishment in years of food scarcity, resulting in poor condition at fledging and poor survival when independent.

Whatever the reason, fewer females than males make it to breeding age. Females also return to the colony at 4.3 versus 4.7 years, breeding for the first time at 5.5 versus 7.1 years, and fledging their first offspring at 7.6 versus 8.7 years. The late start that

RIGHT: Males fight viciously over nesting territories. BELOW: Having failed to learn how to feed for itself, a weak fledgling dies at sea.

ABOVE: An adult returns to the nesting colony. BELOW AND RIGHT: The departing juvenile has much to learn on the open seas.

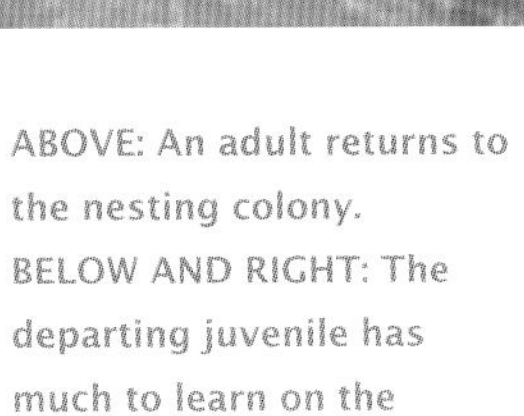

males get on reproduction probably results directly from the biased sex ratio. Once established as a colony resident, males and females survive at about the same annual probability, so the male-biased sex ratio is preserved across all age classes and affects their mating opportunities throughout their lives. With breeding competition imposed by a surplus of roughly one in three males, their prospects of success as young adults are poor against experienced, established, and wily older males.

Now consider natural selection exerted on female breeding behavior under this biased sex ratio. She depends on the male for around half of the egg incubation and nestling care, and for half of her offspring's DNA, so selection should favor females whose mate choice leads them to select males that provide good parental care and good genes. Darwin thought through mate choice, and our thinking today about this topic largely reflects his ideas as expressed in *Sexual Selection and the Descent of Man*. In the intense competition among all those males for access to the limited pool of females, one might logically expect the best males to be chosen consistently for breeding year after year, and the worst ones to have consistently empty dance cards. Females, of course, are in high demand. A female

would have to be a pretty poor performer to fail to breed. Under this scenario, the success of females in securing a partner should be quite uniform compared to males: females should basically all be winners, and males should include many winners but also many losers. The statistical concept of *variance* expresses this quantity of uniformity versus nonuniformity in success. Under standard sexual selection theory and given the particulars of the Nazca booby situation, females should have lower variance than males in the breeding department.

But Nazca boobies do not follow the script. The variance in breeding success is the same for both sexes. Somehow, males are no more likely to be losers (or winners) than females are. The reason is that a male's superior status is not a permanent quality. It varies over time, and females exploit their own status as a limiting resource to track the changing quality of males as they change. Here is how we think this works. A female chooses a male, and they breed. If they raise a nestling to independence, it is probably because both worked hard to find food and perform other care for the nestling. This parental effort costs each of them in physiological terms, making them less likely to be so successful in the subsequent season. Each of them will benefit from changing mates to acquire a 'fresh horse', a partner that has not just raised an offspring. Females are able to exercise this option, because many males are available that had no dance partner recently. They are happy to mate with the female, even though she is a bit depleted. After all, every female breeds, so he has little choice. Contrary to intuition, the more successful a male is, the more likely he is to be shown the door. We call this newly discovered mating system 'mate rotation'. It may be common in cases of two-parent care and sex ratio bias, but has not been looked for specifically.

Successful males do not appear to leave voluntarily, but the divorce will happen nonetheless if the female wants it. Divorce might happen because the female simply leaves the current male's nest site to take up with another male nearby. Females are larger than males, so the original guy cannot do much about that. Sometimes females add insult to injury by divorcing the successful male and trying to keep his site also. In that case the male may resist the ouster, but frequently the female gets her 'fresh horse' to join her at the site and together drive the spent male away. So success begets failure for males under this mating system, but then again failure begets success. Males usually spend only one year on 'time-out' as non-breeders before they join the breeding pool again via female mate choice.

ABOVE: After feeding on schooling fish in offshore waters, often working together with dolphins and other predators, a pair returning to roost in the busy colony exchange loud greetings at dusk.
BELOW: Binocular vision is essential for a plunge-diver.

The Waved Albatross

The Family Affairs of a Critically Endangered Species

Kathryn P. Huyvaert

Department of Fish, Wildlife, and Conservation Biology, Colorado State University, Fort Collins, CO 80523-1474, United States.
<Kate.Huyvaert@colostate.edu>

Dr. Kate Huyvaert began working in Galápagos in 1994. Since then, she has researched a variety of questions on the population dynamics, behavioral ecology, and life history evolution of the Nazca booby and waved albatross. Kate completed her Ph.D. research in 2004 on extra-pair mating behavior of the waved albatross. She is currently Assistant Professor in the Department of Fish, Wildlife, and Conservation Biology at Colorado State University.

AT THE HEART OF STUDIES of evolution by natural selection is the idea that variation exists in life history traits. These are things like the age a frigatebird first lays an egg, the number of seeds an *Opuntia* cactus sets in its lifetime or how long a sea lion lives. In the end, how an ant, a finch or a marine iguana allocates resources to each of these traits is directly linked to its fitness — a measure of how many young it will produce that will produce offspring of their own — and thus, its contribution to natural selection operating in the population. Because resources are finite, organisms also have to make tradeoffs, balancing the requirements of reproduction with the everyday costs of survival. What is fascinating is that neither two species nor even two individuals have the exact same set of traits, and this leaves general theories about life history evolution wide open to exciting new discoveries. Exploring the ins and outs of the life history of the waved albatross (*Phoebastria irrorata*) — and uncovering a rather unusual story in doing so — has played an important part in my own life story.

My very first visit to Galápagos as an undergraduate student in 1994 was to Española Island, the southeasternmost of the islands and home to all but a handful of the world's waved albatrosses, the largest bird in Galápagos. Medium sized as albatrosses go, adults stand about waist high and have long tube-nosed bills, roughly the color of a banana. While magnificent, this species has a goofy persona on land. Prominent eyebrows shade large brown eyes and these masters of the wind and water can, at best, toddle among the rocks and vegetation with a stumbling swagger. Once airborne, though, goofy is replaced by grace as the birds open their wings to their full 2.4 m (8 ft) span and soar above the crests and valleys of the ocean's waves.

BELOW LEFT: A waved albatross on final approach slows down for a controlled landing.
BELOW RIGHT: A mate's-eye view of the gaping display, part of an elaborate courtship dance routine.

ABOVE: Española Island, the oldest and southernmost in Galápagos, is the only sizable breeding colony of the world's only tropical albatross.

A brief life history

As a resident of an archipelago straddling the equator, the waved albatross is the only tropical albatross in the world, but there are many other aspects of its life history in which the word 'only' will appear. They can raise only one chick per year, at best, a process that takes them about nine months. Males return to the colonies along the southern and southeastern coast of Española in late March or early April each year, followed closely by females a week to 10 days later. Established pairs quickly reunite and mate, then the female departs for a few days' respite before returning to lay a single egg, weighing 284 g (9 oz), on the bare ground. Both parents take turns during the 63 days of incubation, making far-ranging feeding trips while the other sits in the sweltering heat of Española for as many as 22 days at a stretch.

By the time the egg hatches, the parents' work has seemingly only just begun. For the next four to five months, they will both dedicate themselves to ferrying meals from the open ocean to their waiting chick. When a hatchling is still a mere fluff-ball, parents continue to take turns watching over it while making short feeding commutes, mostly within the Galápagos Marine Reserve. However, detailed satellite tracking studies of adults caring for larger chicks show they most often travel to the rich waters of the Peruvian coastal upwellings — round trips of up to 3000 km (1860 miles). By December, the surviving chicks lose their brown fuzz and, having acquired the body and plumage of an adult, make the leap from 'landlubbers' to denizens of the air and ocean. Another four or more years will pass before they may return to the colony to make their first attempt at breeding themselves.

COURTSHIP SPECTACLES

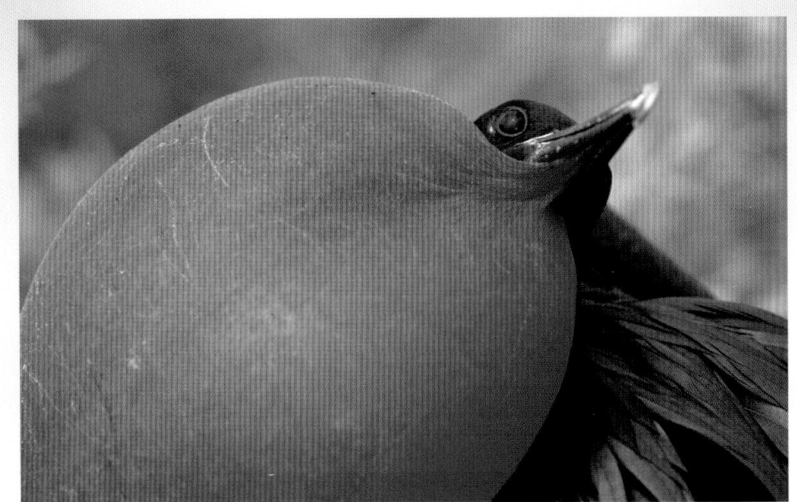

Albatrosses are not the only Galápagos seabirds to engage in spectacular courtship displays. The end of the warm season (February–March) sees colonies of greater frigatebirds, particularly on Genovesa, 'blossom' with the brilliant red balloons of males (above) perched in low bushes trying to impress females passing overhead. Cooing and warbling loudly, they use air sacs known as gular pouches, which they inflate while vibrating their outstretched wings for full effect. Blue-footed boobies, too, gather in raucous colonies to sky-point and dance, lifting their large blue feet rhythmically. Red-billed tropic birds court in flight, emitting ear-piercing screams above their nesting cliffs, while flightless cormorants perform their pirouetting dance in the waters of calm coves.
Tui De Roy

ABOVE RIGHT: The author taking notes amid her study subjects.
FAR RIGHT: The albatross colony at Punta Cevallos is a rowdy place where many unique Galápagos species mingle.

Mating behavior: An albatross's 'one and only'

Walking through the colony during the early breeding season can be something of a sensory overload. The cries of Nazca boobies (*Sula granti*) and swallow-tailed gulls (*Creagrus furcatus*) provide a raucous backdrop for the more subtle sounds of waved albatrosses busily courting among the rocks and shrubbery, a repertoire of greeting honks, sighs, groans, bill-circling, claps and clunks. Arriving adults, after much circling, usually try to make a controlled landing within a few steps of their presumed territory and nesting site, but it seems this plan doesn't always work out.

And that's where the orderly scene ends. Not infrequently, arrivals touch down far from the home territory, with some final approaches ending in ignominious crash-landings. Birds already on the ground often rush over to meet, greet and sometimes attempt to copulate in a rough-and-tumble encounter between individuals who are not typical mates; these are called extra-pair copulations, or EPCs. Given the long-term pair bonds that tend to be the norm among the albatross family, what is going on with the waved albatross? My third and longest expedition to Española in 1996 and 1997 began to uncover a fascinating chapter of this species' unusual story, and later formed the basis of my doctorate.

Life history theory suggests that with only one shot at reproduction per year and the great commitment both parents must make in attempting to raise that single chick to fledging, waved albatrosses will have their 'one and only' mating partner and that this

relationship will be for life. And because adults must make a tradeoff for that reproductive attempt in terms of energy expenditure and future survival, there are strong arguments why individuals should not attempt to breed if even a hint exists that the other partner has mated with a stranger. This means that when we look closely, we should find no evidence of chicks produced outside of the pair bond, so-called extra-pair offspring (EPO).

The first surprise came when we collected blood samples from chicks and their social parents — the term used to describe the pairs that raise them, but who are not necessarily genetically related — and then applied the molecular technique of DNA fingerprinting to evaluate parentage. What soon emerged was that the social fathers of 25% of the chicks in this first sample were not the genetic dads at all!

To better understand this unexpected result, we went straight to the authors of this remarkable story — the albatrosses themselves. Starting in 2000, teams of two to three observers and I recorded all of the 'goings-on' during the copulation part of three breeding seasons in a 3500 m2 (37,500 square ft) portion of the colony, an area about the size of a large playing field, near Punta Cevallos at the eastern end of the island. We banded every bird that entered the area with a large numbered plastic band readable from a distance, and watched their interactions from sunrise to sunset every day for six to seven weeks each year. In all, we documented over 3600 copulation attempts and, very much to our surprise, around 60% of these were EPCs!

Both males and females of all shapes and sizes

ABOVE: Freshly returned from his ocean wanderings, a male feeds the young chick with a rich oily slurry, while the salt-encrusted female prepares to leave after several days on guard duty. Surprisingly, paternity tests show that about one-fifth of the chicks studied were not fathered by the males caring for them.
LEFT: Once the single chick is plump and downy, it will toddle off into the shady bushes, emerging to be fed at the sound of its calling parent.

participated in these EPCs. Some took place apparently peacefully, whereas others were aggressive, causing some participants to quickly run away from the scene. We also tracked the fate of all eggs laid in this study area until hatching, then took a small blood sample from both the chicks and the adults caring for them so we could again test for parentage. In the lab, we found that 14–21% of the chicks from those three years were also extra-pair and, in all of these cases, it was the father who was not related to the chick he was caring for.

BELOW: Many pairs bond for life, rejoining to nest together year after year over three or four decades.

While it is clear that extra-pair copulation and paternity are both important factors in the life history of waved albatrosses, evolutionary explanations for why, exactly, these traits persist are much less clear. From the data collected, we know that a female that switches mates from one year to the next tends to copulate in the first year with her current mate — and, as it turns out, next year's mate too — more than with other males. While this can account for some portion of the huge number of copulations that are outside of the pair bond, life history theory suggests that some other benefit must exist for the relatively large percentage of EPO. One idea that we explored in detail, called the 'good genes' hypothesis, suggests that females will accept and even pursue EPCs with males that are of better 'quality' than their social mates. But we found little support for this: males of all sorts (different shapes, sizes, behaviors and 'qualities') father chicks with females that are not their social mates.

Another set of ideas contemplates aspects of kinship. Because the vast majority of waved albatrosses live on just the one island of Española, and dispersal within the island appears relatively rare, males and females alike could be very close relatives. So, one possibility is that females have EPCs to avoid inbreeding. On the flip side of the coin, a male might, instead, tolerate raising a chick that isn't his own because the true father is a close relative carrying some of his own genes, such as a brother, uncle or father. This makes evolutionary sense, as the caretaking father would benefit because ultimately some of his genes will still make their way indirectly into later generations.

Alas, the data do not support any of these ideas. What we do know for sure is that the waved albatross's extra-pair behavior is an intricate complex of social and genetic dynamics driven not by just one or the

LEFT AND OPPOSITE: Contrary to long-term pair bonding patterns, rough and tumble copulations frequently take place at the landing sites during the brief April mating period.
BELOW LEFT: Unique among birds, the waved albatross moves its large single egg considerable distances during incubation, leading to eggs cracked and lost between lava boulders.
BELOW RIGHT: A female is reluctant to relinquish the egg to her eager mate, taking turns that may last two or three weeks during the 63 days of incubation.

other partner, but by *both* males and females.

As if this were not sufficient, the waved albatross holds other unresolved mysteries. For example, while incubating, an adult may shuffle along the ground with the egg securely wedged between its powder-blue legs and brood patch (a soft featherless patch of skin on the bird's underside), effectively transporting the egg over sometimes quite rough terrain. By the end of the two months-plus incubation, the chick may hatch as far as 36 m (118 ft) away from where the egg was laid. This can have rather detrimental effects on hatching success, with many eggs becoming irretrievably wedged between boulders or falling into fissures. Yet intensive studies examining possible factors driving this behavior, like seeking relief from excessive sun or tick infestations, have failed to cast light on why some birds move more than others and some not at all.

Photo courtesy Mark Jones

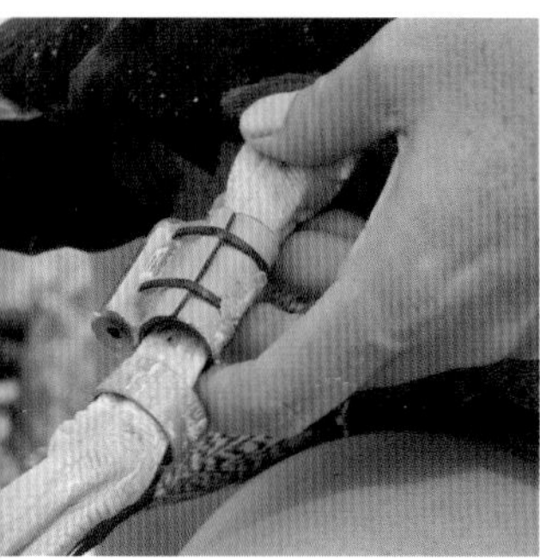

ABOVE: Retrieved at the beginning of the breeding season, a small leg-mounted logging device will reveal invaluable data recorded since the bird left three months prior.

ABOVE RIGHT: Dozens of numbered bands from the author's study site, some still attached to the legs of dead birds, were retrieved from fishermen in Peru, exposing a serious conservation problem threatening the survival of the species.

Survival: Only one chance

One evening, as we sat down to a dinner of tuna burgers and rice, a strange 'clink ... clink ... clink' sound floated out from the bushes behind camp. After several minutes, a male albatross emerged wearing an unusual thin metal band on his leg, not one we had attached. On inspection, we discovered that this bird had been banded at Punta Suárez, at the western tip of the island, by Michael Harris some 30 years earlier, at a time when all of us now gawking at him were, at most, toddlers. In the years since then, we've seen this old-timer wandering in the bushes on several occasions, and his presence reminds us that, while waved albatross reproduction is slow, they can afford to take their time because normally they live very long lives. Indeed, until recently the waved albatross population had appeared stable.

So how do we know whether a population is doing well? One way we measure its 'health' is to estimate rates of change where a measure of 1 means neither growth nor decline, below 1 and the population is declining, above 1, it's growing. In the years since the mating behavior study, we've been carefully documenting the presence (or absence) of all banded individuals in our part of the island each year in May and June. Using these mark-resight data, we can estimate apparent annual adult survival. In recent years, the picture emerging from these analyses is worrisome.

Photo courtesy Jeffrey Mangel

In his time, Harris found that year-to-year survival of adults at Punta Suárez was about 95%. Our estimates at Punta Cevallos nearly four decades later now range from 88% during a mild El Niño to about 92.5% in typical years. While these seem like reasonably high figures, when used to estimate the population rate of change — our measure of its 'health' — in all cases that number falls below 1, highlighting that something extrinsic is contributing to the population's decline. Of the 22 species worldwide, 18 albatrosses are considered threatened to varying degrees, yet until recently the waved albatross was believed safe, so it is clear that something new has begun to happen.

Looking closely for possible sources of mortality,

ABOVE: A fishhook lodged in a nesting bird's throat attests to dangerous encounters on its Peruvian feeding grounds.

RIGHT: Easy to read from a distance, numbered bands allow every individual in the study area to be closely monitored.

waved albatrosses do have a few pathogens, but nothing that stands out as an important cause of death. Plastic ingestion also is not a problem for this species as it is for other albatrosses. However, we have documented accidental and intentional bycatch of waved albatrosses in small-scale fisheries in the cold upwellings off the coast of Peru. In a survey of Peruvian fishing ports conducted by the small NGO Pro Delphinus, 107 albatross bands — some still attached to the legs of dead birds — were recovered from local fishermen, and even though some of the banding records had been lost and thus complete histories did not exist for all, 43 of those bands could be reliably traced to known waved albatross individuals. Although the magnitude of this problem remains difficult to quantify, its negative influence on current population dynamics is starting to show. Thus the waved albatross, unenviably, was recently added to the IUCN's Red List of threatened species as 'critically endangered,' the worst rating before 'extinct'.

While the 'one and only' albatross in Galápagos is in trouble, many hands are poised and eager to help. Current projects in Galápagos include: tracking birds during the breeding season to better understand how often and where they overlap with fisheries, ongoing efforts to document changes in survival and population size, and a project looking at where albatrosses go during the many months they are away from Española after breeding is over. Galápagos is heralded as a living laboratory of evolutionary processes; it is here and almost nowhere else that we can see 'evolution in action.' With some of its glorious flora and fauna at risk, my hope is that someday Galápagos can also become a living example for conservation biologists, and a place that will inspire young people as the epitome of 'conservation in action' — a place where conservation is truly assured.

BELOW LEFT: An albatross returns from a distant feeding foray, a round trip of up to 3000 km (1860 miles).

GALÁPAGOS PETREL UNDER SIEGE

The Critically Endangered Galápagos or dark-rumped petrel, *Pterodroma phaeopygia*, has had the great misfortune of selecting the humid highlands of the lushest islands as its nesting grounds, these areas having collectively been most heavily impacted by both farming developments and the spread of introduced species. Predacious pigs, cats, dogs and both black and Norway rats take a huge toll of adults, chicks and eggs, while invasive plants, such as quinine and blackberry, clog nesting habitat, making it almost impossible for birds to land and take off. Forty years ago the garúa night skies above Santa Cruz resonated with their eerie courtship calls, whereas today to hear this sound is a rare treat, reflecting hope through years of hard work by scientists and park rangers. Ongoing rat control, using grids of regularly stocked bait stations in selected areas of Floreana and Santa Cruz, have given excellent results, with high proportions of chicks fledging every year where previously almost none survived. To help concentrate nesting efforts in such protected sanctuaries, recordings of courtship calls were played to successfully attract prospecting adults at the beginning of the nesting season. Eradication of pigs has also improved the situation on Santiago. But it will require enormous efforts on a wide range of fronts if we are ever to restore this species to its former glory.

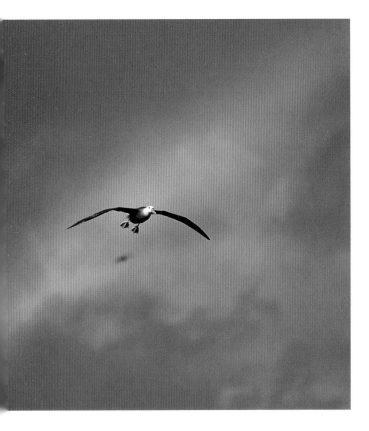

Tui De Roy

Penguins on the Equator
Hanging On By a Thread
Hernan Vargas

The Peregrine Fund (TPF),
Parson 5184B,
Ciudad de Panamá,
República de Panamá.
<pinguinodegalapagos@
yahoo.com>

Dr. Hernán Vargas is Director of The Peregrine Fund's (TPF) Neotropical Science and Student Education Program. His Galápagos childhood originally spurred his interest in biology, and he obtained his first diploma from the Universidad Católica del Ecuador in 1989, followed by an M.Sc. at Boise State University in 1995. Following six years as resident ornithologist at the Charles Darwin Research Station (1995–2001), a research student position at Oxford University (2002–2005) culminated in a Ph.D. in conservation biology in 2006.

WITHOUT A DOUBT, the Galápagos penguin, *Spheniscus mendiculus*, is one of the most unusual bird species found in this archipelago, the family normally being associated with the colder regions of the southern hemisphere. Of the world's 17 species, this Galápagos endemic is the only penguin living exclusively in the equatorial region, and in order to survive in this particular climate, it has developed several unique adaptations.

It is the second smallest of all penguins, with a body length of less than 50 cm (20 in) and averaging just 2.2 kg (4.8 lb) in males and 1.7 kg (3.7 lb) in females. This reduced body size is in fact not surprising, being a trait predicted by Bergmann's law in which the physics of temperature versus body mass dictate that closely related warm-blooded animals increase in size at higher latitudes, as an energy-saving response to cold environments. Another adaptation is to nest mainly inside lava tubes, underneath big boulders or in crevices and

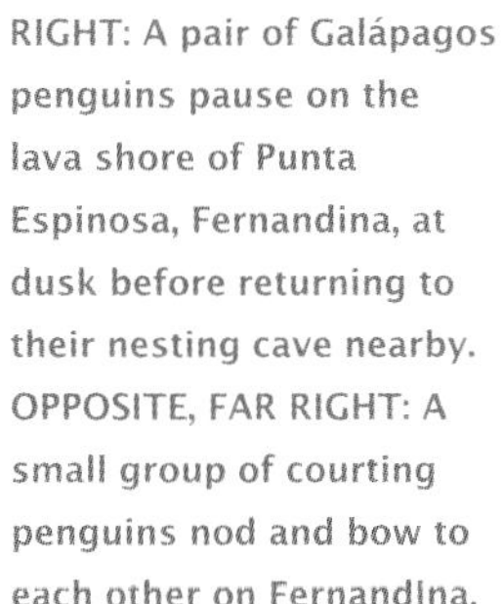

RIGHT: A pair of Galápagos penguins pause on the lava shore of Punta Espinosa, Fernandina, at dusk before returning to their nesting cave nearby. OPPOSITE, FAR RIGHT: A small group of courting penguins nod and bow to each other on Fernandina.

cavities under landslides, offering a cool 'temperate' environment needed during the critical periods of incubation and chick brooding, thus escaping tropical outside air temperatures. The volcanic topography favors this nesting habitat, without which certainly there would be no penguins in the Galápagos Islands.

Although penguins do not normally occur on Santa Cruz Island where I grew up, my involvement with this fascinating marine bird began early in my life. While still a high school student in 1978, I met Hendrick Hoeck, then director of the Charles Darwin Research Station (CDRS), who offered me the opportunity to work as volunteer field assistant with ecologist Hans Kruuk, a visiting scientist studying feral dogs on southern Isabela Island. My first assignment was collecting dog feces to determine diet composition, a task which made me acutely aware of this introduced predator's dire impact when I identified penguin feathers and bones among other prey species. It soon turned out that these dogs had decimated penguin populations throughout the area, while heavily affecting other wildlife such as marine and land iguanas. This eventually justified a successful dog eradication campaign by the Galápagos National Park Service (GNPS).

In 1980, with Sylvia Harcourt and Dan Rosenberg, I again volunteered for fieldwork, this time to participate in a complete penguin census. In due course, my growing interest in this species lead to many years of research and conservation work in Galápagos, and a Ph.D. in conservation biology focused on the effects of climate on Galápagos birds with small populations (penguins, cormorants, flamingos and mangrove finches).

ABOVE: Looking sleek, fat and healthy, a flock of juvenile Galápagos penguins rest along the shores of southern Isabela, an area of particular concern where potentially disease-carrying foreign mosquitoes have been detected.

ABOVE AND BELOW, LEFT AND RIGHT: Swift divers, Galápagos penguins hunt mainly small schooling fish in coastal shallows, sometimes associating with diving brown pelicans, *Pelecanus occidentalis*, below right, to snatch confused prey.

Galápagos penguin ecology

Just as a volcanic habitat on land is essential for nesting success, ocean currents are equally important to make available the amount of food required for penguin survival. The deep, eastward-flowing Cromwell Current, also known as the Equatorial Countercurrent, upwelling where it collides with the Galápagos undersea platform, generates the cold nutrient-laden waters necessary to feed schooling fish consumed by penguins. Hence, 95% of the penguin population occurs around Fernandina and the west coast of Isabela, where sea surface temperatures (SSTs) are lowest. The remaining 5% occurs in small groups on Santiago, Bartolomé and Floreana, where small branches of the Cromwell Current also reach the shore. Sporadic visits of mostly juvenile penguins also occur at Sombrero Chino, Rábida, Pinzón, western Santa Cruz, and even more rarely San Cristóbal, Española, Santa Fe, Seymour and Baltra. I once found a dead penguin on Marchena, perhaps washed up by ocean currents. Since 2003, groups of up to 70 nonbreeding individuals also regularly visit Puerto Villamil, along the southern coast of Isabela. It will remain an enigma why Charles Darwin did not report the presence of penguins in the Galápagos Islands in 1835 having visited locations where penguins are frequently seen today, such as Tagus Cove on Isabela, and Floreana.

The Galápagos penguin is probably the least colonial of all penguin species, with colony sizes usually limited to a handful of nests. This could be influenced by site availability, and by food supplies close to land. Foraging groups usually range from 2 to 12 birds, and rarely exceed 30 individuals. At sea, sharks are probably the only predators of Galápagos penguins. Killer whales and seals, which are known to prey on penguins in southern latitudes, apparently coexist peacefully with them in Galápagos.

Photo courtesy Mark Jones

Photo courtesy Mark Jones

Research

Over the past four decades we have learned much about the Galápagos penguin, starting with Dee Boersma's pioneering Ph.D. research in the early 1970s. She was the first to document the relationship between penguin distribution and upwelling areas, the process of molting as a precursor to nesting, and reproductive success in relation to SSTs [Boersma, 1974]. Studies of foraging behavior conducted by Kyra Mills revealed that the penguins feed mainly in inshore waters in conjunction with other species of seabirds such as brown pelicans, brown noddies, flightless cormorants and Audubon shearwaters [Mills, 1998]. Their food consists primarily of sardines (*Sardinops sagax*), anchovy (*Egraulis* spp.), piquitangas (*Lile stolifera*) and small mullets (*Mugil* spp.). Personally, I have observed an interesting symbiotic interaction in which penguins chasing small fish cause enough confusion to make them easy prey for pelicans, then benefit in turn by stealing the fry from the edges of the pelicans' long bills without much effort.

In further collaborative studies with Antje Steinfurth and Rory Wilson between 2003 and 2005, we continued the application of cutting-edge technology initiated by Kyra Mills in 1994, using logging devices attached to the penguins' backs to record positions, sea temperature and depth [Steinfurth et al, 2007; Vargas, 2006]. This revealed daily excursions along the coast covering between 1.1 and 23.5 km (0.7 and 14.6 miles) from the nest, within less than 200 m (650 ft) from shore, and lasting an average of eight hours. Although the deepest dive was 52.1 m (171 ft), penguins spent more than 90% of their foraging time at less than 6 m (20 ft). These findings suggest that they use a very small fraction of the total upwelling zone, with surprisingly small foraging ranges, which is worrisome because this area coincides with human fishing activities for sea cucumbers, lobsters and mullets. This also contrasts with many other penguin species in other parts of the world, where they may remain at sea for days, or even make long migratory movements covering hundreds or thousands of miles.

During 2003–2006, an intensive monitoring of reproductive activity found that 70% of all nests were located on Isabela, with the highest number of nests at Caleta Iguana [Vargas, 2006]. Other important nesting sites are Playa de los Perros and Puerto Pajas, as well as Marielas Islets, along western Isabela. Interestingly, El Muñeco on northern Isabela is the only place where penguins nest in the northern hemisphere [Vargas, 2006]. We also found nest numbers had declined on Fernandina when compared to data from Boersma in the early 1970s, although we do not understand the reasons for this.

ABOVE, LEFT AND RIGHT: Many small natural caves in friable scoria (left) are used by penguins on the small Mariela Islets in Elizabeth Bay, western Isabela, making this a major nesting site, which has served as a testing ground for rat control. BELOW: A mating pair close to their nesting cave, Cape Douglas, Fernandina.

STORM PETREL MYSTERY

Considered endemic either at species or subspecies level, the white-vented (Elliot's) storm petrel, *Oceanites gracilis galapagoensis*, represents an unsolved Galápagos mystery, having never been found nesting even though it is a year-round resident. A close relative, *O. g. gracilis,* ranges along the coasts of Peru and Chile, where only a handful of nests have ever been located. One of the smallest storm petrels, in Galápagos flocks are commonly seen pattering along the sea surface while feeding very near shore, their dancing, fluttering flight reminiscent of butterflies as they pick tiny food morsels from the water. Two other storm petrels also nest abundantly in the islands, the band-rumped *Oceanodroma castro* and wedge-rumped *O. tethys.* The latter can be seen swarming by the thousands at breeding colonies on Genovesa and Roca Redonda, a very unusual trait since most storm petrels are strictly nocturnal around their nests.
Tui De Roy

RIGHT: A severely heat-stressed penguin gasps on the lava shore near the 1995 Fernadina volcanic eruption, one of several natural and manmade threats that could conspire to wipe out the species.

Censusing penguins

Back in 1970 and 1971 Boersma conducted the first two penguin censuses in coordination with the CDRS, with over 1500 individuals counted each year [Boersma, 1974, 1977]. Subsequent censuses were carried out under the direction of Carlos Valle, Kyra Mills and Isabel Castro. In 1995, I started to work as resident ornithologist at the CDRS and committed myself, 'come rain or shine,' to repeat these surveys every single year, a routine taken over in the past two years by Gustavo Jimenez. Besides invaluable information gleaned from such careful counts, these exercises also serve as a tool to train rangers and students in basic scientific techniques and methodologies, helping to enthuse young islanders — among them Diogenes Aguirre, Carolina Larrea, Monica Soria, Monica Calvopiña, Carlos Carrion, Richard Vokes and Xavier Arturo López.

With crucial GNPS support, methods have been standardized to allow for meaningful interannual comparisons, such as carefully repeated timing and season (early September), itinerary and logistics, field equipment, observer experience, etc. [Mills and Vargas, 1997]. Among the requirements, at least four of the seven participants must come with previous experience, like Hector Serrano, a park ranger, who has participated in every census for more than two decades and knows by heart practically every rock where penguins regularly perch. But still the critical question arises: What percentage of individuals are we counting during each census, and consequently what is the true size of the population? In 1999, applying capture-resight methods, with Howard Snell and Cecilia Lougheed, I estimated that, on average, we counted approximately 57% of the total numbers with each effort [Vargas et al, 2005]. For example, that year we counted 683 individuals, giving us a population size around 1198 individuals, with lower and upper figures ranging between 1054 and 1403.

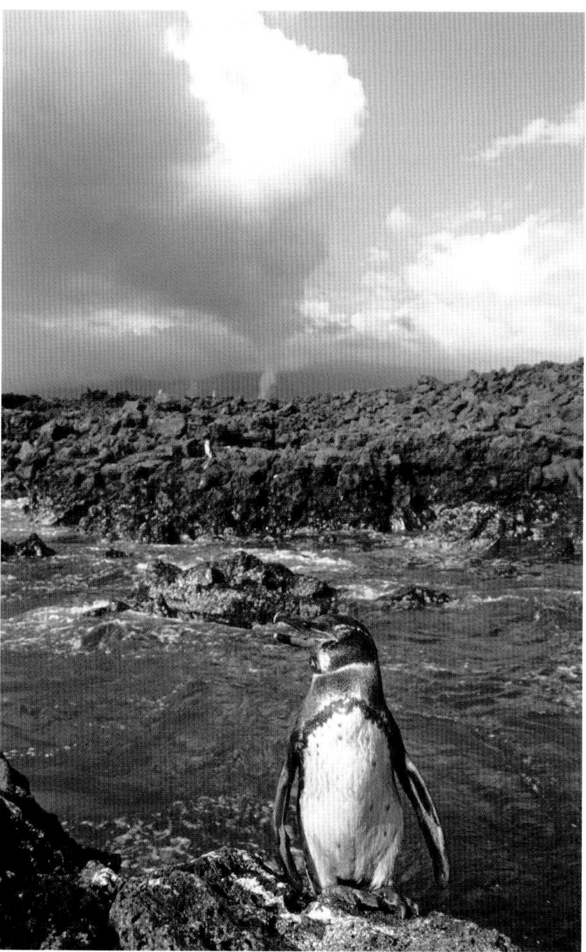

El Niño impact

Applying the same population estimator to Dee Boersma's highest ever count of 1971, the population could have been around 3400 individuals back then, whereas the estimates from the latest report in September 2008 barely reach 1539 individuals [Jimenez and Vargas, 2008]. Thus, it becomes evident that the current population represents less than 50% of what existed in the early 1970s, showing that Galápagos penguins have not been able to recover adequately in over 25 years, after being hard hit by the El Niño events of 1982–83 and 1997–98.

During El Niño, when SSTs rise and upwellings are reduced, fish availability dwindles and penguins simply starve to death. The intense and prolonged El Niño events of 1982–83 and again in 1997–98, were reflected by population crashes of 77% and 65% respectively [Valle and Coulter, 1987; Vargas et al, 2006]. An analysis of the relationship between SST and survival showed that a rise of more than 3°C (5.4°F) above the annual averages caused significant decrease in penguin numbers. Conversely, only with temperatures lower than long-term averages are populations able to recover [Vargas et al 2006]. During cold La Niña periods, penguins can lay two or even three clutches in one calendar year, with concurrent greater survival of both young and adults. With predominantly low sea temperature values and no major El Niño since 1998, they have begun a slow but encouraging recovery.

Correlating such data, we have been able to make tentative predictions of the likely effects of future climate change on the Galápagos penguin, highlighting some somber warnings. One population viability analysis based on the frequency of El Niño events recorded in the Galápagos between 1965 and 2004, concluded that penguins face a 30% extinction probability over the next 100 years, illustrating an important large-scale ecological process [Vargas, 2006; Vargas et al, 2007]. Similar circumstances have doubtlessly occurred since time immemorial, however placing these findings in the context of a 6000-year climatic record gleaned from lake sediments reveals that the weather patterns of the region are increasingly disturbed by more frequent and severe El Niño events.

In 2005, we conducted a population viability analysis of the Galápagos penguin with the participation of international experts in the *Spheniscus* genus. In this exercise, we identified threats to the population, either potential or real. They were: El Niño events, diseases, oil spills, fisheries, introduced predators and even volcanic eruptions [Matamoros et al, 2006]. We concluded that the species may be particularly at risk from combined effects, such as a strong El Niño in conjunction with other dangers, for instance an outbreak of disease and an oil spill all occurring in the same year. In this scenario, the Galápagos penguin could become extinct nearly overnight — a very real possibility.

ABOVE: A penguin emerges from its nest beneath the lava, Cape Douglas, Fernandina. BELOW: The courtship braying call rings out at dusk.

The importance of marking penguins

Because ornithologists working with other species of penguins had recognized that leg bands and flipper tags had negative impacts, it became necessary to design a new marking method for the Galápagos penguin. In 2001, with Cecilia Lougheed, I started marking penguins with microchips (PIT tags), which are known to last indefinitely. Using a 3-m (10-ft) pole with a hoop net, I learned to capture penguins efficiently, an exercise that turned into one of the greatest joys in my life as a biologist. We marked each penguin by injecting a PIT tag in the tarsus region of the left leg. The reason for this placement rather than elsewhere in the body (for example, the back) was to facilitate reading without having to capture the birds again when they are nesting in crevices or lava tubes, where feet are the only parts of the body accessible to a scanner/reader. By threading the electronic reader, mounted at the end of a pole, to where they stand, we can decipher the microchip's code easily without disturbing them at all. Between 2001 and 2008, we marked 875 penguins, the purpose being to conduct a long-term study of demographic factors and population dynamics. This is providing essential data

RIGHT: The characteristic adult plumage of the endangered Galápagos penguin, which is the world's second smallest penguin and the only one living on the equator.

on movement, mate fidelity, reproductive success and survival. In the future we will know the life span of penguins, and how often they reproduce in their lives, and we could also answer the question: what happens when a penguin loses his/her faithful partner during an El Niño year — will the lonely partner remarry or remain widowed for life?

Penguins under siege

With El Niño events increasing, the abundance of disease and parasite vectors may concurrently raise the incidence of pathogens such as avian malaria and others, for which penguin may prove genetically naive, meaning they have no resistance or immunological weapons to defend themselves. As the CDRS ornithologist, I placed emphasis on threatened Galápagos birds and promoted the development of a program to study their pathogens and parasites. The concern regarding the particular risks of avian diseases was based on the case of Hawaii, where avian malaria caused by *Plasmodium relictum* and avian pox have caused the extinction of much of the endemic avifauna. In 1996, Gary Miller and Eric Miller, together with Howard Snell, conducted a sampling of hundreds of penguins in Galápagos, but happily none of them tested positive for malaria. In 2000, Martin Wikelski of Princeton University and myself organized a workshop on diseases, inviting several world experts, including a team from Hawaii, to design an integrated programme of research, and to discuss management measures to protect the native Galápagos wildlife from this potential threat.

RIGHT: A curious penguin meets a snorkeler at Bartolomé.
BELOW: Shark teeth marks in a healed scar indicate of a close call.

Coincidentally, I also came into contact with Patricia Parker of the University of Missouri St. Louis and Eric Miller of the Saint Louis Zoo, as we all shared the same concerns. They offered their expertise and a contingent of students and veterinarians to help the GNPS and CDRS meet the goals we had set out at the international Princeton workshop, and the four organizations started collaborating in 2001 (see also chapter: Parker). As part of this collaboration, between 2003 and 2005 we collected over 500 blood, saliva and fecal samples from penguins to identify potential problems, as well as determine their 'normal' health status. Among other results, I can mention the following:

After visual examination and evaluation of serum chemistry (used as indicator of health) Erika Travis concluded that all tested penguins were in good health, and further laboratory tests yielded negative results for parasites and diseases including: West Nile virus, Newcastle disease, avian paramyxovirus type 1, avian paramyxovirus types 2 and 3, avian influenza, infectious bursal disease, infectious bronchitis, Marek's disease, reovirus, avian encephalomyelitis, western and eastern equine encephalitis, and avian adenovirus type 1 and 2 [Travis et al, 2006]. Although 90% of the tested penguins showed antibodies to *Chlamydophila psittassi*, this was possibly a false positive because no birds indicated symptoms of infection. This study established the basic parameters of the health of penguins, and the news at that point was good.

A truly worrying find, however, is the recent discovery of a *Plasmodium*-type parasite in the blood of penguins sampled between 2003 and 2005, although the birds themselves have so far not shown clinical signs of malarial infection. While writing this article, we are collecting more samples from penguins to determine the prevalence of *Plasmodium* in the population, and to characterize the taxonomy of the parasite at species level to determine the gravity of its potential impact among Galápagos birds.

Introduced mammalian predators

As if all of these dangers weren't sufficient, there's more. Having documented that feral dogs were implicated in decimating penguin populations of southwestern Isabela in the late 1970s, in early 2000 we discovered that feral cats had begun doing likewise. The GNPS dealt to the dogs early on, but could only carry out control of feral cats at selected sites because, being widespread all over the island, they are impossible to eradicate by currently available means. In 1998, realizing that introduced black rats, *Rattus rattus*, are also widespread on Isabela, Santiago, Bartolomé and Floreana — which together hold more than 70% of all nesting penguins (the only rat-free island where they nest is Fernandina) — we decided to test whether they, too, were having an impact. Using brodifacoum, we eradicated rats from two of the three little Mariela Islets off Isabela, where penguins nest in relatively high densities. Luckily, after eight years of close monitoring we can report no appreciable difference in nesting success with or without the presence of black rats, which is just as well considering that eradication of rats, like cats, would be nearly impossible on most larger islands. So far, the second species of introduced rat, the larger Norway or brown rat, *Rattus norvegicus*, which is known to be carnivorous, has not yet reached the penguins' main nesting locations.

Genetic studies initiated by Elaine Akst, later expanded by Jenny Bollmer, documented that the Galápagos penguins have a reduced genetic variability when compared with eight other penguin species [Akst et al, 2002; Bollmer et al, 2008]. They argue that this may be associated with effects of genetic drift (small number of founders) and a number of bottlenecks (significant population reductions) associated with El Niño. Bollmer also analyzed the phylogenetic relationships and found the Galápagos penguin to be most closely related to the Humboldt penguin, *Spheniscus humboldti*, of Peru and Chile, which is not surprising considering their biogeographical proximity aided by the Humboldt Current. This split is estimated at about 4 million years ago. But the human-induced cascade effect of predators *plus* possible diseases *plus* climate change *plus* direct actions of 30,000 human residents — for example, fishing activities causing incidental deaths in fishing nets — and 170,000 tourists visiting the island could end that extraordinary evolutionary journey in the blink of an eye.

ABOVE: Penguins feed in small flocks, most of the time staying within 200 m (650 ft) of shore and only a short swim from their nesting areas.
BELOW: Estimates from the latest yearly census (2008) barely reached a total of 1539 individuals, indicating a population drop of over 50% since the 1970s, and making this the rarest penguin in the world.

The Flightless Cormorant
The Evolution of Female Rule
Carlos A. Valle

Universidad San Francisco de Quito, The Galápagos Academic Institute for the Arts and Sciences (GAIAS), Campus Cumbayá, Diego de Robles y Vía Interoceánica, P.O. Box 17-12-841, Quito, Ecuador. <cvalle@usfq.edu.ec>

Carlos A. Valle is Professor of Ecology and Evolution at Universidad San Francisco de Quito, and Co-director of the University's Galápagos Academic Institute for the Arts and Sciences (GAIAS). He received his Master's degree and Ph.D. in ecology and evolutionary biology from Princeton University. Raised on Santa Cruz Island in Galápagos, he became involved in scientific research of Galápagos seabirds at a very young age. Carlos is a tireless Galápagos conservation advocate, a former Galápagos Ecoregional Director for the World Wildlife Fund, member of the General Assembly of the Charles Darwin Foundation, where he has served on the Board of Directors, and corresponding member of the Pacific Seabirds Group.

BELOW: The Galápagos flightless cormorant has traded the flying abilities of its ancestors for an oversized, muscular body with exceptionally strong legs, as can be seen in the sure-footed leap of a male carrying fresh seaweed for nest-building.

GALÁPAGOS SEABIRDS, much like the fauna of other oceanic islands, are renowned not for their great diversity of species but rather for their uniqueness, brought about by a broad and complex array of ecological, behavioral and physiological adaptations. Such patterns, recurrent among several taxa of the Galápagos Islands terrestrial and marine organisms, result in the Galápagos Archipelago representing a disproportionate contribution in global biological diversity relative to the small number of resident species. Unfortunately, biological uniqueness also correlates with ecological fragility, and in this regard the Galápagos Islands are no exception. A species-poor, isolated ecosystem having evolved free from mammalian predators and far from human intervention, their ecological and evolutionary processes can be easily disrupted beyond the point of no return.

The Galápagos cormorant (*Phalacrocorax* [*Nannopterum*] *harrisi*), more often referred to as the flightless cormorant, exemplifies the evolutionary trend toward the development of astonishing forms. A bizarre and distinctive seabird, it has only recently, yet dramatically, diverged from other cormorants (Phalacrocoracidae) [Kennedy, Valle and Spencer, 2008], both in morphology and behavior. First, it is the only cormorant having developed flightlessness. An extremely rare trait, especially among seabirds, the remarkable reduction of its wings has evolved inversely to body size. However, its breeding system, though less conspicuous, turns out to be even more unusual.

As a graduate student in ecology and evolutionary biology in the late 1980s, I pondered my subjects in order to develop a research proposal worthy of appraisal by a stern academic committee. Most students at Princeton University choose to do their fieldwork in faraway exotic locations, but I immediately decided to return to the islands where I was raised. By the time of this decision, I had been reading extensively about the evolution of parental care and mating systems, and remembered a paper published several years earlier by Bob Tindle, who had first reported the remarkable mating habits of the flightless cormorant. I felt that studying this species would offer not only the perfect topic for my Ph.D.

dissertation, but furthering knowledge of its life cycle would make a useful contribution toward Galápagos conservation management. I will always be thankful to my rigorous thesis committee, in particular my advisor Peter Grant, for forcing me to clarify many questions in my research and seek alternative hypotheses as part of my preparation to become a scientist, sound principles which I adopted and have since applied as a professor myself (see also chapter: Grant).

The breeding system of the flightless cormorant is, without doubt, every bit as extraordinary as its flightlessness, having presumably evolved *in situ* since none of its closest relatives (the double-crested cormorant, *Phalacrocorax auritus*, or the neotropic cormorant, *P. brasilianus*) share any such behavioral traits [Kennedy, Valle and Spencer, 2008]. Firstly, this involves a partial role-reversal, since females initiate and play a more active part in courtship than males. Yet while females aggressively compete for access to available mates, it is the males that actually do the choosing and seem to discriminate strongly among potential partners.

ABOVE: Though seaweed is used most often by cormorants in nest building, individual preferences are reflected in the selection of different items, such as sea urchins or, in this case, starfish.
LEFT: A powerful foot-propelled diver, the flightless cormorant hunts bottom-dwelling fish and has been recorded to depths of 73 m (240 ft), though most dives are less than 15 m (50 ft) deep.

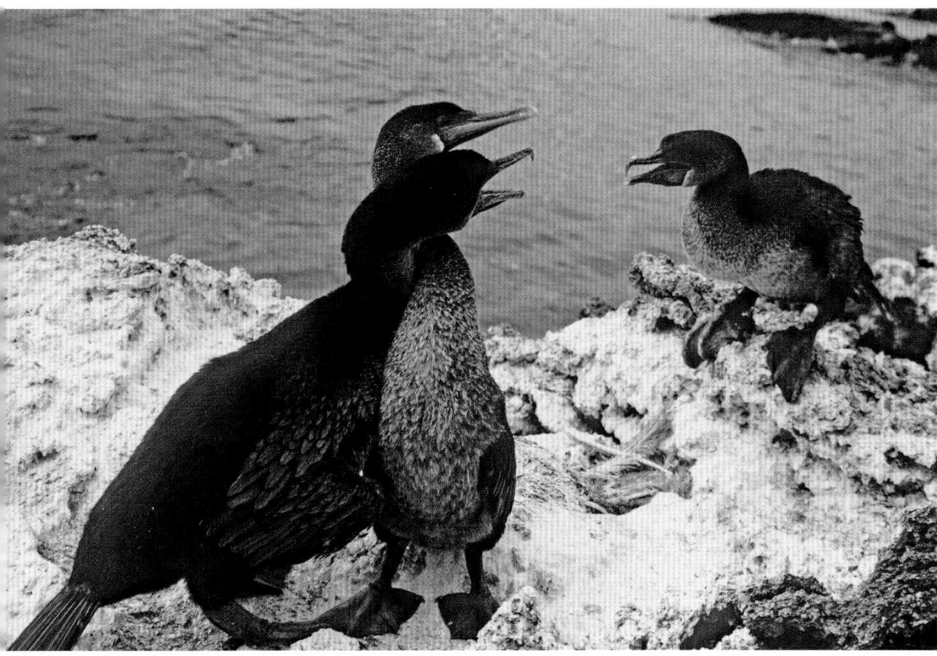

ABOVE: With the assistance of her newly chosen partner, a female (center) at Cape Douglas, Fernandina, rejects her previous mate and takes over their nest site, leaving him to care for their still-dependent fledgling. BELOW: Though useless, wings are dried assiduously after every trip to sea.

Mark Jones

Even more unusual is a mating system termed facultative sequential polyandry, by which the female deserts its mate and offspring — usually when the young fledge at around two and a half months of age — and then goes off to find a new mate, while the male alone raises the young for up to six additional months until they attain full independence [Tindle, 1984; Valle, 1993 and 1994]. These novel traits are not only new for cormorants but exceptional among birds in general.

To explain the terminology: 'polyandry' means a system where one female mates with several males; 'sequential' is because she departs part way through the breeding cycle, effectively mating with one male after another rather than simultaneously; 'facultative' is added because it appears that there are factors influencing not only when she might do so but even whether she deserts at all. Essentially, she appears to time her desertion while still ensuring the survival of her young, which seems to take precedence over the opportunity to remate. This is corroborated by a high variation in the timing of a female's departure, correlated with her mate's parental ability or willingness to care. Females stay longer, even forfeiting opportunities to remate, when brood size at fledging is larger than one, or if the young are maturing slowly.

The evolution of this most extraordinary approach to reproduction raises three fundamental questions, which were to guide my work from May 1988 through April 1994, my last field season, shortly after I completed the defence of my Ph.D. thesis.

What allows one parent to desert?

The shift from monogamy to polyandry has been linked to a relief of the biparental constraint (i.e., one single parent will be as good as two at raising the young) allowing for parental emancipation in association with conditions when food is plentiful or in species where a relatively small clutch size is the norm. Studies addressing this initial concept have revealed that facultative mate desertion tends to be opportunistic and that individuals may take advantage of peaks in food abundance which allow a single parent to raise the brood unaided. These studies also suggest that brood size reduction could be adaptive so as to favor desertion. Mate desertion in the flightless cormorant, however, has been shaped by the combined effects of four main factors all favoring higher rates of reproduction and an earlier desertion by the female instead of the male:

(1) *brood-size reduction to a single young* (from clutches of 2–3 eggs) that releases the necessity of two parents to successfully raise the young;

(2) the *extensive post-fledging parental care* that the young require to become fully independent, during which the female's food contribution to the growing juvenile may be of little ultimate benefit toward offspring survival;

(3) potential *opportunities for remating*, because high rates of nest failure maintain an almost permanent population of potential breeding individuals throughout the year;

(4) *frequent and recurrent nesting opportunities* arising throughout most of the year due to an erratic marine environment that actually triggers mate desertion because females will benefit more from starting another brood sooner rather than helping the male to finish raising a single-young brood.

Why a deserted parent should stay and care for the offspring alone?

This is a question that has been approached by using 'game theory' models [e.g., Maynard Smith, 1982]. These predict that a deserted parent stands to benefit more from its investment in the brood by continuing with lone parental care than if it abandons as well. Deserted males raised their young to independence as successfully as those that were not deserted. However, recently fledged juveniles that are deserted by both parents are unlikely to survive because at that age they are fully dependent on a secure food supply. The willingness of deserted males to care for a chick unaided is thus in agreement with this theory.

Why the parent that would be deserted does not depart first?

This is the most puzzling question in the evolution of sequential polyandry. Given that the flightless cormorant's ancestral condition would have been monogamy with biparental care, and that the opportunity for a parent to desert arises late in the breeding cycle, desertion by a parent of either sex would be expected in equal proportions. That should give rise to ambisexual mate desertion and to the evolution of a sequential polygynandrous system (multiple mates of either sex) in which either parent may desert while the other cares unaided. However, except for rare occasions where the female did not desert and the male ceased feeding the young very late in the breeding cycle, I found that male flightless cormorants almost never desert.

So, having analyzed the questions, I still did not have satisfactory answers to these conundrums. The evolution of sequential polyandry (several males per female) instead of sequential polygyny (several females per male) or ambisexual desertion, should be expected to be favored in any of the following scenarios: the male is constrained from deserting; the female desertion is in the male's interest because he may get a genetic side-benefit; or a male's desertion is not profitable to himself.

Theoretically, desertion by the male could be constrained by several behavioral and ecological factors including, among others: a male-biased sex ratio in either the wider population or at a local scale in the potential breeding population; male territoriality which inhibits his departure because of an allegiance to the breeding site; sex differences in the length of time males and females require to regain reproductive condition after the breeding cycle; poor condition (physiological) of the female that will stop her from further parental care in order to ensure her own survival. Conversely, the male may actually gain a genetic side-benefit from female desertion if the male gets to inseminate the female prior to her desertion, thus either cuckolding her future mate [Valle, 1994]

ABOVE: Massive feet characterize the Galápagos cormorant. BELOW: Nesting on exposed lava just a few paces from the shore, parents take turns guarding the chicks against hawk predation, Puerto Bravo, Isabela.

TWO VERY SPECIAL GULLS

Like so many other unusual Galápagos species, the two endemic gulls present surprising traits in many respects. The lava gull, *Larus fuliginosus* (above), is one of the world's rarest, with a total population estimated at no more than 800 individuals. It is a scavenger found in pairs or very small groups along most shorelines. Breeding only rarely, a pair will aggressively defend its nesting territory, the laughterlike calls a distinctive coastal sound.

Leading a totally different lifestyle, the swallow-tailed gull, *Creagrus furcatus* (right), is the world's only pelagic and nocturnal gull, using its huge eyes to spot prey rising to the sea-surface during the night. Each evening, flocks can be seen leaving their cliff-face nesting colonies to feed far out at sea, returning at dawn. When not breeding, they acquire white heads and range widely throughout the rich waters of the Humboldt Current along the coast of Peru.

Tui De Roy

RIGHT: Twisting, bending and entwining their flexible necks, a pair of courting cormorants pass a booby feather back and forth as ceremonial nest-building material.

or because she is double clutching in a second nest as in some shorebirds [Oring, 1986].

But in the end, the data I amassed lent little or no support to any of the above hypotheses, assumptions or predictions. The breakthrough came from simple field observations of foraging behavior. This showed that concurrent with the flightless cormorant's remarkable sexual dimorphism (males are almost 40% larger than females) mates differ substantially in their average prey size, although they do not vary in their catch rates. Thus, owing to the larger prey that males normally capture, they turn out to be much more efficient food providers than females, especially for nearly fully-grown fledglings. This means that while single males are well able to feed single-young broods even when food availability is low, females alone would not be capable of doing likewise, unless the juvenile was able to efficiently forage and supplement its diet with catches of it own.

Therefore, it becomes profitable for the female to desert as soon as the young leaves the nest and no longer needs guarding against predation from Galápagos hawks (the only, but important, predator of cormorant chicks) or hostile attacks from other cormorants. For the male, on the other hand, desertion at this stage would be disastrous because any fledgling left to the care of the smaller female would have little chance of survival, and thus of furthering his lineage. The larger male must therefore wait longer than females before he can profit from deserting, an opportunity that the female exploits to entrap the male in the 'cruel bind' [Trivers, 1972] of caring alone for the young.

In conclusion, the key factors likely to have favored the evolution of polyandry instead of polygyny is the large practical difference between male and female foraging proficiency. At the same time, the constraint preventing a female from deserting before the young

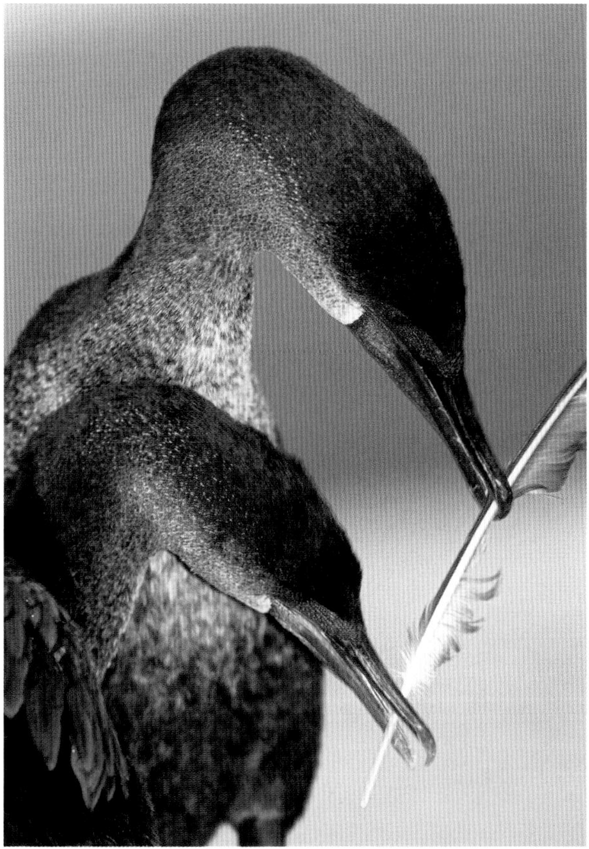

fledge is because these need continuous protection against hawk predation and other dangers during early life. After closer examination, further contributing factors to this remarkable situation also emerged. The high incidence of reduction in clutch and brood sizes, which relax the biparental constraints and allow one parent to desert, mainly result from failure of eggs to hatch and mortality of small chicks due to parental negligence. These aspects cannot be considered as adaptations for desertion since they are major causes of breeding failure and contribute to low reproductive success.

Hatching failure is apparently related to egg infertility that may be a manifestation of inbreeding depression, probably resulting from low genetic diversity in the exceedingly small total population [Valle, 1995] (see also chapter: Parker). When it comes to chick loss, the noticeably large feet of the flightless cormorant, thought to be of major advantage for efficient foraging performance (cormorants are foot-propelled divers), becomes detrimental in the nest as parents (particularly males) frequently squash their small, recently hatched chicks inadvertently under their oversize webs [Harris 1979, personal observations]. The high mortality rate of the last hatchlings from the clutch, another major cause of brood-size reduction, results mainly from parents rapidly increasing the size of prey they deliver, apparently in response to the rapidly increasing ability of the older siblings to swallow such large food items. The late-hatched young therefore often starve to death, or are accidentally thrown out of the nest when the parent attempts to recover from the chick's beak the oversize food it is unable to swallow. Here I can only speculate that such a response may be adaptive because it better benefits the parents to ensure rapid growth of the first-hatched youngsters instead of attempting to raise a larger brood. My field observations corroborated this, revealing that hawk attacks succeeded almost exclusively at nests containing more than one half-grown young because parents were simply unable to efficiently protect more than one chick under their body and wings when these reached half-size or larger.

Thus, the evolution of facultative sequential polyandry in the flightless cormorant differs even more from other species, as food abundance per se is not the factor driving the system. Nor is its evolution the result of direct adaptations favoring mate desertion as such, but rather the side effects of other life-history adaptations.

ABOVE: Spreading its diminutive wings protectively, an adult shields a nearly grown chick from attack by a young hawk, a tactic that almost always fails when trying to raise more than one chick. Cape Douglas, Fernandina.

ABOVE: Using seaweed and sea urchins, a pair begins nest construction along the north coast of Fernandina.
RIGHT: With deep propeller wounds, this cormorant survived only one day after being hit by an outboard motor at Fernandina.
FAR RIGHT: As the largest chick receives priority feeding, rarely do two fledglings survive to this stage.

All of these revelations only helped me appreciate even more poignantly the extraordinary vulnerability of the flightless cormorant as a species. As an undergraduate student back in 1980, I assisted Sylvia Harcourt with a complete census of Galápagos penguins and cormorants. Three years later, I conducted a second census, which revealed a drastic population decline of about 50% for cormorants, as well as a nearly 80% drop for the Galápagos penguin, as a consequence of the severe 1982-83 El Niño conditions (see also chapters: Grove; Vargas). Fortunately, soon afterwards, the cormorants began a quick recovery, reaching their original numbers within a few years, which I estimated to be about 1000 adults. That number was still much lower than the 1600 birds reported more than a decade earlier, although I had long considered this figure to be an overestimation. More recent

censuses by Hernan Vargas and his coworkers have shown that the population has apparently increased further since 1998, with current numbers bordering on 1400 individuals.

Much concern for the species persists, however, because its genetically effective numbers (the number of 'standard' breeding birds within the population) are barely above the theoretical minimum necessary to guarantee long-term viability. Furthermore, besides their remarkably restricted distribution, they face ever-increasing threats from human activities. These include collisions with fast-moving fishing launches operating in the cormorants' near-shore foraging grounds, and the potential danger of diseases spread by recently introduced mosquito vectors (see also chapter: Parker). Drowning in illegal gillnets sometimes set for sharks in coastal waters, plus the possible introduction of black or Norwegian rats from illicit sea cucumber processing camps, are both very real concerns, making the need for tighter controls and effective policing critical for the conservation of this species.

These terrible risks took on a very personal tone for me one day in August 1992. I had just started gathering data on diving performance by attaching recording devices to the backs of several breeding cormorants that I was regularly monitoring. However, that day I wanted to try a subadult cormorant. The only one available being a male I hadn't seen before, it was duly captured, measured, banded, equipped with the device and released. The bird jumped into the water and set off fishing, traveling eastward along the coast, with my assistant and myself following at a distance in our little inflatable boat. After about two hours, the bird, who happened to be one of those that wander among different colonies, had gone several miles when it entered a small cove surrounded by dense mangrove forest. Here we got the surprise of our lives! A squad of about 20 alien invaders and five little vehicles (fast fishing launches) were tucked out of sight, cooking sea cucumbers by the thousands.

That was the day we learned that we were not the only foreign creatures living on Fernandina Island, having stumbled upon a well hidden, illegal *pepinero* camp. That afternoon I sent a message to the National Park and the Darwin Station. Next day the world knew for the first time that Galápagos had a new conservation problem.

ABOVE: Detail of the cormorant's stunted wing.
BELOW LEFT: A small chick struggles with a large mouthful intended for larger siblings.
BELOW RIGHT: Flightless cormorants are strong predators, here taking advantage of waters superheated by a volcanic eruption to overcome a large moray eel.

Sea Lions and Fur Seals
Cold Water Species on the Equator
Fritz Trillmich

Professor, Department of Animal Behavior,
University Bielefeld,
PO Box 10 01 31,
D-33501 Bielefeld,
Germany
<fritz.trillmich@uni-bielefeld.de>

Dr. Fritz Trillmich began work in Galápagos in 1976 and has since spent much time studying the social structure, maternal care, foraging behavior, and population genetics and dynamics of fur seals and sea lions, with a few years' digression on marine iguanas. He is now a Professor of Animal Behavior at the University of Bielefeld.

WHEN I FIRST ARRIVED IN the Galápagos Islands in 1976 I was keen to understand how sea lions and fur seals, so exquisitely adapted to a marine way of life, could deal with the difficulties of a tropical setting on land. Little did I realize then that this is only part of their problem: they must also each deal with a highly variable marine environment that offers boom-and-bust conditions during periods with food abundance, La Niña, and severe food scarcity, El Niño (see also chapter: Sachs). To a large extent, these events drive the population dynamics of both species and influence most aspects of their life histories.

Galápagos sea lion

Originally described as a separate species by Sivertsen, the Galápagos sea lion was later relegated to subspecies status. However, our recent genetic data suggest it has evolved independently of its larger North American relative, the California sea lion (*Zalophus californianus*) over the last 2.3 million years. Thus, *Zalophus wollebaeki* has been reinstated.

Sea lions live in colonies, with males being much larger (up to 200 kg, maybe even 250 kg, or 440–550 lb) than females (mostly 50–80 kg, or 110–175 lb, but up to 120 kg, 265 lb). The large

RIGHT AND BELOW: During a bull's tenure as territory holder, which may last from a few days to six weeks at most, he rarely comes ashore among his sleeping cows, constantly patrolling the shallows against intruders.

territorial bulls are easy to recognize by the marked hump on their heads, due to a prominent bony crest on the skull that serves as the attachment for huge muscles enabling powerful head movements and a forceful bite. Even though sea lions are encountered most easily on sandy beaches, they also inhabit rocky shelves, and actually prefer such sites because these offer access to tide pools in which adults and pups alike can cool off during the heat of the day. They occur everywhere in the archipelago, but with highest densities in the central and eastern islands.

Populations have apparently decreased and are presently estimated as half as large as they were only 30 years ago. The reasons for this decline are poorly understood and may have to do with repeated El Niño events. Severe cases, such as those recorded in 1982–83, 1997–98 and 2002, reduce the survival of pups and yearlings substantially, and may even decrease adult survival. Indeed the 1982–83 El Niño helped me to understand how precarious a life the species leads, always close to the edge of disaster: most years offer them sufficient food due to the high productivity caused by advection of cold, nutrient-rich waters — the Humboldt Current from the southeast and the Cromwell Current upwelling from the west. But when El Niño strikes, these currents fail and surface waters warm rapidly due to the enormous heat of the tropical sun and influx of warm waters from the north. This devastates the sea lion's food resources. Large males are particularly affected since they fast during their time as territory holders. If marine productivity is low when their territorial stint ends, these large emaciated animals do not find enough food to recover, and may starve to death as a result.

In former times, interactions with feral dogs also caused some mortality but due to the controls enforced by the Galápagos National Park (GNP) this problem seems well managed. I am more concerned

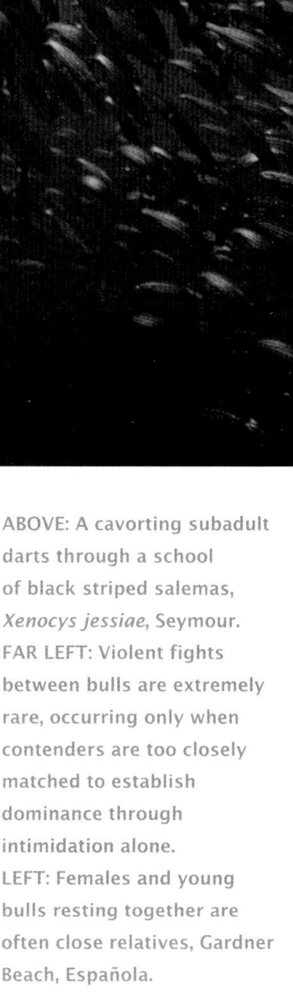

ABOVE: A cavorting subadult darts through a school of black striped salemas, *Xenocys jessiae*, Seymour.
FAR LEFT: Violent fights between bulls are extremely rare, occurring only when contenders are too closely matched to establish dominance through intimidation alone.
LEFT: Females and young bulls resting together are often close relatives, Gardner Beach, Española.

WHALE WATERS

The rich upwelling areas along the shores of western Isabela and Fernandina islands are favored feeding grounds for many marine mammals besides Pinnipeds. Various species of dolphins, plus Bryde's and Cuvier's beaked whales, are seen here regularly, sometimes very close to land. But the most ubiquitous are the large pods of sperm whales, *Physeter macrocephalus*, usually working the deep waters further offshore. Females and their young (above) remain in tropical waters throughout the year, whereas the huge bulls, which are half again as long as cows and may reach 18 m (60 ft), are more solitary and appear to visit seasonally from Antarctic summer feeding grounds, frequenting the female herds only during brief mating forays. Their presence may trigger huge social gatherings sometimes numbering over 100 animals, a fantastic sight. During the 19th century these waters ran red with sperm whale blood as they were ruthlessly persecuted by whaling fleets from both sides of the Atlantic, with attention turning to fur seals pelts as the whales dwindled.
Tui De Roy

about the recent, often negative, interactions with humans. We found animals with large fishing hooks in their skin and mouths and even one with a knife run into its belly up to the handle. The problems arise due to the unfortunate tendency of sea lions worldwide to approach fishing gear and attempt to steal fish, making them very unpopular with fishermen.

When we swam with sea lions we could only marvel at their wonderful diving and underwater maneuvering abilities. As we have no chance of following them into the depths, we equipped a few with electronic instruments that record their feeding forays. We found they can descend to about 300–500 m (1000–1650 ft), but generally search for prey in the upper 50–100 m (165–330 ft) where they find schooling fish like sardines, or organisms that come closer to the surface at night, making them easier to catch. Thus, some individuals dive to the bottom and search for prey there, others dive day and night in open water, and still others prefer to forage during the night when shallow diving brings sufficient food returns. Dive time is usually around 2.5 minutes but may last up to 10 minutes. At a mean swimming speed of 2–3 m/sec (6–10 ft/sec), Galápagos sea lions can cover large areas and long distances while foraging: for example, animals from Santa Cruz have been found close to southeastern Isabela, 70 km (44 miles) away.

Sea lions breed almost all year round, with a concentration of births between September and January. During this long period, males establish and hold territories in areas where many females come ashore. Usually, large bulls patrol the shallows where most activity takes place, announcing their dominance by incessant barking, a sound that characterizes every sea lion colony. Females, juveniles and a few smaller males rest ashore or lie at the water's edge, not paying much attention to the displaying males. Barking makes the male obvious to us as humans, but more importantly to females that come ashore in his section of coastline. Females rarely interact with the dominant bull, and usually react aggressively when he attempts to sniff them. However, when in estrus, a female may approach a bull and induce him to follow. She then crawls around in front of him as he follows her, barking loudly. Near the boundaries of his territory he tries to herd her in, with variable success. Copulation usually occurs in the water, where it can only very rarely be observed.

Bulls fight for prime positions at the waterfront and the most successful males sire most offspring. Holding a territory visibly drains the dominant males' resources. In spite of constant energetic expenses due to chasing and herding, territorial bulls fast most of the time during their tenure, which is limited to two to six weeks as a result. During this time, males shrink in girth as

Photo courtesy Mark Jones

OPPOSITE PAGE BOTTOM: A mother and her newborn pup form a close bond based on smell and the sound of each other's voices.
LEFT: A pup is born tail first at Punta Espinosa, Fernandina.
BELOW: A yearling pup is sometimes allowed to remain, suckling alongside its younger half-sibling, who tends to develop poorly without the benefit of its mother's full milk production.

they slowly lose condition and use up their large fat reserves. Smaller and younger males may sneak into established territories attempting to get in contact with females. Usually, the dominant owner soon discovers these youngsters and chases them out. However frightening and risky this tactic might be, the females seem irresistible, and the smaller males will test their luck over and over again throughout the whole breeding season.

A year after copulation the female gives birth to a single pup weighing about 8 kg (18 lb). This yearly cycle is achieved through a peculiar feature whereby the fertilized oocyte, after forming an initial ball of cells, ceases further development for about two to three months before it attaches to the uterine wall and starts normal embryonic development of a fetus, which is born nine months later. The mother stays with her newborn for five to seven days, protecting it against other sea lions, calling to it, dragging it to the water's edge during the heat of the day and nursing it. This is an important time for female-pup bonding, when they learn to recognize each other's voice amid the general din of the colony. As soon as the pup has become a bit more mobile, the female takes her first feeding trip to replenish her reserves that by now are drained by the production of lipid-rich milk. The first trip back to sea marks the onset of a regular foraging

ABOVE: Playful sea lions ride the perfect wave at Punta Suarez, Española, surfing inside the breaking crest.
BELOW: The tidal cove at Cape Hammond, the author's study site on Fernandina, is an ideal nursery for rambunctious fur seal pups testing their swimming skills while their mothers are away feeding at sea.

cycle, in which periods of absence alternate with periods of pup attendance. As the pup grows older, the duration of absences increases slowly from half a day to about a week. In comparison to other sea lion species, young Galápagos sea lions are dependent on their mothers for an extraordinarily long period. They often suckle for two or three years before they are finally weaned, even though they do some independent foraging from the age of one year.

Lactating mothers are less likely to conceive again and carry the pregnancy to term, as the energetic double-burden of lactation and gestation may also lead to abortion. If a strong female sustains both pregnancy and lactation, conflict arising between siblings often leads to death or abandonment of the newborn. Less frequently, a female successfully bonds to the newborn and defends it against the attacks of its older sibling, in which case they can be seen peacefully suckling together. Sometimes, however, such observations can be of an unrelated pup stealing milk. In an extreme case we saw four young all suckling together! Females are extremely site faithful and daughters seem to inherit the home site from their mothers, so that a Galápagos sea lion colony is actually formed of clusters of closely related subgroups.

Galápagos fur seal

Tropical species are often much smaller than their more temperate relatives and this also applies to the Galápagos fur seal, *Arctocephalus galapagoensis*. It is the smallest of all the seals (Pinnipedia) worldwide, with adult females weighing less than 30 kg (65 lb) on average, and males about twice as heavy. It is closely related to the South American fur seal (*A. australis*) and was even considered to belong to the same species by some.

By the end of the 19th century, Galápagos fur seals had been decimated to near extinction by whalers and sealers who hunted them for their pelts (see chapter: Watkins). When I began my studies, nobody quite knew where I could find an abundant population.

But the then director of the Charles Darwin Research Station, Craig MacFarland, remembered he had seen them at Cabo Hammond on Fernandina Island and suggested we should try there. It was the best tip-off I ever got. What a wonderful, stark place amid lava and, best of all, it was absolutely full of fur seals! As you approach the site it sounds like children yelling in a swimming pool: fur seal mothers call out searching for their young, and juveniles scream for their mothers in return.

Fur seals are not as easily encountered as sea lions because they inhabit more rugged, largely inaccessible coastlines. The jagged, often vertical lava shoreline offers them vital protection from the tropical sun in caves and under large boulders. Here, they climb and jump from rock to rock with remarkable agility. Indeed they are so fleet on difficult surfaces that territorial fur seal males find it easy to chase away sea lion bulls when the latter venture into fur seal habitat, even though the trespassers may be three or four times larger. They occur predominantly in the western region of the archipelago, which is cooler and more productive due to local upwelling of the Cromwell Current bringing cold, nutrient-rich waters to the western flanks of Fernandina and Isabela.

Your best chance to meet a fur seal ashore is around full moon. Why so? It took me about 15 years to figure this out. Fur seals exploit the most productive marine areas by feeding on deep-sea species like lanternfish (Myctophidae). Being so small, they cannot dive to the great depth at which these prey organisms occur during the day. The deepest dive we recorded was by one adult female who reached 170 m (560 ft), but most range between 20 m and 30 m (65–100 ft) only. At night, the prey species migrate close to the surface and that is why Galápagos fur seals forage almost exclusively during the hours of darkness. On bright moonlit nights, however, the prey remains deep. Therefore the period around full moon does not offer good feeding opportunities, and so many animals spend a few days ashore during this phase.

The fur seal's choice of sites with difficult access makes it hard to obtain reliable population estimates. In the late 1970s, when we sailed around the islands with Fritz and Carmen Angermeyer in their beautiful little boat, the *Nixe*, we estimated around 30,000 fur seals for the entire archipelago. More recent counts, however, suggest that numbers have dropped to about 15,000, raising concerns, as we do not yet understand the reasons for this decline.

Fur seals reproduce on land, always close to the sea, as they need access to water for cooling off during the heat of the day. They often jump in briefly and return to land until they have dried and need to cool themselves once more. Bulls establish terrestrial territories in sites harboring many females, and

ABOVE AND LEFT: Mother and pup fur seals reunite with shrill calls audible above the sound of the surf.

BELOW LEFT: A vestige from their cold-water origins, fur seals have thick coats of dense wool protected by stiff guard hairs, trapping a layer of insulating air, as can be seen by the escaping bubbles in the wake of a diving bull.

BELOW: A mother and her newborn pup relish close tactile contact that helps them bond.

ABOVE: During warm weather, fur seals sleep at the sea surface, unable to keep cool on land.

viciously defend their area against other males. In contrast to sea lions, copulation takes place on land, each bull mating with all the females that come into estrus in his territory, a trait that makes it easy for us to measure male reproductive success. Unlike other fur seal species, Galápagos bulls are very tolerant of humans in their territory, becoming non-aggressive after only a short period of habituation. This allows close observation of all their breeding activities, as the observer is soon completely ignored. In fact, I got a terrible fright once when a territorial male ran right across my lap in pursuit of an intruder he had seen behind me.

RIGHT: Interactions with people are a mutually enjoyable experience, but could threaten the animals with outside contamination. BELOW: A breeding bull fur seal, considerably heftier than the females, guards his territory on land, taking frequent dips to cool down.

Females mature when about five years old. Interestingly, after her first mating a female will remain simultaneously pregnant and lactating for the rest of her life. Becoming receptive shortly after giving birth in what is termed postpartum estrus, she will carry the next fetus while rearing the previous offspring. Like many marine mammals, delayed implantation in fur seals means that the pup is born a year after copulation.

During September and October, pregnant females come ashore about two days before parturition. A mother stays with her newborn for almost a week and then is impregnated by whichever male happens to be the territory holder where she chose to give birth. Immediately after copulation she becomes highly aggressive against males and chases them away. After that, she alternates spending a day or two on land nursing her pup, and foraging at sea for one to four days. This routine is strongly influenced by moonlight as described above, but interestingly, many females return to shore to rest in daytime during longer foraging trips yet hide away from their pups by coming ashore elsewhere.

Fur seal mothers may care for their pups for up to three years, or in rare cases even longer. As with sea lions, if the previous pup is still alive and has not been weaned at the end of its first year, a female may find herself with both a newborn and its older, still dependent sibling, a situation often leading to bitter conflict between the two over access to their mother and her milk. Although she tends to defend her newborn, usually the older offspring wins, and the smaller one eventually starves to death. It is hard to observe such a perfectly viable pup slowly dying while struggling for access to its mother. Only when food is very abundant, particularly when cold waters in highly productive La Niña years cause nocturnal organisms to migrate close to the surface, are yearlings weaned without putting up much resistance. In such plentiful years, young fur seals grow exceptionally well, reaching sufficient mass to become self-sufficient divers by the end of their first year, but more often they are still partly dependent on maternal milk and therefore fight for their survival against their newborn half-sibling [Trillmich and Wolf, 2008].

The sea lions and fur seals of the Galápagos have adjusted to life in a tropical environment and its enormously variable marine productivity in many different ways. This is most obvious in their long periods of maternal care, but also in the slow growth rates of their offspring, which allow them to survive when food is scarce. But these features, combined with fluctuating climatic influences on their population dynamics, can lead to disaster when El Niño and adverse anthropogenic effects combine. Most of all, I am fearful of the contact between sea lions and humans and their pets in the larger settlements (see chapters: Parker; Watkins). Such interactions can potentially lead to the spread of disease with unforeseeable consequences for these unique and beautiful animals. Avoiding a manmade catastrophe is one of the greatest challenges for Galápagos conservation managers today.

Parasites and Pathogens
Threats to Native Birds
Patricia Parker

Dr. Patty Parker is the Des Lee Professor of Zoological Studies at the University of Missouri-St. Louis, and Senior Scientist at the Saint Louis Zoo. In 1990, she began long-term research in Galápagos, and has led the UMSL/Saint Louis Zoo/CDF/SPNG efforts to understand disease threats to Galápagos birds since 2001. For this work, she received the Order of Scientific Merit from the people of Santa Cruz in 2008, and has published numerous papers on this and other Galápagos topics with her students and collaborators. She is a fellow of the Animal Behavior Society, the American Ornithologists' Union, and the American Association for the Advancement of Science.

Department of Biology,
University of
Missouri-St. Louis,
8001 Natural Bridge Road,
St. Louis, MO 63121,
United States.
www.umsl.edu/~parkerp

WildCare Institute,
Saint Louis Zoo,
1 Government Drive,
St. Louis, MO 63110.
<pparker@umsl.edu>

IN THE FALL OF 2000, Hernan Vargas and Martin Wikelski organized an international workshop to understand the threats posed to Galápagos birds by the arrival of new diseases. Shortly before, I sat in the office of Eric Miller at the Saint Louis Zoo and we brainstormed how my new joint position between the University of Missouri-St. Louis (UMSL) and the Saint Louis Zoo might best facilitate research between the two institutions in light of their strengths, interests and institutional missions. Since I had already been working in Galápagos for several years, and also because the Saint Louis Zoo has a strong veterinary staff with interests in wildlife medicine, one of the ideas was to study the threats posed by the arrival of new avian diseases in these islands. When we realized that the Galápagos National Park (GNP) and Charles

LEFT: The nocturnal swallow-tailed gulls and Galápagos marine iguanas are but two of the vulnerable endemic species that may be immunologically naïve, meaning that, in the absence of native diseases, their defences against foreign pathogens are much weaker than in continental counterparts.

ABOVE: Darwin's finches, such as this small ground finch, *Geospiza fuliginosa*, nesting in spiny *Scutia pauciflora*, may be particularly at risk from introduced pathogens, as is demonstrated by the demise of avian fauna on other isolated islands, such as Hawaii, most of which has become extinct.

Darwin Research Station (CDRS) were considering exactly the same thing, we offered to help. We began working together in 2001.

The motivation behind this collaboration was to help Galápagos avoid becoming Hawaii all over again. The Hawaiian honeycreepers have undergone an adaptive radiation similar to the Galápagos finches, but one striking difference separates the two groups today: almost half of this Hawaiian group has gone extinct within the last 200 years since the colonization and commercial development of the islands by humans, whereas to date no Galápagos bird species has been lost. The Hawaiian case represents some of the best-documented examples of extinctions caused by the arrival of new diseases to which the endemic birds had no resistance. Chief among the pathogens responsible in Hawaii were the blood parasite *Plasmodium relictum*, that causes avian malaria, and the avian pox virus. Birds of some species dropped dead from malaria alone, others from pox alone, and yet others from combined infections of both pathogens.

We knew that symptoms looking like pox were already present in Galápagos in domestic chickens and in some wild bird species, and we were also aware that the arrival of the *Culex quinquefasciatus* mosquito had been documented in the mid-1980s. The significance of the latter arrival is that this particular mosquito species was the vector for the avian malaria parasite *Plasmodium relictum* which had caused the inroads in Hawaii. So our first efforts were to confirm that what appeared to be infections of the avian pox were due to that virus and not some other cause, and to corroborate that the *C. quinquefasciatus* mosquito had indeed established itself. Unfortunately, both were true.

These two early studies required close cooperation between partners. The pox study utilized the virus expertise of Terry Thiel, and molecular characterization of the pox virus by my lab, both at UMSL, in collaboration with Virna Cedeño and her students at the GNP molecular lab on Santa Cruz Island, with sample collection by Gustavo Jimenez of the CDRS and Tim Walsh of the Saint Louis Zoo.

The mosquito study involved the entomological expertise of Noah Whiteman at UMSL, in association with Simon Goodman and Andrew Cunningham of the Zoological Society of London, also through their connection with the GNP laboratory. This form of multipartner interinstitutional collaborative spirit would characterize all of our work to follow from 2001 until now.

Avian pox

Let me explain a little more about these two important diseases. A virus in the Avipoxvirus group causes avian pox. It can be highly pathogenic depending on the strain of the virus (since viruses do not reproduce sexually, their different forms are usually referred to as 'strains' of a particular group, rather than 'species'), and also on the susceptibility of the bird that becomes infected. When an infective virus particle, called a virion, enters a bird through a break in the skin, it immediately sheds its outer envelope, enters the cells, and within 24 hours is replicating itself thousands of times within the cells near the site of entry. Within a day or so, the bird shows pustules at these sites, and these may grow extremely large. Those pustules may become secondarily infected with bacteria. They eventually crust over to form a scab, and when this falls off, virions are released into the environment, where they can remain in the substrate for long periods until coming in contact with some break in the skin of the next hapless bird. UMSL graduate student Jenni Higashiguchi is studying the relative importance of this form of passive environmental transmission and the alternative transmission by biting insects. Any insect that bites a bird at the site of an infection will pick up virions on its mouthparts, which it can then mechanically transmit to the next bird it bites, facilitating transmission by delivering both the infective virion particles as well as the break in the skin. We think that this form of transmission is probably particularly important during wet seasons, when blood-sucking insect populations flourish.

During the 1982–83 El Niño, Hernan Vargas documented the extreme mortality impact of this virus on mockingbirds on Santa Cruz (no juveniles infected at the outset of the study were alive at the end, while 72% of the uninfected juveniles survived), and Bob Curry and Peter Grant had similar findings in the mockingbirds on Genovesa. Our ongoing studies are focused on trying to understand why some species seem particularly susceptible (several finch species and mockingbirds) while others are less so, and to better understand species-specific as well as island-specific patterns of prevalence of the virus.

ABOVE: A small ground finch, *Geospiza fuliginosa*, approaches death from an introduced avian pox virus.

BELOW: This microscope test slide, produced from a museum specimen of a finch collected more than a century ago, shows that the pox virus disease arrived with early Galápagos colonists.

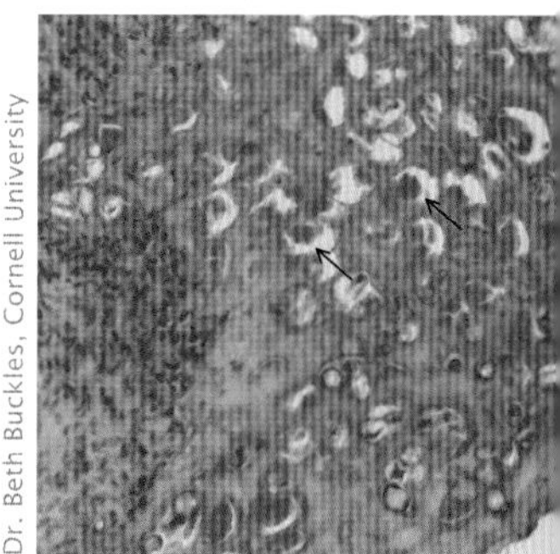

Dr. Beth Buckles, Cornell University

Malaria

In contrast to the pox virus, the blood parasites in the Apicomplexan group that can cause malaria require passage through the digestive and circulatory systems of a particular insect vector, in which they undergo sexual reproduction. Because of this, the insect is considered the primary host of the parasite, and different parasites in this group have developed fairly specific parasite-insect relationships (such as *Culex quinquefasciatus* mosquito as vector for *Plasmodium relictum* in Hawaii, mentioned earlier). Following sexual reproduction in the insect, the parasite is passed to a secondary vertebrate host through the saliva of the biting insect. The parasite then goes through a massive multiplication phase in the liver of the animal before sending forms called merozoites to the bloodstream, where they enter red blood cells and undergo further development. From there, the next biting insect that takes a blood meal picks them up, and, if it is the correct host insect for that parasite, it can complete its life cycle. Because of these life cycle requirements, the parasite tends to have a very specific relationship with a particular insect species, but be fairly broadly infective in the taxonomic group of vertebrates that are bitten by that insect. Which vertebrate any particular parasite infects thus depends mainly on the affinity of the biting insect for taking blood meals from reptiles, mammals or birds, and on the victim's susceptibility.

Two of the major genera of Apicomplexan blood parasites infecting birds are *Haemoproteus* and *Plasmodium*. In general, those that cause the sometimes devastating disease symptoms that we call

RIGHT AND BOTTOM: Blood tests on various bird species revealed that endemic Galápagos doves, *Zenaida galapagoensis*, and the great frigatebird, *Fregata minor*, both carry the blood parasite *Haemptereus*, seen in the microscope slide below, although neither show overt signs of illness.

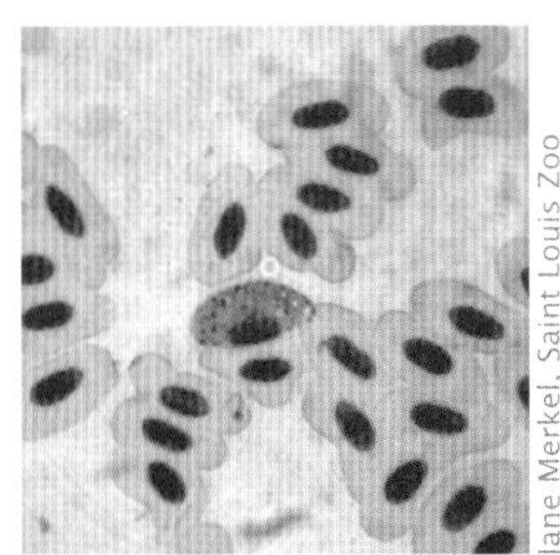

Jane Merkel, Saint Louis Zoo

'malaria' are in the genus *Plasmodium*, although this genus also includes some forms that are less harmful. Closely related parasites in the genus, *Haemoproteus* are generally less harmful, even though some studies have measured fitness consequences of *Haemoproteus* infections such as producing fewer offspring. The point is that *Haemoproteus* infections rarely are the direct cause of death of a bird, while some species of *Plasmodium* can cause rapid death in many bird species. In studies led by Saint Louis Zoo veterinarian Luis Padilla, and UMSL graduate students Diego Santiago and Iris Levin, we have found several forms of *Haemoproteus* parasites in Galápagos birds, including almost all of the Galápagos doves sampled, and a number of great frigatebirds, magnificent frigatebirds, Nazca boobies and swallow-tailed gulls. Our preliminary analysis suggests that the form of *Haemoproteus* in the doves is distinct from that in the frigatebirds and boobies, but that swallow-tailed gulls may be infected with whichever type they encounter (the dove-type or the seabird-type). In no case does it seem that this parasite causes them great harm, but this remains to be studied in detail.

Of greater concern is the *Plasmodium* that we discovered in the Galápagos penguins (see chapter: Vargas) in June 2008 from samples taken in 2005. While the genus is definitely *Plasmodium*, its exact identity (species) is not yet clear, as it seems genetically distinct from the other previously described *Plasmodium* parasites of birds. Whatever species it turns out to be, it may not be a terribly harmful variety, as we found it in several penguins that appeared generally healthy (extremely pathogenic varieties may cause very quick death, and may be difficult to detect because few infected birds live long enough to be selected in a random sample). In addition, between 2005 and 2008 the penguin population appears to have been stable or even slightly increasing. Nonetheless, the GNP launched an immediate follow-up survey, sampling penguins at several sites and focusing particularly on locations where the infected individuals were found earlier. If the parasite is found currently, intensive mosquito sampling will be undertaken along the shorelines with the greatest number of infected penguins to identify which mosquito is the vector. Sampling of terrestrial birds in the immediate vicinity will be carried out as well, to see if the parasite may have been spread to other species. We have not detected *Plasmodium* in any other Galápagos species so far, having tested many individuals from 16 species on 16 islands, but we have not yet sampled terrestrial birds in the immediate areas of the penguin colonies.

Botfly

The botfly *Philornis downsi* is another particularly harmful parasite that currently resides on the islands. The work of Birgit Fessl, Sonia Kleindorfer and Rachael Dudaniec has revealed the devastating impact that this fly can have on nestling finches and other small birds. The female fly deposits an egg in the nostril of a newly hatched chick. When it hatches, the larva may kill its small host outright by consuming the tissues surrounding the nostril. Alternatively, in a short time the larva settles in the substrate of the nest, moving up to take occasional blood meals from nestlings. Eventually, the larva pupates within the nest floor itself, and emerges as an adult. These researchers have documented the mortality of nestlings related to number of parasites infecting a nest (death is more likely as the number of parasites increases), and body size of the bird (the smaller finches are more likely to die than the larger-bodied terrestrial birds, like the mockingbirds).

Other potential disease-causing agents

Aside from these three known threats (pox, blood parasites and the *Philornis* fly), our work has described many other potential disease-causing agents on the islands. We group these into three general categories: (1) Those that appear to have arrived with the colonizing birds that were the ancestors of

Jody O'Connor, Flinders University, Australia

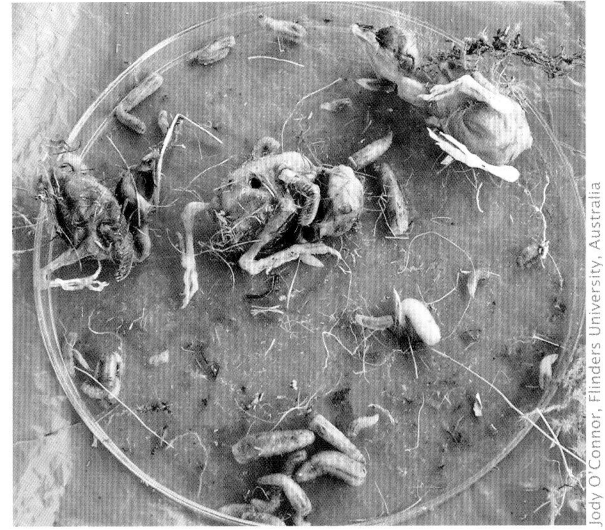

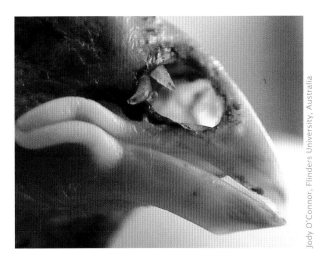

Jody O'Connor, Flinders University, Australia

LEFT, TOP AND BOTTOM: The introduced botfly, *Philornis downsi*, causes devastation in nesting Darwin's finches and other small land birds, sucking blood from the nestlings and invading their nasal cavities (left), eventually causing death, as seen from the nest content remains in this petri dish.

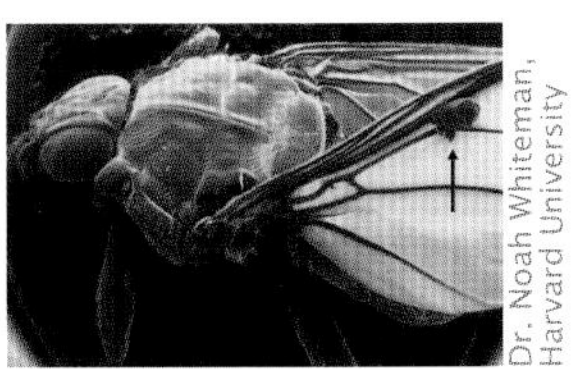

Dr. Noah Whiteman, Harvard University

ABOVE: A minute mite attached to the wing of a parasitic fly that is evolving along with its host, the Galápagos hawk, all help to reveal their histories of interactions on the islands.
FAR LEFT: Galápagos hawks apparently diverged into an endemic species only relatively recently, showing weaknesses in their immune system due to genetic isolation.

ABOVE AND BELOW: A microfilarid nematode parasite (below), infecting both Galápagos penguins and flightless cormorants that share the same habitat, is an interesting example of a parasite naturally jumping species.

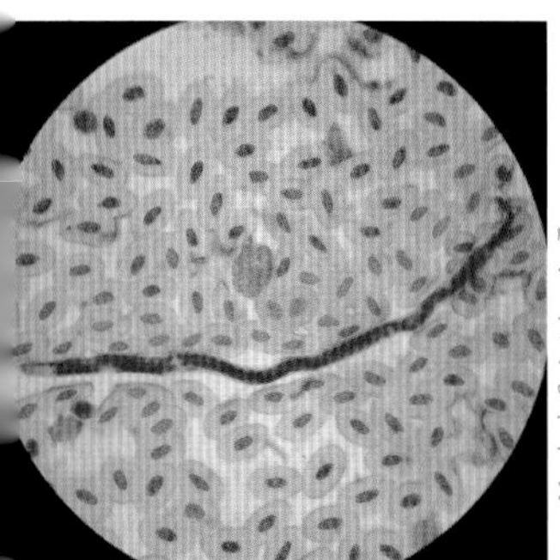

Jane Merkel, Saint Louis Zoo

today's endemic species. Parasites in this category evolved alongside their hosts, and appear to represent several examples of new species of their respective types of parasites. The lice and hippoboscid biting flies that are specific to a Galápagos bird species, or group of species, are important examples of this category. Because of the long evolutionary history that these parasites have with their hosts, they have likely arrived at a sort of equilibrium where the birds can easily manage the infections under ordinary circumstances. They thus provide very interesting opportunities to study the evolutionary process of coadaptation, as in the work of Noah Whiteman on the Galápagos hawk and its ectoparasitic lice, flies and mites, all of which occur only on this one species and therefore must have arrived with the ancestral hawks.

(2) The second category includes parasites that may have made their entry into Galápagos naturally with one ancestral bird, and 'jumped' to another native species since that arrival. Some of these may be of some concern, depending on how recently that host-switch occurred. For example, Jane Merkel and Erika Travis of the Saint Louis Zoo discovered that the Galápagos penguin and the flightless cormorant harbor a microfilarid nematode; genetic studies by Noah Whiteman and morphological research by Hugh Jones independently confirmed that this is a single species that cross-infects both bird species that share the same habitat in almost completely overlapping ranges. Further studies should reveal the direction of this transfer.

(3) The final category represents those pathogens that arrived within the last 200 years, coinciding with human colonization, and are likely to be a consequence thereof. This group certainly includes the avipox virus, and perhaps the *Plasmodium* in penguins as well. It is these more recent arrivals that are of greatest concern, as the endemic birds will be immunologically unprepared against classes of pathogens to which they have never previously been exposed.

Pathogens can arrive naturally on a migrating bird or windblown biting insect, or perhaps more commonly on an animal brought in by humans, or an infected insect stowed away in the baggage compartment of a ship or plane. Modeling by Marm Kilpatrick suggests that the most likely route of arrival of the dreaded West Nile virus, for example, would be through an infected mosquito that hitches a ride on an airplane. It is these possibilities that have led to the recent fumigation protocol for both baggage and passenger compartments of all planes headed for Galápagos from mainland Ecuador (see also chapter: Merlen).

We have also taken a close look at the parasites and pathogens in the most common barnyard bird, the domestic chicken. Studies led by veterinarians Nicole Gottdenker and Catherine Soos have revealed massive evidence of current or past infections, including a wide variety of pathogens known to have caused significant declines in wild bird species in other parts of the world, some with close Galápagos relatives. We do not yet know whether any of these agents have made the switch to wild Galápagos birds with which chickens interact in the agricultural zone, but studies are underway. Meanwhile, this information has led to important recommendations for the management of chickens to minimize opportunities for transfer by keeping backyard birds enclosed. Our genetic studies showed that the pox virus in the chickens is not the same as that infecting the wild birds, but the other chicken-borne pathogens may be more broadly infective and are a source of great concern.

Finally, we are always testing for pathogens of recent worldwide concern, such as West Nile virus or avian influenza. However, it is important to remember that one reason these pathogens have received such broad media attention is their potential to infect humans, particularly with recent records of human mortality. While this is certainly an important public health concern, there are plenty of other pathogens that can cause birds at least as much harm, without having a history of 'jumping' to humans.

West Nile virus and avian influenza have not reached Galápagos, and we hope they never will. Meanwhile, we and our partners will remain vigilant for the arrival of any such potentially devastating agents, and work with the GNP and the CDRS to erect action plans in preparation for such an event. The immediate response of the Park to the news that *Plasmodium* had been detected in penguins was to assemble people, materials and a boat to mount a new survey, all within six days of the information being received, with plans for an immediate follow-up trip to identify the vector and check for infection of land birds. This rapid response suggests to us that the situation is well in hand. Disease studies are now considered in new management plans, such as a project to reintroduce the critically endangered Floreana mockingbird to its former home island from two small islets where it is currently restricted. Disease surveys led by veterinarians Marilyn Cruz of the GNP and Sharon Deem of the Saint Louis group will document the pathogenic agents that may face those repatriated birds on their ancestral island. The new GNP laboratory, built with funding from the Darwin Initiative via partners Virna Cedeño, Simon Goodman and Andrew Cunningham, means that much testing can now be done on site without lengthy delays for exportation and test results.

Arrival of new diseases will remain a major threat to the insular fauna, but the people of Galápagos are arming themselves with the latest technologies for rapid detection and response to such perils.

SEA LION MYSTERY POX

A sea lion with terminal 'pox' symptoms languishes near death.

In the 1980s a wave of an unknown disease swept through the sea lion population. The disease was called 'pox,' even though the infective agent was never determined. This disease caused major losses, particularly among pups and juveniles, with several localized later recurrences. Symptoms included an outbreak of boils all over the body, and gradual paralysis setting in from the hind flippers forward. It was last observed in 2004 on Fernandina, where many pups were infected. Canine distemper, which in recent years has affected many pet dogs in Galápagos, could also become a source of infection for sea lions. Even though known cases of this disease are usually caused by a slightly different virus, called 'phocine distemper virus,' the canine distemper virus has infected and killed seals by the thousands at Lake Baikal. Similar dangers may come from domestic cats as carriers of a protozoan parasite causing toxoplasmosis, which on occasion is capable of jumping to other mammals and even birds, with disastrous consequences. On Isabela, 63% of feral and domestic cats tested positive in a recent study, prompting urgent plans for further tests on sea lions, penguins and cormorants.

Fritz Trillmich & Patty Parker

Success in Biological Control

The Scale and the Ladybird

Charlotte Causton

Charles Darwin Foundation,
Puerto Ayora,
Santa Cruz Island,
Galápagos Islands, Ecuador.

Dr. Charlotte Causton has been working in the Galápagos Islands since 1997 and helped create the Terrestrial Invertebrates Department, which she led for five years. Her primary research interest is invasive species management. She has extensive experience developing methods for controlling insects in areas of conservation importance, with particular expertise in biological control. She is also technical advisor to the Galápagos Quarantine and Inspection System.

I ALWAYS KNEW THAT the post that I had accepted at the Charles Darwin Research Station (CDRS) would be exciting, but little did I realize that I would be involved in one of the most challenging and contentious conservation research projects ever carried out in Galápagos. Or that I would ultimately be responsible for intentionally introducing an insect predator into the islands!

Insects are among the lesser known and less visible of the Galápagos species. But when the endemic *Scalesia* trees (see also chapters: Hamann; McMullen) and other threatened plants started to die, a small, white insect came to the attention of conservation workers and the Galápagos community alike. *Icerya purchasi*, commonly known as the cottony cushion scale, feeds on the sap of plants and causes fruit and leaves to fall prematurely, resulting in dieback of branches and even death in some Galápagos species. Plants that have been attacked can be easily seen from afar because they are often covered by black sooty molds that thrive on the large quantities of honeydew produced by the scale insect. Entire plants can be smothered by these molds and as a consequence are unable to carry out photosynthesis necessary for survival.

BELOW: Originally from Australia, the cottony cushion scale, *Icerya purchasi*, is an insect pest causing major damage to citrus orchards worldwide. Accidentally introduced to Galápagos in the early 1980s, it gradually spread to all islands, with devastating effects on many native and endemic plants.

Icerya purchasi is an insect pest originating from Australia, but it has been transported around the world on fruits and plants, and thus has established itself in many places, including the isolated Galápagos Islands. It was first reported on San Cristóbal Island in 1982 and is thought to have been inadvertently introduced on ornamental plants that were imported to decorate the streets. In a matter of 14 years, wind currents had carried the pest to most islands in the archipelago, making it hard or impossible to control by conventional means.

Soon after I was hired to run the Terrestrial Invertebrates Program, I was asked what I thought the chances were of controlling this insect using its natural enemies — a technique known as biological control. I replied that I thought the prospects were good, although we would be in for some challenges. The cottony cushion scale is a pest of citrus orchards worldwide and has been successfully controlled by releasing its predator, an Australian ladybird beetle, *Rodolia cardinalis*. However, it wasn't just a simple matter of bringing this predator to Galápagos and turning it loose. Because of their isolation, Galápagos ecosystems are highly susceptible to alien organisms and disease (see also chapters: Parker; Merlen), so special precautions would need to be taken to determine that this introduction would not unintentionally harm any of the unique fauna. Furthermore, this was the first time that biological control had ever been considered for Galápagos, so we had no prior experience on which to base our work.

The possibility that the ladybird might turn out to be a problem rather than a solution for ecosystem conservation provoked much debate among scientists. For the better part of a year and a half, emails went back and forth, strongly expressing differing views on whether biological control should be permitted for the Galápagos Islands at all. Some were in favor if appropriately safe procedures were employed, while others were emphatic that it should never be attempted because they considered it dangerous and thought that this would set a precedent for introducing other species. My role during this time was to come up with an impartial review of the benefits and risks should this project go ahead. Although biological control had had some bad press worldwide, most of it was based on early efforts when little groundwork was carried out to evaluate its safety. In some cases the introduction of animals such as mongooses in Hawaii and cane toads in Australia became far more of a problem than the species that they were intended to control! Nowadays, pre-introduction screening of the candidate biocontrol agents is extremely rigorous to ensure that those types of disaster are not repeated. In the end, it became clear that biological control was a viable option.

In 1998, a committee composed of resident and external scientists, and senior Galápagos National Park Service (GNPS) staff gave the go ahead to carry out an in-depth research program to assess the safety of introducing the Australian ladybird beetle to Galápagos. This programme wound up taking six years to complete, as we had to investigate virtually every facet of the beetle's introduction, searching for any potentially negative impacts. These were then weighed against the damage caused by the cottony cushion scale to Galápagos flora in order to decide whether the benefits of this type of control were worth the risk of deliberately introducing an alien species.

For financial and logistical reasons we decided to conduct the studies in Galápagos, within purpose-built facilities at the CDRS on Santa Cruz Island. The onus was now on me to design and oversee the construction of a low-budget, yet highly secure building that would prevent the ladybird from escaping. A huge responsibility anywhere, but especially so in Galápagos where building even a regular house can be logistically challenging! Fortunately, I was able to rely on the help of my colleague Lazaro Roque-Albelo and on technical expertise from leading institutions in biological control (CABI in the U.K. and CSIRO in Australia). I also had a very patient and innovative local contractor. This proved of great importance as we deliberated how best to design water and air-cooling systems that couldn't serve as an escape route for 3-mm (0.1-in) beetles. We also needed to design a system of rooms within rooms in order to create as many barriers as possible between the beetle and the Galápagos environment. Construction took the better part of a year. I then had the task of hand-carrying 96 live ladybirds, kindly donated by CSIRO in Brisbane, halfway across the world and making sure that my charges survived in spite of travel delays.

ABOVE AND BELOW: *Scalesia* and *Darwiniothamnus*, both endemic plant genera, became victims of the cushion scale attack during the 1990s.

The first few months were mentally exhausting as my colleagues and I began figuring out the safest way of keeping the beetles alive and healthy while studying them. Veronica Brancatini from CSIRO–Brisbane provided invaluable advice on rearing techniques. We also had to develop protocols to make sure that they did not accidentally leave the building with us on either our clothing or equipment. It took a while

RIGHT AND FAR RIGHT: The natural predator of the cottony cushion scale, the small Australian ladybird beetle, *Rodolia cardinalis* (left), was the only viable option for controlling this pest in Galápagos, but before it could be deliberately released, its potential effects on other species, such as the spotless ladybird (right) and many other Galápagos native species, had to be ascertained.

BELOW: Ladybird beetles have been released on the larger volcanoes of Fernandina and Isabela where the scale affected endemic species such as *Darwiniothamnus tenuifolius* shrubs, seen here growing on the summit of Wolf Volcano.

Photo courtesy Heidi Snell/Visual-Escapes.smugmug.com

to convince some people that it really was necessary to don shower caps, lab coats and slippers before working with the beetles. To top it off, all this gear had to be kept in the freezer to make sure nothing could survive on it. All eyes were on us, and we were only able to breathe more freely once things were up and running smoothly.

A wonderful team of research associates and Ecuadorian undergraduate students worked with me on these exhaustive studies. Piedad Lincango, Tom Poulsom and Carolina Calderón in particular put in long hours inside the insect containment facility, observing the beetle's behavior and determining whether it fed on any Galápagos insects. We tested its voracious larvae on 16 insect species of conservation value presumed to be at high risk of predation, whereas adult beetles were tested on eight. We also watched for interactions between the Australian ladybird and four Galápagos insect predators, including two native Galápagos ladybirds.

One of the most exciting parts of this project was that in our quest to verify whether Galápagos insects might be adversely affected, we had to visit parts of Galápagos rarely frequented by people. I can't begin to describe how it felt to be camping on the rim of the rumbling Fernandina volcano surrounded by bright orange land iguanas, or sitting inside the crater of Eden Island looking out at the shimmering turquoise sea while being circled by Galápagos hawks. Lucky and privileged are two adjectives that I would use.

We spent many days in the field searching for endemic scale insects to test against the ladybird. Nothing was known about the natural history of these insects so we had to rely on old collection records to find them. One particularly elusive insect was a species known as a ground pearl, which lives inside wax shells in the soil. We spent days checking for ground pearls among the roots of all 'yellow plumed' plants where they had once been recorded in 1920, only to discover many days later that we were surrounded by them! We hadn't noticed them because they didn't look like any scale insects we had ever seen before.

Considerable time was also spent in confirming the degree to which the cottony cushion scale was having an impact on Galápagos plants. This involved carrying out inventories of attacked plant species throughout the archipelago and quantifying the extent of damage. We also carried out controlled experiments with potted endemic plants to compare the growth rates of plants with and without scale insect infestations.

Our research showed that the cottony cushion scale was clearly a threat to endemic plants as well

Photo courtesy Heidi Snell/Visual-Escapes.smugmug.com

as to certain native insects that feed exclusively on these plants. We recorded at least 62 native or endemic plant species on 15 islands that were attacked. According to the Red List of Threatened Species kept by the IUCN, six of these species are listed as endangered or critically endangered, and 16 are threatened. It was also determined that the scale damages some commercially important crops, in particular citrus trees.

On the other hand, our studies strongly suggested that releasing the Australian ladybird beetle should not be a threat to Galápagos invertebrates. Our experiments demonstrated that in the islands it should only be able to complete its life cycle if it can feed on the targeted species. The ladybird appears to have formed such a close relationship with this scale insect and its near relatives, that its larvae will sooner starve than eat other insects. Further experiments suggest that adult ladybirds are likewise specialist feeders.

Because of its specific feeding habits, it is unlikely that the Australian ladybird would ever compete for food with Galápagos insect predators, particularly as Galápagos ladybirds also appear to be picky eaters, and none were recorded feeding on the cottony cushion scale during extensive observation. Only two insects were found feeding on the cottony cushion scale: an endemic green lacewing insect and an introduced moth larva, but not at levels sufficient to effectively control its numbers. The moth was not affected by the Australian ladybird. On the other hand, the lacewing might in fact diminish its efficacy as a control agent because, when we placed them together in a small petri dish, it attacked the ladybirds, often killing them.

Our research clearly showed that the ladybird satisfied most requirements for its introduction to the Galápagos Islands. Then, however, the concern was raised that, like some other beetles, it might be toxic to insect-eating birds. So we went back to the drawing board, but this time to design and carry out experiments to see whether Darwin's finches would feed on the beetles, and if so, whether this might be harmful to them. We tested two species of finch and neither showed adverse reactions after being fed the beetle. In fact, they seemed to find them distasteful.

As a result of all our studies, the GNPS concluded that the benefits of introducing the beetle far outweighed any potential negative impacts, and its release was authorized.

Because this project was something new conceptually, with a high visual impact in Galápagos, we wanted to make sure that everyone fully understood the extensive groundwork that had been carried out before this decision was reached. The best way to do this was to actively involve the community in the release and monitoring of the beetle. This aspect became another highlight for me. What could be more fun and satisfying than turning conservation science into a 'hands on' activity that everybody could take part in.

Photo courtesy Heidi Snell/Visual-Escapes.smugmug.com

ABOVE, LEFT AND RIGHT: A before-and-after comparison of the same location shows the dramatic effect on mangroves and *Maytenus* trees growing near the shore in Puerto Ayora, dying in 2001 (left) and their subsequent full recovery in 2008 (right). BELOW: The scale's honeydew also generated heavy fungal growth that further smothered plants such as white mangroves, *Laguncularia racemosa*, preventing photosynthesis. BELOW LEFT: This humble but efficient building at the CDRS was built especially to carry out the exhaustive tests needed to verify the potential consequences of using the ladybird as biological control agent.

Photo courtesy Heidi Snell

A B C D E F G H I J K L

TOP TO BOTTOM: A. Native orb spider *Neoscona oaxacensis* with prey, Alcedo; B. native painted lady butterfly *Vanessa virginensis*, Wolf Volcano; C. Galápagos centipede *Scolopendra galapagoensis*, Española; D. endemic painted grasshoppers *Schistocerca melanocera*, Santa Cruz; E. endemic coastal grasshopper *Schistocerca literosa*, Española; F. female endemic carpenter bee *Xylocopa darwinii*, Cerro Azul; G. endemic longhorn beetle *Eburia lanigera*, Santa Cruz; H. native manzanillo hawk-moth larva *Erinnyis ello encantada*, Cerro Azul; I. introduced paper wasp *Polistes versicolor*, Santa Cruz; J. endemic nocturnal flightless long-horn grasshopper, *Liparoscelis cooksoni*, Alcedo; K. beach gnats swarming around marine iguanas, Santiago; L. native diurnal ornate moth *Utetheisa ornatrix ornatrix*, Alcedo Volcano; bottom row, left to right: native hawk-moth *Manduca rustica colopagensis*, Santa Cruz; wolf spider *Heteropoda venatoria*, preying on introduced cockroach *Periplaneta americana*, Santa Cruz.

TERRESTRIAL INVERTEBRATE FAUNA

Surprisingly perhaps, invertebrates dominate Galápagos terrestrial ecosystems, far outnumbering all other animal species. They form an important part of the food chain, play key roles as pollinators, and act as soil builders by decomposing organic matter. Nearly 3000 species of land invertebrates have been reported, 51.7% of them endemic. The largest group is insects, with 1555 species dominated by beetles and moths.

This microfauna is rich in endemic species but poor in diversity when compared to the South American mainland. Most of their ancestors probably arrived as winged adults, helped along by wind currents. Fewer traveled by sea on rafts of floating vegetation and some rode on other animals. At least 50 groups of invertebrates have undergone further speciation since their arrival, with many flightless species such as ground-dwelling beetles, darkling beetles and issid plant-sucking bugs, plus, notably, the endemic land snail family Bulimulidae, which has undergone a most spectacular adaptive radiation process: about 80 known species and subspecies originating from a single ancestor (see Parent and Coppois).

In the last few years, one of the priorities of the CDF has been to evaluate the conservation status of endemic land invertebrates according to IUCN criteria. So far, 48 *Bulimulus* land snails and 13 Lepidoptera (moths and butterflies) have been identified as threatened with extinction, 26 of these being critically endangered, due primarily to a combination of introduced species and habitat loss. To date, 543 invertebrates have been inadvertently introduced to Galápagos, of which six are highly invasive: two species of fire ant, two paper wasps, the cottony cushion scale (this chapter) and a parasitic fly that causes mortality in nestlings of small birds (see Parker). Another 55 species are potentially invasive. However, much more research is needed to understand the full ecological impact of most introduced invertebrates.

The new Terrestrial Invertebrate Reference Collection at the CDRS, with over 400,000 specimens of both introduced and native species, is the most comprehensive collection of Galápagos invertebrates in the world. This reference material is an indispensable tool for ecosystem research and for detecting new introductions to the islands. CDRS scientists provide technical assistance to the GNPS and SICGAL, the Galápagos quarantine and inspection system, by identifying any new invertebrates discovered in imported goods. Recent specimens have included a human botfly and spiders collected from a stuffed snake!

Charlotte Causton and Lazaro Roque

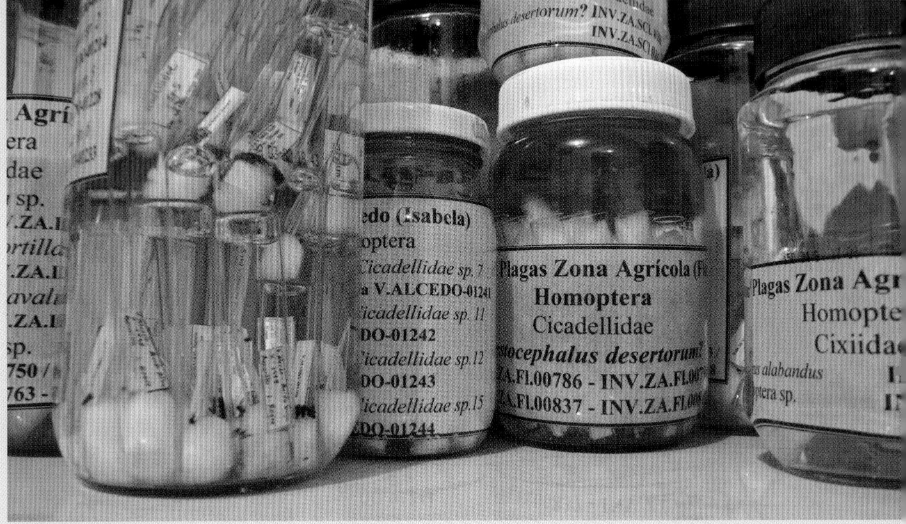

RIGHT: The CDRS invertebrate museum, seen with curator Lazaro Roque, contains a vast insect collection used for identification and comparative analysis.

KILLER ANTS
Source: Charles Darwin Foundation

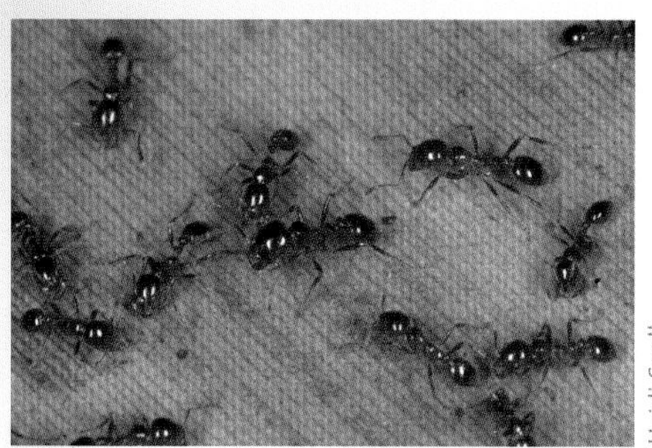
Heidi Snell

The tiny red fire ant, *Wasmannia auropunctata* (left), barely 1 mm (0.04 in) long, can bring down large prey by its concerted venomous attacks, whereas the large nocturnal endemic *Camponotus macilentus* (below), over 10 times its length, is harmless. A total of 44 ant species have been recorded in Galápagos. Of these, four are endemic, 30 are known to be introduced, and 10 are of undetermined origins. *Wasmannia*, which arrived around 1910–1920, is considered one of the most aggressive invertebrate species ever introduced to Galápagos, followed closely by the more recent (1981) tropical fire ant, *Solenopsis geminata*. Having first established themselves on inhabited islands, they have since spread to other islands — either by flying or through human transport — such as the recent establishment of the tropical fire ant on tiny Champion Island near Floreana, where some of the critically endangered Floreana mockingbirds survive in very small numbers. Both these ants prey heavily on native invertebrates, such as land snails and insects, as well as young birds. They even affect nesting success of larger species, including land iguanas and giant tortoises. After *Wasmannia* fire ants were found on Marchena in 1988, the specific nonresidual insecticide AMDRO was used between 2000 and 2002 to successfully clear the 22-ha (54 acre) area, with surveys in four succeeding years showing no recurrence. However, clearing all islands of these pests by currently available means remains impossible.

Tui De Roy

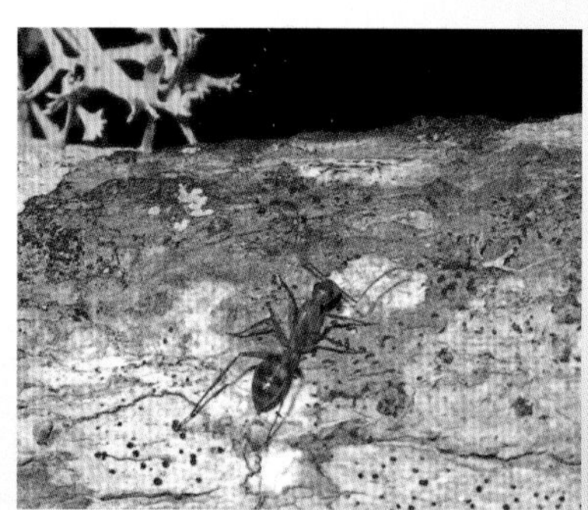

The release program was a joint effort between CDRS and the GNPS and followed an intensive six-month community educational campaign. During this time CDRS technicians were busy rearing large numbers of beetles in preparation for their release. In January 2002, town officials, with the help of high school students, simultaneously set beetles free on the inhabited islands of Santa Cruz, San Cristóbal, Isabela and Floreana. Since then, over 2000 beetles have been released in priority areas including the islands of Marchena, Fernandina, Pinta, Pinzón and Rábida, where endangered species of plants were seriously affected by the scale pest.

Monitoring shows that the beetle has successfully established on most islands and has even reached some other islands unaided — a graphic example of how exotic organisms can easily spread throughout the archipelago. Once more funds are raised, we will identify which areas require additional releases. It is not anticipated that the beetle will completely eradicate the cottony cushion scale from Galápagos, but rather will keep it at nonpest levels, thereby reducing its impact on Galápagos plants. This certainly seems to be the case in several areas that we have kept under observation. It was extremely gratifying to witness the blackened and dying mangroves of Puerto Ayora become lush green plants once again.

What is even more satisfying is that the local people could also see these changes and were able to experience first-hand the practical results of cutting-edge conservation research. The benefits of this project are numerous: healthier plants no longer ravaged by the scale insect; an infrastructure developed and procedures in place for future biological control testing; and a community able to appreciate that the research that we conduct improves their livelihoods as well as the conservation value of the Galápagos Islands.

RIGHT: School children and the local community participated in the ladybird's release.
FAR RIGHT: Media events were conducted by the GNP director (left) together with the author (center) and her team to announce the biological control measures.

Photo courtesy Heidi Snell/Visual-Escapes.smugmug.com

Photo courtesy Heidi Snell/Visual-Escapes.smugmug.com

Saving 'Lost' Plants

Finding and Nurturing the Survivors

Alan Tye

Secretariat of the Pacific
Regional Environment
Program (SPREP),
PO Box 240,
Apia, Samoa.
<alant@sprep.org>

Dr. Alan Tye left his native United Kingdom for his first post-Ph.D. job in Africa, and has worked in tropical ecology and conservation ever since. His longest residency was in the Galápagos Islands, where he was head of the Botany Department of the Charles Darwin Research Station (CDRS) for 11 years. During this time he also established and built up the Terrestrial Invertebrates Program, and became Director of Sciences before leaving for his current position with SPREP, based in Samoa and serving 21 countries and territories of Oceania. His work has focused on bird and plant research, conservation of tropical forests and island ecosystems, and the management of both threatened and invasive species.

THROUGHOUT THE WORLD, plants have always been the Cinderellas of conservation, neglected in favor of the more charismatic pandas and whales, or in the case of Galápagos, the giant tortoises and Darwin's finches. However, Galápagos plants demonstrate, to at least the same degree as these famed animals, the unique characteristics for which the archipelago is renowned. Evolution in isolation has produced many oddities, and a high proportion of these plants are found nowhere else in the world. Such endemics create the main structure of most Galápagos habitats, and many of the animals depend on them, whether for food, water, nesting places or shelter. So, from a conservation perspective, the Galápagos flora is just as valuable as its better known animals.

On many of the islands, the native vegetation has suffered dramatic destruction through clearance for farming and development, the introduction of voracious herbivores such as

LEFT: Endemic tree ferns, *Cyathea weatherbuyana*, here with a vermilion flycatcher, were once typical of the Santa Cruz highlands but are becoming rare, out-competed by a growing list of invasive species introduced ever since farming began.

ABOVE: Even the most unobtrusive endemic plants have key roles to play in the ecosystem, such as *Tiquilia nesiotica* growing in the volcanic cinders of Bartolomé, whose minute flowers are eaten by the lava lizard *Microlophus albemarlensis*.
BELOW: The beautiful *Lecocarpus darwini* of San Cristóbal belongs to an endemic genus with three species limited to the oldest, southern islands. The Floreana species is considered vulnerable, whereas the other two are Endangered.
RIGHT: Teetering on the brink of extinction, the delicate San Cristóbal rock purslane, *Calandrinia galapagosa*, is threatened by a variety of factors in spite of conservation efforts.

goats, alien plant species invading and transforming habitats, and insect pests brought inadvertently with goods from the mainland. Because of the key role of plants in the Galápagos ecosystem, botanists have devoted much effort to measuring such habitat changes and the loss of some high-profile species such as the *Scalesia* trees. However, with some 225 endemic plants (as compared with just a few dozen endemic birds and reptiles, for example) research efforts have not managed to keep track of them all.

This was the situation that struck me when I first began working at the CDRS in 1996. We had a reasonable idea of major changes in the vegetation, but very little knowledge of the current status of most of the individual endemic species — the components of that vegetation. We did not know the extent of decline in particular species caused by goats or by forest clearance. We therefore could not establish which species were the most seriously threatened, and thus which should receive the most urgent attention. Calls had been made for action to save a number of iconic species, but since we lacked comprehensive knowledge, we had no real idea whether those that had captured the attention of botanists were truly the ones most in need of conservation and restoration, or what to do about them if indeed they were.

This realization was the reason I made it a priority to start a major effort for the strategic conservation of Galápagos plants, beginning with a search for crucial funds. With only four staff in 1996 and no budget for rare plant work, the CDRS Botany team was not in a position to have much of a positive impact. The program received a boost a year later with a grant from the United Kingdom's Darwin Initiative, which focused on vital survey work to establish the status and distribution of Galápagos endemics.

Which plants are threatened or even extinct? Obtaining the crucial information

Going back through the literature showed that botanists visiting the islands in the past had tended to return to the same places — generally spots where it was fairly easy to land from a boat — and had not penetrated very far inland except along a few well-known trails. Consequently, there was no information at all for huge areas of most of the larger islands, and what was available was often more than 30 years old. Since many of the dramatic changes to the Galápagos ecosystem have occurred quite recently, the old information was not adequate for conservation planning.

We therefore embarked on a comprehensive survey of the archipelago — an enormous task that is still not finished more than a decade later, but which we recognized from the outset would take many years to complete. Each field trip usually lasted a week or so, using a small boat to gain access at different spots along the coast, and deliberately targeting areas where it was difficult to land and from where consequently there was very little previous information. In the first 10 years we investigated Santiago and Española almost completely, plus most of San Cristóbal, Pinta and large tracts of Santa Cruz, along with most of the offshore islets associated with these larger islands.

The second thrust of this program was to produce a complete Red List of Galápagos plants, evaluating all of the endemic species by the IUCN criteria. The first draft Red List for the flowering plants and ferns was completed in 2002. This has since been kept up

Photo courtesy Heidi Snell/Visual-Escapes.smugmug.com

to date by periodic revision, while additional groups such as seaweeds and lichens are being added. The IUCN criteria depend heavily on knowledge of the abundance and distribution of a species and on information about change (declines, reductions in area): the surveys feed information into a Geographic Information System (GIS), which provides the basis required for Red-Listing.

The Red List is our prioritization tool, for the first time allowing us to identify the most seriously threatened species, objectively and comprehensively, and, as a result, to focus our conservation efforts where they are most needed. From this exercise, about 20 species have emerged as critically endangered (CR), the highest threat category, and thought to be in immediate danger of extinction.

However, one unforeseen outcome of the Red-Listing was the emergence of a group of an additional 10 or so apparently 'lost' species, most of them not seen for 30 years or more. Together, these and the CR species became the targets of our most intense research and conservation efforts. Obviously, in the case of the lost species, this translated into more survey work — to try to find them.

Rediscovering 'extinct' plants

Searching for 'lost' species can be a disheartening task when scouring hostile terrain with no idea whether survivors exist, but conversely locating even the smallest relicts represents a huge morale boost. Such a breakthrough first came in 1995, the year before my arrival, when the Santiago Island endemic *Scalesia atractyloides* was rediscovered by National Park rangers who found five trees growing on a crater wall, out of the reach of goats. This gave hope that other species, which had not been seen for many years, might also still exist. In 1997, searches on Floreana paid off when we rediscovered the Floreana flax, *Linum cratericola*. This small shrub was only ever known from two small craters and, as previous quests had failed to find any live plants since 1981, it was about to be classed as extinct. A scientific paper recording its demise was already in press when we encountered 13 live plants.

This new discovery gave added hope and impetus to our search plans for other lost species. Just like the Santiago *Scalesia*, most of the remaining Floreana flax plants were found clinging to rock faces out of the reach of goats, with a few seedlings that had sprung up on the ground below due to recent rains, right alongside a goat trail. By the time of our next visit a few weeks later, most of these young plants had already disappeared. Direct action was obviously necessary, so the third aspect of our rare plants program was immediately activated — protection.

SCOURGE OF THE HILL BLACKBERRY
Source: Charles Darwin Foundation

Originally from the Himalayas but cultivated worldwide, the hill blackberry, *Rubus niveus* (left), was first introduced to San Cristóbal in the 1970s, and then to other inhabited islands as a living fence and for its fruit. It has become rampant in the highlands of San Cristóbal and Santa Cruz, and to a lesser extent on Isabela (Sierra Negra and Cerro Azul), Floreana and Santiago. A terrible weed on farms, it also invades natural highland areas, where it quickly develops into impenetrable thickets, smothering all else. Fruiting within six months of germination, seeds are easily transported by birds, especially the introduced ani, and can last up to 20 years in the ground. Intensive control work is being carried out on Santiago and Floreana by cutting and then spraying resprouts with herbicide, a process that must be repeated every three months in established patches. Pre-emergent herbicides are being trialled, as well as experiments to reduce germination through shading and root competition by other plant species. Developing a biological control agent is a high priority.

Tui De Roy

Photo courtesy Alan Tye

LEFT: Until goats were exterminated on Santiago, the endangered *Scalesia atractyloides*, rediscovered in 1995, survived only on cliffs and crater walls.

MARCH OF THE RED QUININE
Source: Charles Darwin Foundation

Introduced from tropical South America in 1946 for attempted cultivation as a source of malaria medication, by the 1970s the red quinine tree, *Cinchona pubescens* (left), began spreading rapidly throughout the highlands of Santa Cruz. It now covers at least 11,000 hectares (27,000 acres) between 500 and 860 m (1640 to 2820 ft) elevation and invades all humid habitats, from *Scalesia* forest through *Miconia* belt to open fern and sedge zone. At about two to three years, saplings begin producing huge quantities of winged seeds easily dispersed by wind. Fast growing and taller than native species, a dense forest eventually develops that shades out almost everything else and also affects the soil. Dense groves impede the takeoff and landing of the critically endangered Galápagos petrel, while roots invade nesting burrows. The CDRS and GNP have successfully cleared large areas after developing effective control methods that involve application of picloram herbicide into cuts made in the trunk of larger trees (stumps otherwise regrow from root stock) and manually uprooting juvenile plants. Total eradication is a high priority but it is estimated that it would take 10 to 15 years and US$6–8 million. Because it is likely that quinine also suppresses the invasion of other pest plants like blackberry and guava, its control needs to factor in subsequent consequences for management of the ecosystem.

Tui De Roy

An emergency fund was used to build goat-proof chain-link fences around most of the flax plants, and within six months of their rediscovery, the remaining individuals were secure. Regular monitoring since then has seen their population increase to around 400 plants, while goat control in the surrounding area allowed more plants to become established outside the two fenced areas. However, the species still qualifies as CR and its situation will remain precarious as long as goats are present, dependent on continuous fence maintenance and monitoring. National Park plans to eradicate goats completely from Floreana are indeed very good news for the flax and other endemics there.

ABOVE RIGHT: Agile and ravenous, feral goats quickly strip the native vegetation, which has no defense against such onslaughts.
RIGHT: Palo santo trees, *Bursera graveolens*, dot a goat-ravaged landscape on Santiago, now rapidly recovering in the wake of their eradication.

But Floreana flax plants are small, and are also threatened by a new scourge — the spread of the introduced scrambling shrub *Lantana camara*, which is invading the natural vegetation, creating dense thickets where native plants cannot survive. *Lantana* already covers one of the two former locations where the flax was recorded, and is currently encroaching into the site of the remaining plants. Continual management of *Lantana* will be needed for the foreseeable future if the Floreana flax and many other endemic plants are to survive there. The threat from invasive *Lantana* is even more insidious than that of goats, and not so easily conquered.

Rediscovery of the Floreana flax and Santiago *Scalesia* left us with just two more plant species, one for each of these same two islands, whose status we were eager to confirm as they were both considered almost certainly extinct. The Santiago species is *Blutaparon rigidum*, which was only ever collected twice, in 1895 and 1906, when it was fairly common in at least one area. Belonging to the plant family Amaranthaceae, many of which are favourite foods for goats, it may have been exterminated when the goat population on Santiago exploded in the mid 20th century. On Floreana, the enigmatic *Sicyos villosa*, a relative of cucumbers and pumpkins, was described by its discoverer, Charles Darwin, as 'forming great beds injurious to vegetation,' but strangely no one has recorded it since. Despite lack of success so far, searches for both these species continue.

To those two widely recognized possible extinctions, we added a third species when investigating other plant specimens collected by Darwin. Again on Floreana, Darwin had discovered a small shrub that was named *Delilia inelegans*, but later botanists had assumed that the specimen was simply an odd example of a more common species of *Delilia*, rather than a separate species found only there. However, when I examined Darwin's specimens

Photo courtesy Alan Tye

ABOVE: A goat-excluding fence around a small rocky hill on Santiago's west coast.
BELOW: Help may have come too late for *Blutaparon rigidum*, last seen alive on Santiago in 1895 and 1906.

ABOVE: Floreana's rolling interior is dotted with extinct volcanic cones, providing tiny hiding places for plants decimated by introduced grazers — and hope for rare-plant seekers. BELOW: Charles Darwin's own specimen is all that remains of the enigmatic *Sicyos villosa*, which he discovered on Floreana and described as 'forming great beds injurious to vegetation,' but which has not been recorded since.

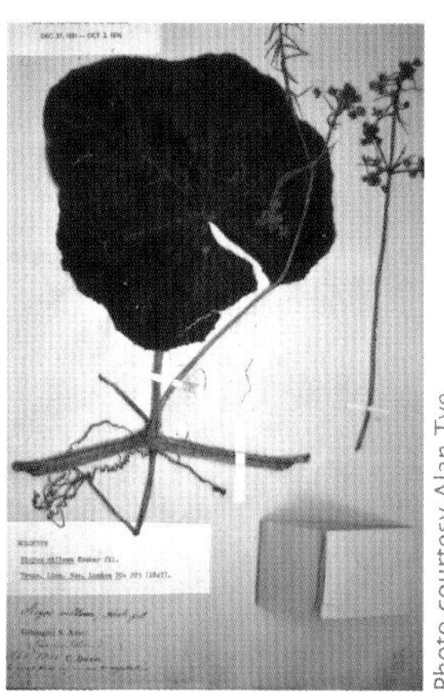

Photo courtesy Alan Tye

in Cambridge and Kew, it became obvious that *D. inelegans* is genuinely something different — a second species endemic to Floreana that has only ever been collected by Darwin.

The infamous position of Floreana as the home (or former home) of so many endangered and extinct species probably has a lot to do with the fact that it was the first island in Galápagos to be settled by people in the early 19th century, and that it has supported huge herds of feral cattle, pigs, donkeys and goats for many decades.

Since 2007, Floreana has become the subject of an ambitious island restoration program, with plans to control all major introduced pest species and encourage regeneration of natural plant and animal communities. With this, further searches for more flax populations, along with the 'extinct' *Sicyos* and *Delilia*, should gain new impetus. On Santiago, now that feral goats have been eradicated, there is renewed hope that the search for *Blutaparon* in and around its last known site might reveal that a few seeds or plants survived in crevices out of the reach of the voracious goats. Further searches for lost species are required on other islands too, especially the four main inhabited ones, as well as surveys for assessing changes in the distributions and abundance of other endemics threatened by continuing environmental degradation. Meanwhile, some other lost species are (or were) denizens of remote areas on uninhabited islands, which remain to be fully investigated.

Turning declines around — managing threatened plants

The main purpose of survey work is to provide the basic information necessary for evaluating status and threats. Armed with that knowledge, practical management plans can be drawn up for threatened species, and effective action taken to halt and reverse declines. The first step in that direction is often a biological study to identify precisely how a particular species is being affected. Threat factors can be many, including seed predation, browsing by introduced mammals or insects, competition from introduced plants, or causes that are even more difficult to detect, such as lack of pollination due to reduced plant density, or reduced fertility due to population size restriction and consequent genetic incompatibility. Studies can sometimes reveal answers quickly, or they may take many years before it becomes clear how to intervene to reverse a decline. Biological studies of the Floreana flax contributed invaluable understanding of its population dynamics and revealed that, if the main known threats of feral goats and invasive *Lantana* could be controlled, there would be no other evident barrier to full recovery. The plants produce plenty of fertile seed, have a reproductive strategy that enables the species to survive droughts, and can support attacks by native snails and competition with native plants. In contrast, intermittent studies of the San Cristóbal rock purslane *Calandrinia galapagosa* have so far failed to resolve conservation issues completely, and have revealed a complexity of threat factors, including not only grazing by goats but also attacks by stem-boring insects, possible diseases and lack of clarity over the genetic, taxonomic and conservation status of different forms of the species.

Biological studies such as these have the disadvantage of requiring an investment in research, often over several years, and there have consequently been few of them, even though they represent the primary tool for conservation management. This is doubly regrettable because, above and beyond generating the necessary data, detailed plant studies offer training opportunities for a new generation of young Galápagos researchers to learn about, value and eventually fight for the conservation of 'their' species and islands.

Sadly, the work achieved thus far pales compared to what is still urgently needed, while consistent funding remains the critical factor to stabilize and sustain an effective plant conservation program. Of the 32 CR and 'lost' Galápagos endemic plant species and subspecies (as at the end of 2008), only two have received reasonably detailed study to determine the reasons for their declines and to plan their recovery; only nine have been more or less regularly monitored for more than one year (and in seven of these cases that monitoring is now in abeyance); only three have current management programs in place. And 14 have still not been seen for some 30 years. In conclusion, thus far only about 15% of the 32 most at risk Galápagos plants have been given a more secure future through direct conservation action, a disappointing figure for one of the world's topmost natural icons and conservation flagships.

SCALESIA ADAPTIVE RADIATION

Scalesia crockeri,
Baltra Island

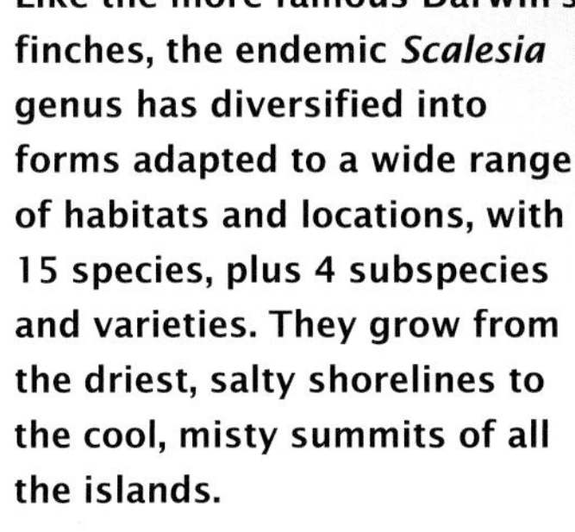

Like the more famous Darwin's finches, the endemic *Scalesia* genus has diversified into forms adapted to a wide range of habitats and locations, with 15 species, plus 4 subspecies and varieties. They grow from the driest, salty shorelines to the cool, misty summits of all the islands.

***Scalesia microcephala*, Wolf Volcano summit, Isabela Island**

***Scalesia affinis*, Alcedo Volcano shoreline, Isabela Island**

***Scalesia pedunculata*, highlands, Santa Cruz Island**

***Scalesia atractyloides*, eastern lowlands, Santiago Island**

***Scalesia vilosa*, northern coastline, Floreana Island**

***Scalesia helleri*, southern shoreline, Santa Cruz Island**

For me personally, working in Galápagos has been a bittersweet experience, with a few outstanding conservation successes set against a backdrop of overall insufficiency of effort, coupled with sustained environmental degradation, particularly on the inhabited islands. The successes show us what can be achieved — the challenge is to multiply those efforts and build on the results.

LEFT: The introduced scrambling shrub *Lantana camara* is invading the natural vegetation of several islands, creating dense thickets where native plants cannot survive.

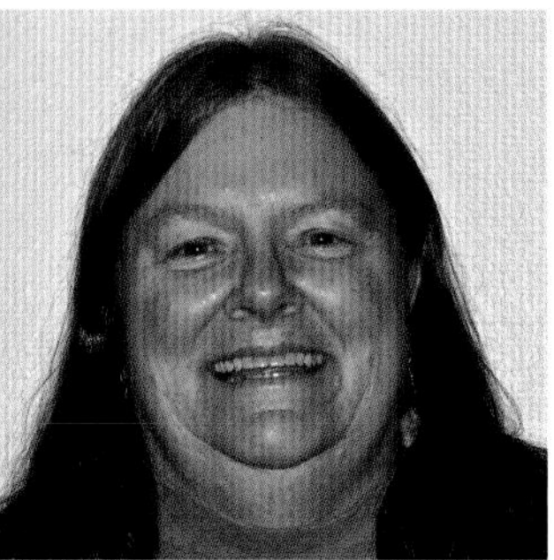

Reign of the Giant Tortoises

Repopulating Ancestral Islands

Linda J. Cayot and Washington (Wacho) Tapia

259 Karen St,
Quincy, CA 95971,
United States.
<lcayot@galapagos.org>

Dr. Linda Cayot earned her Ph.D. on the ecology of Galápagos giant tortoises and worked at the Charles Darwin Research Station (CDRS) from 1988–98, successively heading the Departments of Herpetology and Protection of Native Vertebrates, and later as the first coordinator of Project Isabela. She continues as a consultant for the Galapagos Conservancy and others, and is a member of the Charles Darwin Foundation, currently serving on its Governance Committee.

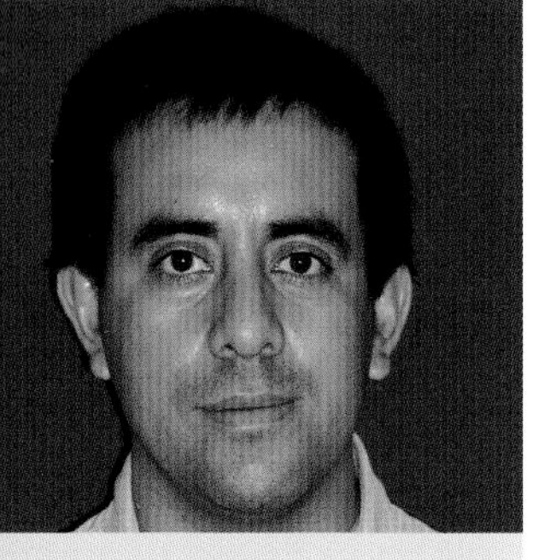

Technical Department,
Galápagos National Park,
Santa Cruz, Galápagos,
Ecuador.
<wtapiaa@gmail.com>

Washington Tapia was born in Galápagos, and has been working for the Galápagos National Park Service (GNPS) since 1998, where he started as leader of the introduced mammal control team and currently heads the Department of Conservation and Sustainable Development, integrating scientific research with protected area management. After finishing high school on Santa Cruz Island, he attended the Universidad Técnica del Norte in Ibarra, Ecuador, on a CDF scholarship, and completed his Master's thesis on giant tortoises in 1997.

Linda's experience

WHEN I FIRST TRAVELED to the Galápagos Islands in 1981 to begin research on giant tortoises, I never dreamt that Galápagos itself would become my life. But it did. When I arrived, the tortoise breeding and rearing program had been underway for 15 years, run jointly by the GNPS and the CDRS. Yet there was still so much to learn and do. My original research centered on the behavior and feeding ecology of the giant reptiles, comparing the smaller saddle-backed tortoises on Pinzón Island with the large domed tortoises on Santa Cruz. My two-and-a-half year introduction to these animals and their islands would serve me throughout my professional life.

By coincidence, I arrived at the time of a severe drought, followed by the incredible 1982–83 El Niño (see also chapter: Sachs). Living in the field through these two extremes gave me a profound understanding of what drives the Galápagos ecosystem. During the drought, most land animals suffered due to lack of food and water, while marine life flourished in the cold, nutrient-rich seas. Nesting of most land bird species stopped, mortality of both birds and reptiles increased, and vegetation died or went into deep dormancy. Tortoises seemed to be the fulcrum on which this teeter-totter rests. They just waited, some hardly moving for weeks at a time. Their size and slow metabolism — those same characteristics that made them ideal food sources for sailors of past centuries on their long voyages — allow them to simply wait out a drought.

RIGHT: A young tortoise gorges on lush annual vines after heavy rains, Urbina Bay, Isabela.

Then the rains came. The entire ecological situation flipped, and marine species began to die as the seas became too warm and food species disappeared. Terrestrial plants and animals, on the other hand, thrived. Again, tortoises did well, yet also seemed to be waiting for better times. They had plenty of food, but their normal habitat in the highlands of Santa Cruz was flooded and the resulting dense jungle-like vegetation made movement difficult. They responded with a mass migration to the lowlands — males and females alike.

In February 1983, I was settled in on the second day of a two-day watch of one large Santa Cruz male. He had spent the day resting and feeding in the same general area. Then, at about 3 p.m., he unexpectedly began his migration to the lowlands, and he was in a hurry. The route he chose was the stream — yes, the months of heavy rainfall had created streams flowing from the highlands to the sea, a rare sight in Galápagos. I stashed my regular notebook, grabbed my waterproof one, slung my daypack onto my back, and in rubber boots, rain pants and jacket, I crawled into the water after him. On hands and knees, the water washed along my midriff. With the force of the current we both moved quickly. The only problem was that we had to maneuver around rocks, and the dense *Clerodendrum* thickets overhead caught on his carapace and nearly tore my daypack off. We moved on, both in and out of the river, getting farther and farther from my study area. Then after a little over an hour, he just as unexpectedly pulled up onto the bank, began to feed and finally found the perfect spot to settle in for the night. I headed home.

When I learned to follow the many trails the park rangers had cut through the vegetation, I soon realized they had followed ancient tortoise trails. What I learned during El Niño was that the tortoises had also chosen the easiest routes, along the often dry but more open riverbeds. The climate, the islands, and the tortoises were in balance — that is, until humans arrived.

In 1988, after completing my Ph.D., I returned to the CDRS as Head of Herpetology and have remained involved ever since. My main responsibility during the initial years was the tortoise program, including supervision of the Tortoise Center, where eggs from captive-bred Española tortoises, along with those collected from wild nests of other threatened populations, are incubated. The hatchlings are then reared here for their first few years of life, until they are big enough to survive in the wild — usually by age four — even if introduced predators are still present.

By the time I came on the scene, the scientists and park rangers working in the center — among them

ABOVE: Giant tortoises are active ecosystem managers, creating lasting rain ponds and maintaining heavily grazed open meadows, as seen on the caldera floor of Alcedo Volcano, Isabela. LEFT: A tortoise sleeps in a dusty hollow, capable of waiting out prolonged droughts without much hardship.

ABOVE: A very old male saddleback, *G. ephippium*, on Pinzón. Hatchlings from this island must spend their first years in captivity to survive predation by black rats, whose eradication is being planned.
RIGHT: Sadly, Lonesome George from Pinta, the world's most famous tortoise, is the last survivor of his species, *G. abingdoni*.
FAR RIGHT: One of 12 original Española females used for captive breeding since the 1960s, whose offspring have helped repopulate the island.
BELOW AND OPPOSITE CORNER: Two famous faces at the Tortoise Center: Diego, the Española male returned from the San Diego zoo in 1977, and Lonesome George, in residence since 1972.

Craig MacFarland, Howard Snell, and the tortoise caretakers Guillermo Jaya from CDRS and Fausto Llerena and Cirilo Barrera from the GNPS, plus many of their colleagues and predecessors, had carried out experiments and field research to determine best practices for the successful nesting, incubation, hatching and rearing of the tortoises. It fell to me to implement many of their findings. We built more outdoor corrals, decreased living densities, and created sufficient feeding and watering sites to enhance survival. The results soon becoming apparent as survival of hatchlings in the center shot up from an average of 82% in the 1980s to over 97% in the 1990s. We also expanded nest searches on their parental islands to obtain the highest possible genetic mix. Repatriation trips were the highlight of this work.

With our daypacks full of small tortoises, we headed inland to prime tortoise habitat. There, we pulled the youngsters out, checked their markings and released them. Within minutes they disappeared among the rocks and vegetation. Home at last.

Working alongside park ranger Fausto Llerena was inspirational. With many years both at the center and on all of the islands where tortoises survive in the wild, his knowledge and experience were legendary — and I reaped the benefits on many occasions. One day we were hiking along a narrow rocky trail on Pinzón, searching for tortoise nests in barren country where everything looks much the same, when Fausto suddenly stopped. He knew the nearby muyuyo tree (*Cordia lutea*) was a regular nesting site, and sure enough, there was a nest only a few yards away.

Part of my job at the CDRS was to educate and mentor young Galapagueños (local residents) who would become tomorrow's conservation leaders. During my first year as Head of Herpetology, a bright high school student joined my crew as a volunteer. His name was Washington Tapia, or Wacho to his friends. He was 17 years old and passionate about reptiles. Wacho worked tirelessly and later went on to head the Department of Ecosystem Management at the GNPS.

Wacho's experience

Since childhood, I was enraptured by the unique nature of my home islands, so when I finished high school, I volunteered at the CDRS. I wanted to contribute to the conservation of this earthly paradise. I started with marine reptiles, then did some work with invertebrates, introduced mammals and plants, but when I arrived at the Breeding and Rearing Centers for giant tortoises and land iguanas, I knew that of all the unique biodiversity of Galápagos, it was reptiles — and most especially giant tortoises — that I would dedicate my greatest efforts to conserve.

Thanks to a CDF scholarship, I was able to attend university on the continent, and in 1995 I began fieldwork for my thesis on the population status of the giant tortoises of Cinco Cerros on the slopes of Cerro Azul Volcano on southern Isabela. This population fascinated me because it includes two very different giant tortoise morphotypes living in sympatry, one of which is an enigmatic form with a flattened shell we refer to as *aplastados* ('squashed' in Spanish), whose ancestry remains unclear. Later DNA analyses have shown that the genetic relationships among the various subpopulations of southern Isabela tortoises are very complex, including some groups that look similar but have been genetically isolated, probably through recurring volcanic eruptions.

At the Universidad Técnica del Norte in Ibarra, Ecuador, I had studied sympatry at a theoretical level, but the opportunity to work in a place where this unique and complex ecological relationship occurs naturally was perhaps the most enriching experience of my life. Not only did I learn a great deal about giant tortoises and their ecological significance, I also grew to understand the complexity of the archipelago's ecosystems, especially the influence that the climatic cycles have on their delicate balance. Through my work I gained a deep appreciation of the functional role of tortoises as ecosystem engineers — how they mold their habitat by their activities and thus contribute to the development of other species of fauna and flora.

In October 1998, one year after completing my thesis, I was working for the GNPS when a parasitic crater on Cerro Azul erupted, right in the highland tortoise habitat at Cinco Cerros where I had done my

ABOVE: Bred at the Fausto Llerena Tortoise Center, a 20-year-old male Española tortoise, *G. hoodensis*, forms the core of the reestablished wild population.

ABOVE: Tortoise droppings form an integral part of the cycle of life.

FAR LEFT AND LEFT: On Cerro Azul Volcano, the flattened *aplastado* tortoises (far left) living alongside more normal domed type *G. vicina* (left) were the subject of one of the authors' study.

Number of giant tortoises repatriated by species and decade — 1970–2007

POPULATION*	DECADE				TOTAL
	1970	1980	1990	2000	
Española	79	208	696	499	1482
Pinzón	182	86	244	40	552
San Cristóbal	42	13	0	0	55
Santa Cruz	0	67	28	269	364
Santiago	115	90	282	129	616
Cerro Azul (Isabela)	103	102	8	371	584
Sierra Negra (Isabela)	0	51	61	938	1050
Wolf Volcano (Isabela)	14	23	3	0	40
TOTAL	535	640	1322	2246	4743

* Population names are island names except in the case of Isabela Island, where populations are limited to single volcanoes (Cerro Azul, Sierra Negra and Wolf).

Source: CDF and GNP files.

Lonesome George — Last Pinta Tortoise

DATE	EVENT
Dec 1971	Live tortoise sighted on Pinta by snail biologist Joseph Vagvolgyi.
March 1972	Tortoise located by park wardens during a goat hunting expedition and brought to the Tortoise Center on Santa Cruz.
Soon after	U.S. media began to refer to the Pinta tortoise as Lonesome George (LG), after comedian George Gobel.
1972–92	LG located in a large exclosure near the sea, usually with two female tortoises of unknown origin; over time, with excessive love and feeding by his keepers, LG became overweight.
1992	New corral for LG constructed adjacent to new visitor trail; two female tortoises brought from Wolf Volcano (northern Isabela) to accompany George; morphologically, Wolf tortoises considered the most similar to Pinta tortoises; George put on a diet to improve his health and his potential for reproduction.
1993	Efforts at sexual stimulation of George during a four-month period resulted in some behavioral changes but no reproduction.
1994–95	General evaluations of LG by animal nutritionist and veterinarians result in some improvements in his diet.
1999	Results of genetics analyses of Española tortoises show them to be the species most closely related to Pinta tortoises.
2007	Genetics group from Yale University discover a hybrid tortoise on Wolf Volcano that is half Pinta tortoise; the original blood sample was collected in 1994 but complete analysis of museum specimens was required before being able to identify the hybrid.
July–Aug 2008	Two females housed with LG nest in July and August, but eggs are infertile; a thorough genetic sampling of Wolf tortoise in December searches for more Pinta genes.

Española Tortoise Restoration Program

DATE	EVENT
1963–74	All tortoises found on Española transferred to the Tortoise Center on Santa Cruz — a total of 14 animals (2 males and 12 females) — initiation of a breeding program for giant tortoises.
1968–78	Goats eradicated from Española (introduced there prior to 1905); a total of 3355 goats removed.
1970	First successful nesting areas created after much trial and error; tortoises nest.
1971	First hatchling tortoises from the captive adult population.
1975	First repatriation of young tortoises back to Española — a total of 17.
1977	Return of Diego — the Española male that had been in the San Diego Zoo since the mid-1930s.
1990	First nests of repatriates encountered on Española; also two hatchlings eaten by hawks.
1991	First live hatchlings resulting from reproduction by the repatriates encountered on Española.
1994	Census of repatriates on Española — minimum survival estimated at 55%.
2000	1000th tortoise repatriated to Española.
2003	Hybrid Pinzón-Española tortoises found on Española suggesting that a Pinzón tortoise was mistakenly repatriated to Española during the early years of the program.
2007	Plan approved to suspend repatriations on Española and place the Española tortoises reared at the Tortoise Center on Pinta as part of the Pinta Restoration Project.

History of Giant Tortoises in Galápagos

ERA	EVENT
2-3 mya	Giant tortoises arrival, dispersal and evolution.
1700s	Exploitation by buccaneers and early visitors — low levels.
1800s	Exploitation by whalers and fur sealers — estimated to have removed between 100,000 and 200,000 tortoises.
1800s & 1900s	Introduction of rats, goats, pigs, dogs and donkeys to Galápagos.
1900s	Decimation of habitat of tortoises by introduced mammals on many of the islands.
1900s	Exploitation for oil — tortoises killed on site and their carcasses rendered for collection of heating oil to be used on the continent.
1959	Establishment of the Galapagos National Park and the Charles Darwin Foundation and the initiation of systematic reviews of the tortoise populations.
1965	Initiation and implementation of the captive rearing program for giant tortoises.
1970	Initiation and implementation of breeding program for tortoise populations with few remaining adults.

Milestones of Giant Tortoise Restoration Program

PERIOD	ACTION
Early 1960s	Evaluation of tortoise populations.
1965	Collection of first eggs from nests on Pinzón (natural recruitment zero due to black rats); solar incubators built near the sea.
1963-74	Collection of Española tortoises and initiation of a breeding program.
1969-71	Field studies on breeding, nesting and incubation; results incorporated into management methodologies in the Tortoise Center.
1969-77	Incorporation of additional populations into the program (Wolf Volcano, Santiago, Cerro Azul, San Cristóbal, Santa Cruz and Sierra Negra).
1970	First repatriation — 20 tortoises released on Pinzón.
1970	Construction of the Casona — a building open to visitors that would house the young tortoises.
1972	First outdoor corrals built for young tortoises.
1979	Solar incubators constructed inland near the Casona.
1985	Electric incubators designed for experiments to enable temperature control.
1985-87	Experiments on rearing conditions for young tortoises; results indicated that tortoises survive best outdoors in corrals with a soil substrate.
1987-90	Incubation experiments on temperature-dependent sex determination and hatching success; based on results eggs now divided with two-thirds incubated at 29.5°C (85.1°F) to produce females (approximately 120-130 days) and one-third incubated at 28°C (82.4°F) to produce males (up to 175 days).
1990	New outdoor corrals built and all tortoises removed from the Casona; improvements in rearing techniques decrease mortality of young tortoises in the center from an average of 18% per year in the 1980s to less than 3% per year in the 1990s.
1990	New visitor trail constructed to provide better viewing of young tortoises; corral for LG constructed along this trail.
1990s	Genetic analyses of all tortoises in the center able to determine origin of the majority of the unknown tortoises.
1994	Tortoise Center opened on Isabela.
2003	Tortoise Center opened on San Cristóbal.
2008	Two females living with LG lay eggs (three clutches) for the first time. Although these were infertile, it is hoped that this marks the beginning of more breeding activity.

Tables: Linda J. Cayot and Washington Tapia

ABOVE: Two legendary figures: Lonesome George and his caretaker, park ranger Fausto Llerena, after whom the Tortoise Center was named.
BELOW: The growth progress of baby captive tortoises is closely monitored.
RIGHT: Another Española hatchling emerges from its egg in simple but reliable incubators, where trial and error over the years has lead to a doubling of the hatching success, from under 25% in the first decades of the program to over 50% today, with hatchling survival above 97%.

thesis work. Given that this population had already been greatly reduced by predation and competition with introduced organisms, and was still subject to clandestine hunting, the GNPS decided the time was right to remove some individuals to initiate a captive breeding program. Thanks to my knowledge of the area and of the individual *aplastado* tortoises and their seasonal movements, I was fortunate to be named leader of the team of park rangers and scientists that travelled to Cinco Cerros to transport them to the Tortoise Center in Puerto Villamil. In two days, we found and evacuated 17 tortoises to form the nucleus of a new breeding program; these same animals have already produced hundreds of hatchlings.

Past, present and future

The history of giant tortoises since the arrival of the first humans in 1535 is perhaps the saddest of all Galápagos species. Reports from the first navigators to visit in the 17th, 18th and even 19th centuries described the giant tortoises as extremely abundant and widely distributed. The Spanish called them 'galápagos,' after a type of horse saddle because of the shape of their carapace — thus the archipelago was named for the tortoises.

But the arrival of humans initiated a series of events that would result in the extinction of some species of tortoise and the decline of many others. Buccaneers, whalers and fur sealers gathered tortoises as live food for their long voyages. Scientific expeditions collected specimens for their museums, then later live tortoises for zoos. Entrepreneurs from the continent came to Galápagos to render the tortoises for their oil. And perhaps most threatening of all, the numerous mammalian and other species introduced to the islands, either intentionally or accidentally, wreaked havoc on tortoise populations — through predation (pigs, rats, dogs, cats, fire ants) and competition/habitat destruction (goats, donkeys). The tortoises had lived on these islands for hundreds of thousands of years, safely isolated from the rest of the world, but they could not adapt to such sudden change. By the mid-20th century, three tortoise species were extinct, two were near extinction, and several others were endangered. But the longevity of tortoises turned out to be their saving grace, allowing some to simply outlive the horrific times and survive to an era when humans with a changed view of the world would arrive and put things right.

The establishment of the CDF and GNP in 1959 marked the tangible beginnings of the desire to reverse that history. A review of the status of the tortoise populations began almost immediately. The conditions on Pinzón, where black rats had thwarted any tortoise recruitment since the late 1800s, stimulated the immediate start of the rearing program in 1965.

The most glowing success in the nearly half-century of giant tortoise conservation work is the resurrection of the Española tortoise population. Once doomed to extinction, the last 14 individuals (12 females and two males) were rescued from their goat-ravaged island and bred at the Tortoise Center, along with Diego, a third male kindly returned from the San Diego Zoo. Twenty years after the first captive-bred hatchlings emerged and 15 years after they were first returned to their ancestral island, the repatriates started breeding on their own. Today, with the extermination of the feral goats in 1978, the gradual subsequent recovery of the vegetation, and nearly 1500 tortoises released on the island, the island is slowly undergoing a period of restoration towards a more pristine condition. While work in genetics to determine the health of the population continues, it appears that such intensive assistance is no longer required.

The story of the Pinta Island tortoise took a much more tragic turn. Lonesome George, the only representative of his species, survives alone at the Tortoise Center where, for 35 years, efforts to breed him with some of his closest relatives have persistently failed.

In the last two decades, new technologies have allowed us to make major advances in the eradication of introduced

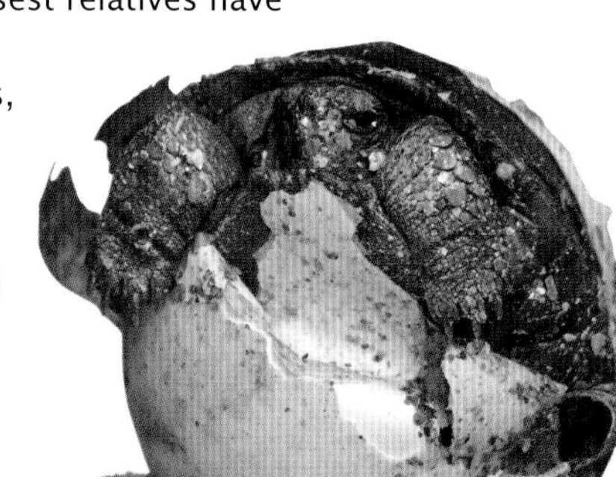

mammals, as well as in our understanding of tortoise ecology and genetics. The advent of DNA analyses of Galápagos tortoises, led primarily by a group of scientists from Yale University, has been particularly enlightening (see also chapter: Caccone & Powell). For years, the taxonomy of Galápagos tortoises had been debated and repeatedly rearranged, but today each island population (or the five individual volcanoes in the case of Isabela) is considered a separate species.

Genetic analyses have also been used in the search for a solution to the Pinta problem. The island, free of goats since 2000, is undergoing rapid regeneration of its vegetation, but something is missing — tortoises! The vegetation is coming back so densely that many of the light-loving species are disappearing. The island needs its major herbivore back — to eat the vegetation, to open areas through its movements and resting habits, and to disperse seeds. DNA results have shown that Española tortoises are most closely related to those of Pinta (DNA samples were taken from George and from the few museum specimens of Pinta tortoises). With the Española population well on the road to recovery, we can now divert the current and future Española hatchlings at the Tortoise Center to repopulate Pinta Island. The first release of 120 young Española tortoises was planned for late 2008.

However, nothing can ever be that simple. No sooner had these plans and policies been laid out, than two events occurred which could someday result in the return of a close to pure Pinta tortoise, putting the current repopulation plans for the island on hold.

In 2007, blood samples from tortoises on Wolf Volcano revealed the presence of a Pinta-Wolf hybrid, indicating that at some point in the human history of Galápagos, buccaneers, whalers or others accidentally or purposely caused Pinta tortoises to land on northern Isabela. A more complete search of the volcano is currently being carried out with the hope of finding more hybrids or perhaps even an original Pinta tortoise. Could the possibility now exist of resurecting the species through selective mating of these hybrids?

Latest twist — Wacho recounts

The plan for the ecological restoration of Pinta was developed in part because the GNPS had resigned itself to the fact that Lonesome George would never reproduce. So on 21 July 2008, when Fausto Llerena informed me that a freshly dug nest had been discovered in Lonesome George's corral, laid by female #106, one of the two females from Wolf volcano sharing his corral, I was incredulous. My first thought was that this was a mistake, but I knew that was impossible, as there is no better expert in recognizing tortoise nests than Fausto. I headed to the Tortoise Center thinking that it would turn out to be a false nest, but upon opening it we discovered nine eggs, though only three were intact. Even so, I doubted their fertility, or thought perhaps the female had stored sperm from before her time with Lonesome George, which tortoises are capable of. But then on 1 August female #107, the other Wolf volcano female living with Lonesome George, also nested. And female #106 nested again on 8 September. I was finally convinced that George had become reproductively active, stimulating both females to nest for the first time in their 16 years living with him. Fourteen eggs were placed in the incubators. Since the sex of Galápagos tortoises is triggered by the temperature at which the eggs are incubated, we set 10 at a female-producing 29.5°C (85.1°F), and six at half a degree lower to obtain males. Sadly, but perhaps not surprisingly, anticipation turned to disappointment as we waited impatiently during the four months incubation for the most-awaited baby tortoises on earth. As it turns out, none of the eggs were fertile. Still, with a potentially reproductively active Lonesome George, and the possibility of more Pinta genes in the Wolf Volcano population, there may still be hope for the recovery of the Pinta tortoise species, currently considered Extinct in the wild.

The GNPS, together with the CDF, and their many collaborators around the world continue to advance the conservation of Galápagos, with the goal of returning all of the uninhabited islands to near pristine conditions, and especially ensuring the survival of their namesake — the Galápagos giant tortoise.

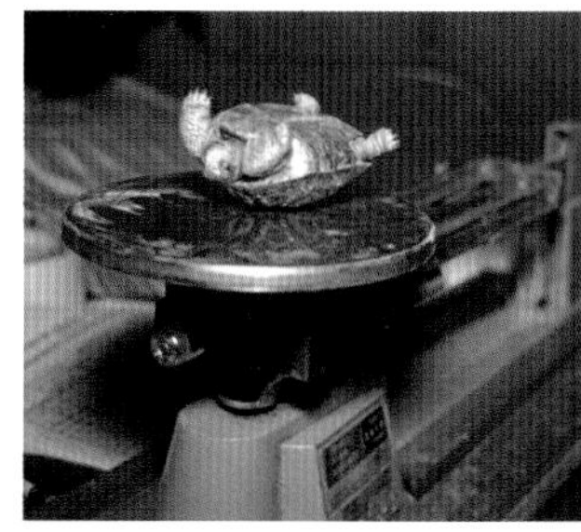

LEFT: In March 2000, under the glare of media attention, the 1000th baby Española tortoise was repatriated to its island of origin.

ABOVE: A hatchling is weighed before moving to the outdoor pens, where it will spend the first few years of its life to increase its chances of survival in the wild.

BELOW: Staff from the Galápagos National Park and Charles Darwin Research Station routinely review tortoise breeding procedures, a long-standing bi-institutional project.

Project Isabela

Ecosystem Restoration through Mega-eradication

Karl Campbell with input from Linda Cayot (Galápagos Conservancy), Gonzalo Banda (Dublin College), Felipe Cruz (Charles Darwin Foundation) and Victor Carrion (Galápagos National Park)

Island Conservation,
100 Shaffer Rd, LML
Santa Cruz,
CA 95060, United States.
<karl.campbell@
islandconservation.org>

Karl Campbell has worked in the Galápagos since 1997. He came up the ranks of Project Isabela, from volunteer to manager/coordinator of field operations and strategies. During this time, Karl gained his Ph.D. in refining and developing Judas goat methods to enhance the efficiency of detecting the last goats in eradication campaigns. Karl still lives in the Galápagos, but now works for Island Conservation, a US-based nonprofit group that conducts eradications of invasive mammals to restore island ecosystems and prevent extinctions on islands around the globe.

THANKS TO THEIR EXTREME isolation, as a rule oceanic islands have developed unique ecosystems that often exclude the presence of most terrestrial mammals. Over millions of years, other types of animals tend to fill the ecological roles more traditionally occupied by large mammals, particularly grazers. The ancient islands of New Zealand had their flightless birds, from kiwi to moa, while in Galápagos reptiles took up the challenge. The giant tortoises are particularly striking examples, with some forms adapted to graze the highland meadows and others to browse the arid lowland thickets. So it is not surprising that when humans came on the scene with their bevy of domesticated farm animals, the ecological changes that followed were as profound as they were devastating.

BELOW: The sun rises over a timeless scene as giant tortoises begin to stir in their caldera homeland, where the largest population survives. Thanks to their long lifespan and slow metabolism, the tortoises were able to endure the goat invasion on Alcedo Volcano until Project Isabela restored peace to the ecosystem.

Domestic goats, *Capra hircus* — hardy, adaptable and fast-breeding — headed the list of menaces that began to invade Galápagos over two centuries ago. Some escaped when set ashore temporarily to fatten up, others were deliberately released so they would multiply and provide meat for visiting ships. Invariably, the result was the same: on one island after another their numbers increased within a few decades, until the land became denuded at the expense of the reptilian grazers. This process was still going on even when the first conservation efforts were set in motion in the 1960s. While early hunting campaigns were busy removing goats from smaller islands, other populations were still exploding on larger ones, such as Santiago and Isabela.

Linda Cayot, who first came to Galápagos to study the tortoises, recalls her dismay when she became aware of the vertiginous speed of sweeping habitat loss as goats invaded the relatively pristine northern half of Isabela Island. She first visited Alcedo Volcano, where the largest concentrations of giant tortoises remain, in 1983, before the goats had arrived from the southern, inhabited part of the island. She remembers walking into a prehistoric landscape of dense, mist-shrouded forest crisscrossed by low tortoise trails. But when she returned 10 years later, their world had completely changed. She writes, 'Where dark forests had once stood, patches of dry grass dotted the sun-baked dirt. The life-giving pools of water had dried into dust bowls and the trunks of the once lush trees lay strewn across the ground. With no place to go, the tortoises, also sun-baked and dusty, still congregated in the dry 'pools.' Horrified at what I saw, I wanted to run from the place. But I stood frozen, unable to look away. A tear slipped down my cheek. Tortoise heaven had become a living hell and we, the human race, were to blame. Our method of destruction was the introduction of exotics — this time, goats. They now outnumbered tortoises nearly 20 to one.'

In 1997 Linda became the first coordinator of the newly formed Project Isabela (PI), a bi-institutional endeavor mounted jointly by the Galápagos National Park (GNP) and the Charles Darwin Foundation (CDF). Its aim was island restoration on a scale never before undertaken anywhere in the world. This was also when I first arrived in Galápagos as a volunteer. As Linda's assistant, one of my first activities that year was to participate in a workshop that laid out a template for the eradication of goats from all of northern Isabela Island. This workshop brought together local knowledge and international experts. Through these discussions a plan was developed that would assemble the most advanced methods from around the world in order to rid such a vast area of goats.

But first and foremost we had to establish what the realistic prospects might be for such an ambitious objective to succeed. So, after carrying out a survey trip with the workshop members, the first milestone was reached when everyone agreed that this may indeed be *possible*, though all noted that it would be extremely challenging. Isabela was over 20 times larger than the largest island where similar work had ever been successfully carried out anywhere on Earth. To further complicate matters, there are no roads or other facilities over the entire northern half of the island, and it is riddled with caves and lava tunnels for goats to hide in.

ABOVE: While Alcedo Volcano (foreground) was the center of operations, Project Isabela covered the entire island, including Darwin, Wolf and Ecuador volcanoes in the distance. BELOW FAR LEFT: A giant net for mustering goats on foot is drawn across denuded Santiago. BELOW MIDDLE AND RIGHT: Proud animals very much in the wrong place, goats ravaged giant tortoise habitat on Alcedo.

Photo courtesy Project Isabela archive

Photo courtesy Project Isabela archive

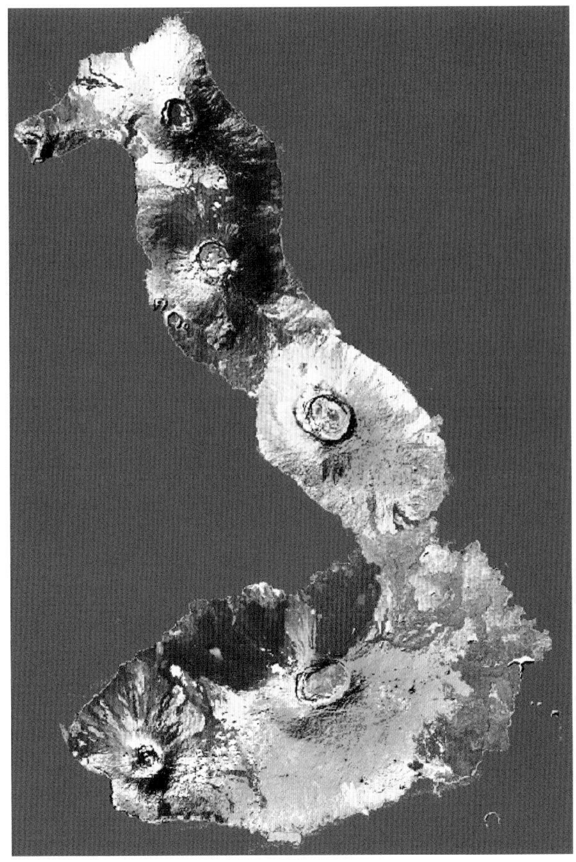
Courtesy Project Isabela archive/NASA

ABOVE LEFT: A large Santiago herd corraled in a crater.
ABOVE RIGHT: With a total surface area of 584 sq km (225 square miles), Isabela is by far the largest island ever successfully tackled for feral mammal eradication, a task complicated by its extremely rugged terrain and absence of logistical facilities such as roads, footpaths, runways or basic supplies like fuel and water. Enhanced relief on a true colors NASA satellite image shows the complex topography.
RIGHT: During the drought of 2000, tortoises struggled to find shade after trees in large areas of Alcedo's caldera rim were toppled by goats.

I was an undergraduate at the University of Queensland in Australia at the time. Over the next 12 years, this workshop and the project that ensued propelled me into the highly specialized field of island restoration via mammal eradication. Felipe Cruz (CDF) and Victor Carrion (GNP), both Galapagueños passionate about conservation, would soon take over from Linda as co-directors of PI, and in the process would become my supervisors, mentors and lifelong friends.

Work on the ground started in 1998, although the first six years were dedicated to building up our technical capacity and securing funding. Gradually a team was assembled by transforming a rag-tag group of rough and ready local hunters who worked for the GNP into an elite squad that was highly skilled in firearms, hunting tactics, GPS use, radio telemetry, dog handling and local logistics. This transformation was accomplished by a combination of intense fieldwork and training, while being driven hard by the project management, whose vision was to achieve results that others thought unobtainable. We used Santiago Island as our training ground, where the feral pig eradication efforts had been ongoing for 30 years using traditional ground-hunting means, and the island was now down to the last few individual animals. Applying the new methods we were developing, their final extermination was swiftly completed, making Santiago the world's largest island in the world successfully freed of pigs. Likewise on Pinta, another goat-infested island where prolonged hunting efforts were approaching the final push, we rapidly concluded the operation by trialing 'Judas goats,' a system that will be described later.

These major accomplishments, both on Pinta and Santiago, happened simultaneously, inspiring great confidence, particularly among our funding bodies, such as the Friends of Galápagos organizations and the Global Environment Facility from which we received a major grant via a complex hierarchy of UN bilateral programs.

By 2001 we went into high gear. We had to import specially trained goat hunting dogs from New Zealand, along with their trainers. Local dogs were selected and crossbred to build up our kennels. We also invited international bidding for a massive aerial hunting contract that was eventually won by a private helicopter firm, also from New Zealand, involving two helicopters and their pilots, a mechanic and several expert aerial hunters. We initiated the complicated importation process for custom-made semi-automatic rifles as well as over half a million rounds of ammunition into Ecuador. Eventually, our ground hunting team grew to 55 local hunters, while we bred and trained a world-class team of 110 hunting dogs.

Almost every step suffered major delays, yet we managed to keep things moving nonetheless. Importing ammunitions and firearms was a frustrating process that took more than two years. Getting helicopters into the country was equally fraught, and likewise obtaining the Ecuadorian civil aviation permits for pilots, mechanics and aircraft operations. Every element was complex to implement, but we knew we were charting new ground, so we called in many favors, used every trick in the book (as well as many that weren't), and above all we carried a

Photo courtesy Project Isabela archive

Photo courtesy Project Isabela archive

Photo courtesy Project Isabela archive

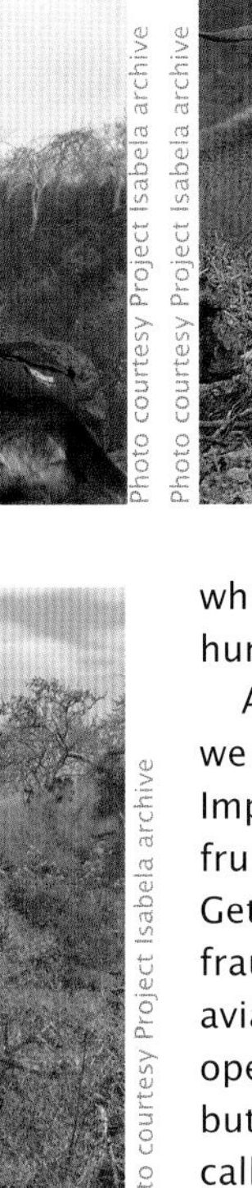

Photo courtesy Project Isabela archive

ABOVE: Goats on Alcedo stripped vegetation far beyond the tortoises' reach.
LEFT, TOP TO BOTTOM: Hunters working from a deftly positioned, low flying helicopter; trained dogs fitted with special booties to protect them from sharp lava; a hunter and his dogs track a radio-collared Judas goat.

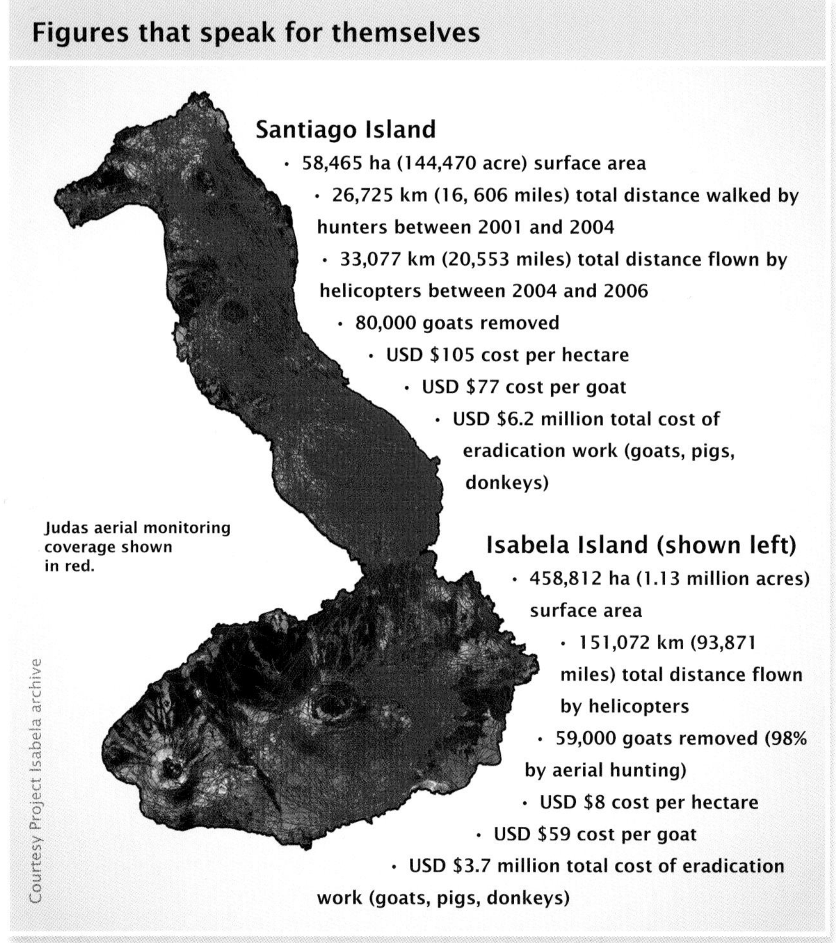

can-do, will-do attitude that made things happen. It was a great team to be a part of.

Figures that speak for themselves
Santiago Island
• 58,465 ha (144,470 acre) surface area
• 26,725 km (16, 606 miles) total distance walked by hunters between 2001 and 2004
• 33,077 km (20,553 miles) total distance flown by helicopters between 2004 and 2006
• 80,000 goats removed
• USD $105 cost per hectare
• USD $77 cost per goat
• USD $6.2 million total cost of eradication work (goats, pigs, donkeys)
Judas aerial monitoring coverage shown in red.
Isabela Island (shown left)
• 458,812 ha (1.13 million acres) surface area
• 151,072 km (93,871 miles) total distance flown by helicopters
• 59,000 goats removed (98% by aerial hunting)
• USD $8 cost per hectare
• USD $59 cost per goat
• USD $3.7 million total cost of eradication work (goats, pigs, donkeys)
Courtesy Project Isabela archive

Once helicopter blades started turning in early 2004 it was all go. We decided to continue fine-tuning our expertise by tackling the goat problem on Santiago first. As we waited for the firearms to arrive we trialled the use of miles-long netting, which was easy to string up across the denuded landscape. This proved hugely successful, enabling us to round up thousands of goats in a very short time by using mustering teams of men on foot, with mules to transport the gear. The rifles still hadn't arrived so we also worked with shotguns from the helicopters, which resulted in spectacular-colored bruises radiating from the center of everyone's chest to their forearms — we were all padding the shotgun butts and taking anti-inflammatories with every meal. Three weeks later the semi-automatic rifles arrived. By the time we finished with Santiago in 2006, roughly 80,000 goats had been eliminated, and we could already see the long-suppressed native vegetation beginning to rebound.

Finally, in April 2004, we were ready to tackle Isabela. Our teams were divided based on expertise. Galápagos is a hard place to maintain good communications, but we had worked with each other so closely for so long that most things were intuitive. Our 'beer o'clock' sessions at the pizza restaurant Media Luna in Puerto Ayora, Santa Cruz Island (our second office) had been a part of the group's culture and provided a forum for ideas to be tabled, information to be passed on, strategies to be developed and everyone to stay on the same wavelength. When fieldwork became intensive, our team was infrequently in town and Media Luna closed down.

We built our operational base at Punta Alfaro, on a sun-baked slab of barren lava along the eastern shore of Alcedo Volcano, a Spartan cluster of tin-roofed equipment sheds, fuel and water tanks, and basic accommodation. From here, we established temporary fuel depots for local operations strategically placed around the northern half of the island. We began to fly systematic tracks and clear blocks that divided the volcanoes into manageable units. At the same time, we were dropping materials (lumber, roofing metal and plastic water tanks) inland at regular intervals to construct rainwater collection points, which would supply the ground hunters that were to follow once the aerial hunting had reduced the goats to very low numbers.

RIGHT AND FAR RIGHT: Through the 1980s, moss-clad trees on the rim of Alcedo performed a crucial role in capturing fog-drip into puddles used by tortoises (left). The same place photographed in 1991 and again in 1995 reveals drastic habitat destruction (right).

Our first surprise came because we were simply not prepared for how effective aerial hunting would be in the open Galápagos environment. I remember calling Felipe after less than two weeks based in Punta Albermarle, the most northerly point on Isabela where we initiated the aerial campaign. That call was a request to get our camp moved because we'd put all of Wolf and Ecuador volcanoes and half of Darwin Volcano down to low levels ready for the 'Judas goat' operation.

Judas is a term given to a goat that is used as a hunting aid when the population has reached extremely low densities. These are goats that are captured, fitted with a radio telemetry collar and released. Being a social species, the collared animals go in search of other remnant goats. They are then tracked down periodically and any associating uncollared goats are shot. Eventually only Judas goats remain, which can be detected and removed.

Traditional Judas goats, as they had been used on eradication campaigns elsewhere, were considered 'dead' because they could be tracked down and shot any time, so there would be no danger of them escaping. But all previous campaigns had used less than 30 Judas goats, while we had plans for deploying at least 600 on Isabela. If we had even a 5% failure rate on collars we'd have goats getting around that would be able to breed, thus leading to failure of our ultimate goal. Plus, even goats with functioning collars would be breeding during the time span needed to complete our work — too many things could go wrong, especially on the scale we were planning. We very soon realized we would be opening a whole new field of operation, one where we couldn't simply import the best techniques from elsewhere, as was the case with the aerial hunting and the use of dogs. Eradication is a demanding field as there is simply no margin for error. Just one pregnant surviving goat can represent total failure. What we needed were 'super' Judas goats that we could rely on 100 percent — we needed goats that were sterile, and that would intensively search out and be searched for by other goats.

I left Galápagos after the Pinta eradication to conduct trials back in Australia, and returned 10 months later with the means to make these super Judas goats a reality. I applied techniques from veterinary medicine to conduct sterilizations — tubal ligation for females and epididymectomy (similar to vasectomy but in a different spot) for males. We could do these surgeries fast and in the field, and we could terminate pregnancies with a hormone injection. I'd also worked out how to put female goats into a near permanent estrus (or heat) using hormone implants. This was the ticket because females in estrus search intensively for males and males will track down females in heat and stay with them. We later coined the term 'Mata Hari' for these Judas goats, after the irresistibly seductive World War 1 double agent. We first conducted a large-scale experiment with Judas goats on Santiago as part of the final sweep there to see whether sterile estrus-induced females worked more effectively than sterile males or normal females. As suspected, male goats couldn't resist associating with our estrus-induced Judas goats, and were found in their company twice as often as with the untreated females. With this tool, we removed males from the remnant population faster, and consequently stopped reproduction in the last remaining females.

Work on Isabela was intensive. We had two helicopters working flat-out for three months, until one day we suffered a 'heavy landing' when we lost power high above a volcano and had to glide down to a near-crash landing on a small patch of flat lava. Early on we had prepared everyone for the likelihood of fatalities at some point during the project, yet fortunately we had none. Only thanks to our pilot's extraordinary skill did we all walk away from the wreck unharmed, but the helicopter was a total write-off. This nearly crippled the project. Five months after the incident we restarted helicopter operations, completing the remainder of the project with just a single machine. Dog teams and ground hunters were used strategically in thickly vegetated parts of the island, where they carried out systematic GPS-guided sweeps. The dogs were trained to locate the goats, then quickly round them up and keep them bailed up so they wouldn't scatter until the hunters arrived.

On 30 March 2006, Felipe broadcast these electrifying words to an incredulous world: 'As of

ABOVE LEFT AND RIGHT: Photos taken 12 years apart show deep erosion and dying stumps replacing a tree fern-sheltered pool between 1983 and 1995.

BELOW: By the time the last goat was gone 11 years later, only one healthy specimen of the endemic *Cyathea weatherbyana* survived inside a protective enclosure. The revegetation process is advancing in leaps and bounds.

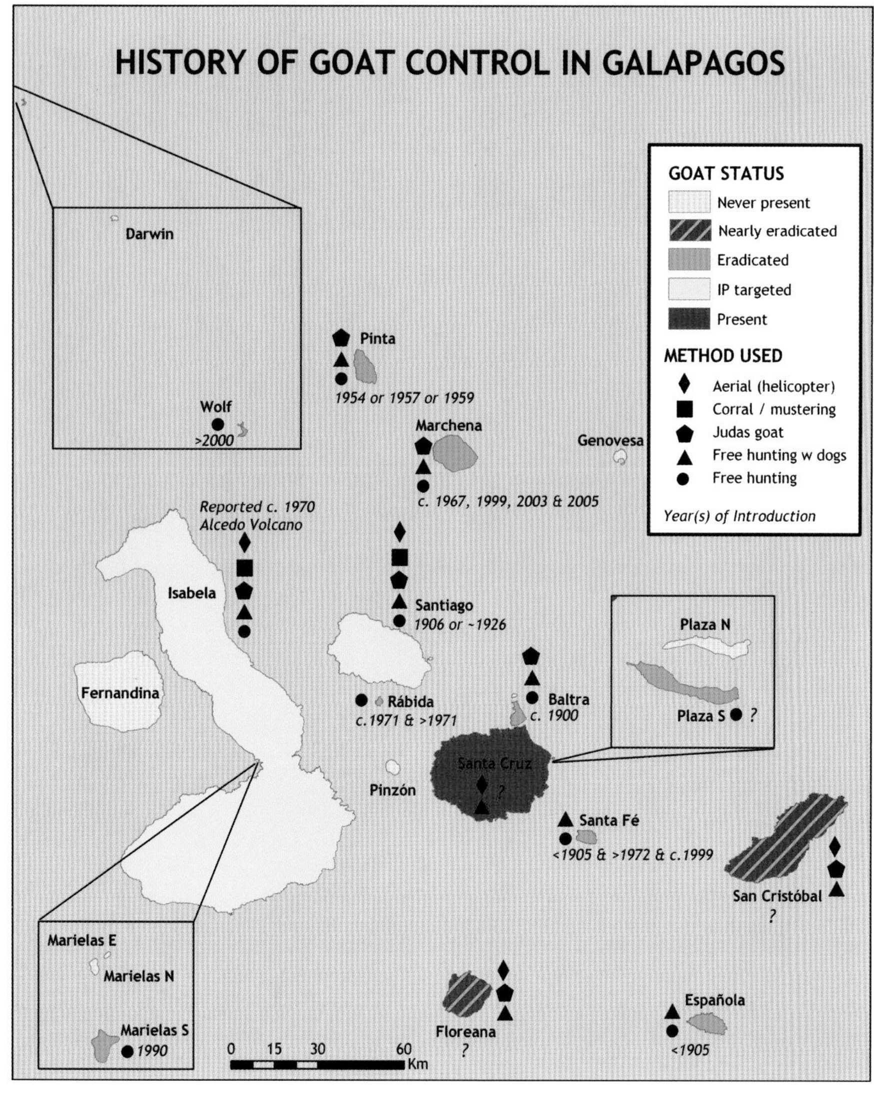

BELOW, FROM LEFT: On Isabela, predatory feral cats remain, but not so the tortoise-nest-trampling donkeys. Endemic rails, *Laterallus spilonotus* (below far right), are rebounding on Santiago.

today (2 p.m. Galápagos time), we have closed down all field activities on Santiago and Isabela islands. The structure and facilities at Cowley camp are gone, all the other smaller camps have been evacuated and all the personnel and dogs are safely back home. The helicopter is in the process of being packed in a container to go home too. All the objectives and challenges planned many years ago have been fulfilled and proudly we can report that we have gone beyond what was planned and expected. Therefore, I would like to say that we have done what the World believed was impossible. Pigs, goats and donkeys on Santiago as well as from the whole of northern Isabela are now history!'

For me, the most gratifying part of doing eradication work has been seeing the spectacular recovery of suppressed native species. On Santiago, Galapagos rails (*Laterallus spilonotus*) were approaching local extinction in the 1980s with pigs, donkeys and goats present. Today, the population of these tiny endemic birds resembles what Darwin described when he visited here over 170 years ago, 'The upper region being kept damp by the clouds, supports a green and flourishing vegetation. So damp was the ground, that there were large beds of a coarse cyperus, in which great numbers of a very small water-rail lived and bred.'

Vegetation everywhere has demonstrated an even more dramatic response. While we were still removing the last goats on Alcedo, we saw *Scalesia* forest immediately beginning to recover from the dormant seedbank, and *Darwiniothamnus* seedlings recolonizing the grasslands outside of craters that had been their refuge from goats. Even slow-growing trees like guayabillo (*Psidium galapageium*), cat's claw (*Zanthoxylum fagara*) and *Opuntia* cactus were popping up in the Santiago highlands, quickly reversing the deforestation process caused by the goats.

Now the GNP is leveraging the capacity built during Project Isabela to remove large feral mammals from all islands — even those with human settlements — the ultimate goal being a goat-free archipelago. Floreana Island is the first in this new effort, and goat eradication there is near, while feral donkeys and cattle are already gone. Knowing that action can be taken in time and that I have been a part of such a positive conservation process provides me with an enormous amount of motivation to see this tool implemented on other islands in need of similar work, even beyond Galápagos. Few, if any, other conservation tools are this powerful.

Reports from the Front

Personal Accounts from National Park Field Staff

Wilson Cabrera and Omar Garcia

Wilson Cabrera and Omar Garcia are both long-time Galápagos Park Rangers, with over 10 years experience. Wilson works with the Introduced Animals Control Unit, where, in between carrying out arduous expeditions to clear feral animals from sensitive areas, he manages Geographic Information System (GIS) data and radio telemetry operations, and also designs field strategies for pest control. He was acclaimed as the Galápagos National Park's (GNP) crack aerial and ground hunter during the Isabela Project. Omar is an accredited sea captain who commands the GNP's flagship patrol vessel, the *Guadalupe River*, and has carried out many of its most ambitious law enforcement operations. Both are Galápagos natives whose families were among the early settlers before the GNP came into existence half a century ago.

Galápagos National Park Service,
Isla Santa Cruz,
Galápagos, Ecuador.
www.galapagospark.org

Wilson's Diary: Tracking Down the Wolf Island Goat

It all started as a perfectly normal day in the office, writing reports and planning our next operations, when one of my colleagues returned from the field with stunning news: a goat had been spotted on one of the most distant and pristine islands in the GNP, the northern outpost of Wolf Island. Immediately we sprang into action, putting in motion a well-practised routine at the Introduced Animals Control Unit, where I am one of the park rangers specialized in hunting.

A trip had to be planned at short notice, which meant chartering a private vessel, not always easy as they are usually busy fishing or carrying tourists. We did not ponder the disturbing questions of how or why this goat had been released on the island,

LEFT: With a total area of 1.3 sq km (0.5 square miles) and a maximum height of 253 m (830 ft), the precipitous cliffs of Wolf Island, along with the rich waters that surround them in the far north of the archipelago, are a familiar arena to both authors, whether hunting down a rogue introduced goat, or chasing high-seas fishing poachers.

DOG DAYS ON ISABELA

Although limited in range by the availability of fresh water, small numbers of feral dogs occurred for a time on both Floreana and Santa Cruz, with dire impacts on native wildlife. Nowhere, however, did they become established with such deadly proficiency as on southern Isabela. Their history seemed to go back to the Spanish colony, whose records hold that the viceroy of Peru dispatched a vessel to release dogs on Galápagos in order to eliminate the goats upon which pirates and buccaneers relied in between plundering raids on the colony's riches. Whatever their origin, besides hunting in packs among large herds of feral cattle, they effectively decimated a long list of Galápagos species — land and marine iguanas, penguins, cormorants, sea lions, fur seals, boobies, tortoises and many more. In 1980, the GNPS launched a successful campaign to finally eliminate these dogs, a tricky operation requiring innovative application of deadly 1080 poison to minimize incidental kills of native species. Crisscrossing Cerro Azul volcano on foot, park rangers shot wild cattle at regular intervals and injected bite-size baits of fresh meat with a minimum deadly dose of poison sufficient to kill a dog, but not harm a raptor or reptile who might later feed on its carcass. The baits were then hidden in grass or under lava ledges all around the exposed, poison-free carcass. The result was that scavenging birds such as hawks or lava gulls could safely feed on the dead beast while dogs readily sniffed out the hidden, poison-laced baits. With the dogs gone, many species rebounded — tortoise nests could once again hatch safely and penguins have re-established an important breeding colony around Caleta Iguana on the southwest coast. Interestingly, feral dog genes live on in a few domestic dogs as local hunters occasionally captured puppies, favouring their hardy nature when breeding them with their own goat-hunting dogs. Spindly legged black and white descendants can still occasionally be seen on highland farms.

Tui De Roy

our mission being to eliminate it as soon as we could, before it was able to inflict too much harm on this otherwise untouched habitat. We knew the geographical conditions of the island would make this expedition challenging, but we were determined — the job had to be done, and quickly.

That evening we boarded the local fishing boat *Oberlus II.* We were an eclectic group consisting of myself and another park ranger, Fidelino Gaona, Pamela Martínez, a technician from the GNP's veterinary lab (to examine the goat carcass and try to determine its origin), and an officer from the Ecuadorian Marines as a firearms handler (a requirement of the Joint Command of the Ecuadorian Armed Forces). Along for a rare chance to visit the island were Henry Herrera, an entomologist from the Charles Darwin Research Station (CDRS), and photographer and local conservationist Tui De Roy.

After a 30-hour crossing we sighted the island in the first light of dawn, having made a brief intermediate stop at Pinta Island the afternoon before. A hurried breakfast, and we were off to find a landing spot, leaping ashore between waves onto a giant boulder at the base of the island's massive cliff face. This is one of only two spots where it is possible to get ashore, as there are no beaches here at all. Luckily, the huge swells that normally pound this exposed shoreline were manageable, and soon we were looking for a way to scale the formidable rampart toward the summit plateau, 215 m (705 ft) above. We found a sort of chute where the vertical face eased off to about a 75-degree angle at best, and picked our way between unstable boulders and loose, sliding rocky debris mixed with broken vegetation.

RIGHT: Dwarfed by jagged crags, senior hunter Wilson Cabrera (a speck of red in the middle of the photograph) leads the team nearing the summit plateau of Wolf Island by following a precarious chute riddled with loose and falling rocks.

It took us nearly two hours to reach the top, much of the time fearing for our safety as we took calculated risks dodging falling rocks, clinging to unstable faces or scrambling up slippery scarps. But we were determined, aware of the paramount importance of our mission. We felt the responsibility heavily, not only to find and eliminate this one goat, but to make absolutely sure that there were no others lurking around. All along the climb we saw numerous birds, both terrestrial and marine, such as nesting swallow-tailed gulls on cliff ledges and red-footed boobies in the small shrubs clinging to the face.

Just as we scaled the final wall and crested the summit ridge, we suddenly encountered the goat quite near to us, but before we were able to get her in our sights, she dashed away like lightning, jumping from ledge to ledge, much more agile than ourselves. We tried to discover where this wily animal had taken refuge along the precipice by walking out on narrow lava cornices, following the contours of the cliff as far as we could go, but did not spot her again. All day we inspected the entire upper reaches of the island some 200 m (650 ft) above the ocean surface, following the goat's spoor and finding its favorite feeding and resting spots, but all in vain.

Curious Galápagos doves and mockingbirds followed us along, reminding us of the wonders that we were here to protect. By 4 p.m. we knew it would

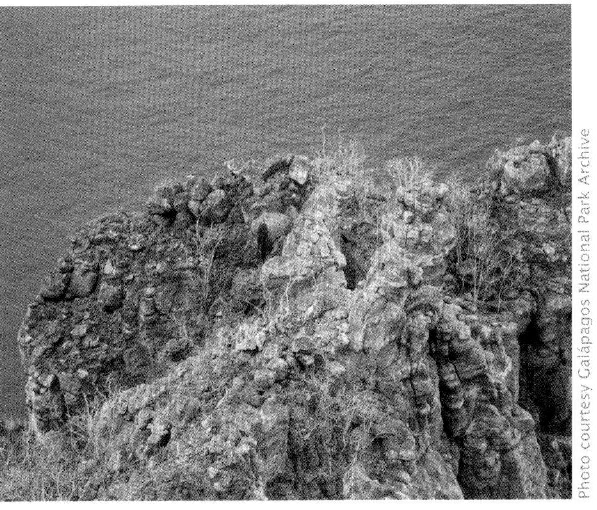

Photo courtesy Galápagos National Park Archive

ABOVE: After tracking the elusive goat all day, the hunters (near top of photo, one in red and one with a blue hat) wend their way back down amid a dense colony of nesting Nazca boobies, disappointed and empty-handed for now. The chartered fishing boat *Oberlus II* (white, at right), serving as support vessel, waits at the anchorage shared with the National Park's larger craft, the *Sierra Negra* conducting a shark research expedition. LEFT AND FAR LEFT: Visible for only a brief instant, the hunters' quarry — a single female goat set ashore by unknown perpetrators (center of photo at left) — takes refuge along narrow cliff ledges contouring the vertiginous cliffs, where the hunters follow her spoor unsuccessfully.

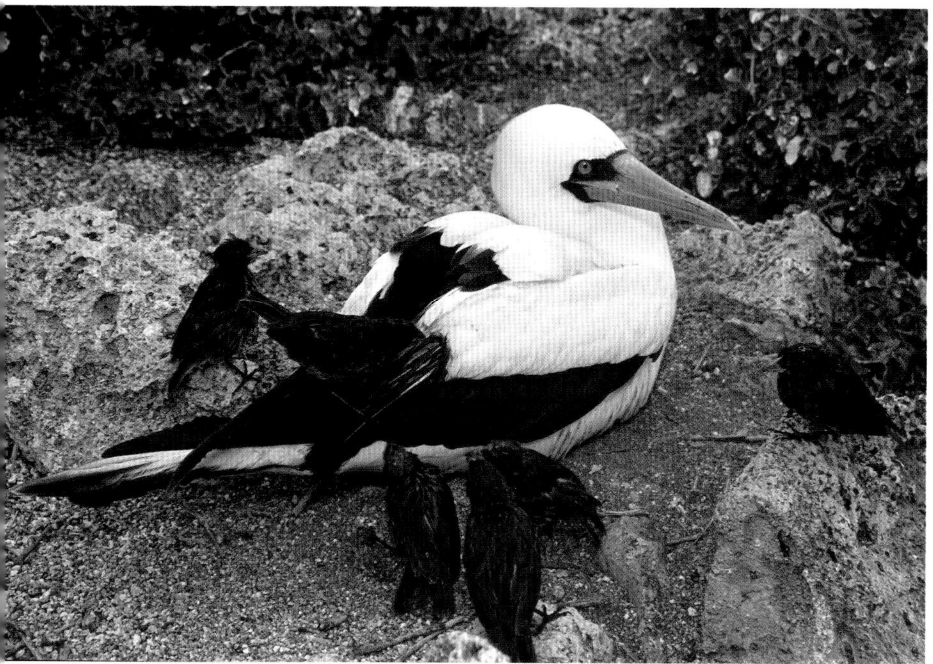

ABOVE: The bloodthirsty habit of vampire finches, who drink booby blood in dry years, is noted as the team crosses the island. BELOW FAR RIGHT: The fast park patrol launch *Guadalupe River*, commanded by Captain Omar Garcia.

take us at least another hour and a half to regain the shore, so we headed to the northern end to complete our tour of the island, and also to avoid the dreaded slippery ascent face. We had to traverse a hair-raising rocky knife-blade some 5 m (16 ft) long and ending in a vertical 2 m (6.5 ft) spur, without the help of a rope or other safety gear. But again, we were committed, and there was simply no turning back. We eventually made our way onto a lower plateau where large numbers of Nazca boobies were sitting on eggs, and here we saw the vampire finches busily taking blood from the boobies' wing feathers, an amazing behavior that occurs only here.

We worked our way down a final steep slope, always cautious to avoid knocking loose rocks onto our companions ahead, and eventually arrived, exhausted and battered, onto a barnacle-covered terrace from where it was possible to leap between crashing waves into the small dinghy that had come around to pick us up.

The next day we were supposed to head back to port, but frustrated by our lack of success, we decided to give it another go. At 6 a.m., before the sun had broken over the horizon, we were already ashore, and by 7:15 we'd made it back up to the top, where we renewed our search for the elusive goat. All day we tracked our quarry intently, urged by an intense determination to come out triumphant in ridding this island of an animal that we know can cause irreparable damage. But time was our enemy, and once again the goat eluded us. That night we undertook the journey home, which would take us almost 40 hours against wind and current, deeply disappointed at our failure. We knew this trip represented an enormous expense for the GNP, not to mention the high risks taken by its staff, all for the sake of a goat set ashore by unknown persons who haven't got the slightest appreciation of what it means to protect this unique environment for the future of humankind.

It took a second expedition, one month later, for my colleagues, using the experience we had gained, to finally eliminate this one goat from Wolf Island. As dedicated park rangers we can now sleep easier knowing the island is safe again, for the time being.

On Patrol: Captain Garcia Recounts

We had just set out on a routine patrolling tour of the western waters of the Galápagos Marine Reserve (GMR) aboard the fast GNP vessel *Guadalupe River* under my command. That evening, at 11 p.m., we picked up several suspicious-looking boats some 50 km (30 miles) offshore of Fernandina Island, well within the protected waters where only certain types of Galápagos fishers are legally permitted to operate. We spent all night observing them, noting much activity with intermittent lights, so just before daybreak we moved in, hoping to catch them all. But as soon as we were seen, the fishing boats began to flee at full speed, leaving behind all their gear. We made several radio calls on VHF channel 16, but there was no response, so we gave chase. About 20 minutes later, we closed in on the nearest vessel and could clearly read its name, *Triunfo*, and its home port, Punta Arenas in Costa Rica. Two more similar boats were about a mile ahead of it, all doing around 10 knots. We continued trying to make radio contact, but it was clear they intended to leave the GMR.

Our onboard Ecuadorian Navy representative offered to open fire, but I suggested he might injure or even kill someone, so I ordered our fire-fighting water cannons switched on instead. My crew awaited orders as we pulled up alongside. Twice I attempted to ram the side of the poachers' hull to make them

Photo courtesy Galápagos National Park Archive

stop, but each time my maneuver was evaded. Only after we aimed our water jets at their engine exhaust stack did the vessel come to a halt. Seeing what had happened, the other two boats also turned around and gave themselves up. We inspected all three ships, then escorted them back to where their longlines were still set. We began pulling these in at 11 a.m. but did not finish the operation until 7:30 the next morning, during which time we were able to rescue a number of sharks that were still alive on the hooks. In the end, we ascertained that each boat had had around 80 km (50 miles) of longline in the water, and we could only imagine how much destruction these lines would have continued causing had we left them behind while delivering the three ships to port authorities where they would be judged.

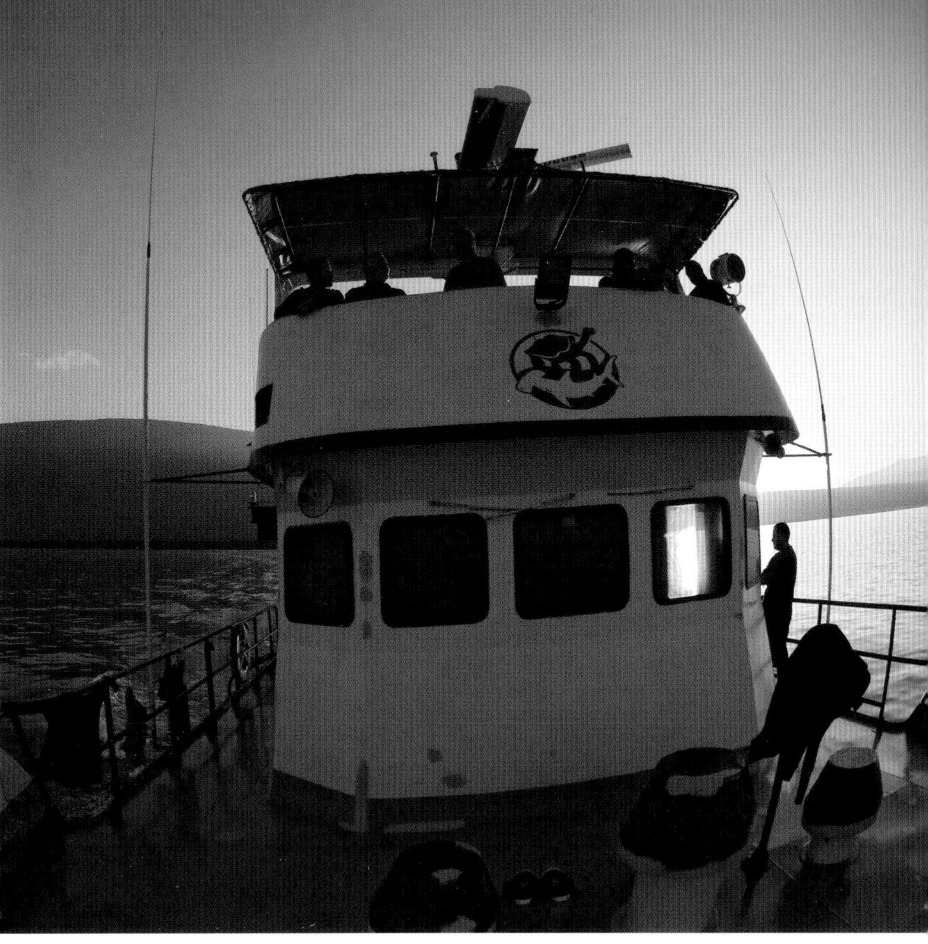

The *Tatiana II*

In my eight years as patrol boat captain, I had many other tense moments, for example the case of the *Tatiana II*, an Ecuadorian flagged vessel. It took us two years of surveillance and covert information gathering before we could catch these poachers in the act.

It was late in the day and we were anchored at Genovesa when our patrol aircraft, the *Sea Wolf*, buzzed us. The radio came alive, 'Poaching vessel sighted north of Pinta Island.' They gave us the coordinates. We knew the *Tatiana II* was illegally taking sharks in the marine reserve but we needed to catch her in the act, otherwise the wily fishermen would simply say they were passing through after operating in waters outside the reserve, more than 65 km (40 miles) offshore. So the GNP pilot had climbed to high elevations where the plane was unlikely to be noticed in the late afternoon sky, until he could spot the vessel lying far below, getting ready for the night's hunt. We weighed anchor immediately but approached the area very slowly under the cover of night, all lights switched off. Soon we had her on the radar, and could detect two smaller tenders working nearby, so we were sure they were setting their longlines. We didn't want to wait much longer as this would mean the death of an additional couple of dozen sharks — a normal night's catch — so we launched our own fast tender and moved in at once.

Now everything went very quickly, using the element of surprise. As part of our cooperation treaty with the Ecuadorian Navy, I sent the armed marine who was working with us along with several of my crew, to swiftly board the mother ship. Opening the hold, immediately they could see one freshly dead shark lying on ice, so I ordered one of my men to

ABOVE: On patrol, the park launch *Sierra Negra* sets out at dawn.
BELOW LEFT AND RIGHT: The poacher *Tatiana II* was intercepted through close air-sea coordination between the park's seaplane (left), relaying timely information to the patrol vessel *Guadalupe River*.

Photo courtesy Galápagos National Park Archive

Photo courtesy Galápagos National Park Archive

ILLICIT WILDLIFE TRAFFIC

GNP Archive

Dried genitals from three bull sea lions, seized after an informant tip-off, worth USD$300 each to a poacher and far more for traffickers.

The two products most persistently poached from the Galápagos Marine Reserve are sharks and large numbers of sea cucumbers, harvested out of season and surpassing the annual quota. All sharks are protected throughout Ecuadorian waters, yet both longlines and occasionally gillnets are set illegally. Often just the fins are sought, sometimes cut from still-living sharks, a practice condemned by the United Nations Food and Agriculture Organization (FAO). Sea cucumbers have not only become seriously depleted, the illegal catch is often processed in secret camps hidden in mangrove areas, with serious associated dangers of introducing alien species to otherwise pristine coastlines. A further attack on marine wildlife is the taking of sea lion genitalia (penises and testicles) for the Asian aphrodisiac market. Prime breeding bulls are targeted for their purported greatest sexual potency, posing great dangers to the reproductive potential of the sea lion population, undermining their social fabric. Little is known about the volume of this trade, which appears to be handled by underground networks often exiting via Peru, but at a going rate of US$300 per animal, incidences are on the rise since the decline of other lucrative catches. In November 2008, an informant led to the arrest of one trafficker, Zhang Youpeng, owner of a Chinese restaurant in Guayaquil. Three dried penises with testicles, along with a stash of sea cucumbers, were seized on the premises.

Godfrey Merlen

stand on the hatch while the captain was brought over for questioning. When I confronted him he tried all sorts of tricks to soften me up and come to an 'agreement.' First, he broke down in tears, said he only had three sharks in the hold and asked me to just allow him to toss them overboard and forget the whole thing. Then he admitted there were actually 65 sharks on ice and offered me money, even proposing to provide me with a tuna fishing vessel of my own. When I refused, he became sullen and asked me for permission to return to command his ship as we escorted it back to port.

Meanwhile, my other crew were tracking down the tenders manning the longlines. They were very nervous as they moved in unarmed in pitch darkness (park rangers are not sanctioned to carry firearms), remembering one of their colleagues a few years ago who was grievously wounded by gunshot while approaching an illicit sea cucumber fishing camp. As my men now closed in on their quarry, suddenly one of the fishers revved up and bore down on their inflatable dinghy, certainly no match for the fast fiberglass launch used by the poachers. But one of my crew had taken it upon himself to slip his own licensed handgun in his pocket, so he fired in the air and the collision was averted. However, they abandoned their lines in the darkness so we were unable to retrieve them and stop the baited hooks killing more sharks. Later, we found they had a firearm on board but as we caught them by surprise they hadn't had time to dig it out.

By now dawn was breaking and the armed navy guard called over the radio saying there was trouble

Photo courtesy Galápagos National Park Archive

RIGHT: The *Tatiana II*'s sickening haul consisted of several dozen shark carcasses, laid out on the Baltra dock the day after her capture. Taking so many large predators is like beheading the marine ecosystem, yet the *Tatiana II*, like many other captured vessels, got away lightly.

Photos courtesy Galápagos National Park Archive

aboard *Tatiana II*, with her captain running round trying to open the sea cocks and scuttle his ship. Finally, we got everything under control and later that day handed over our prize to the navy authorities at the nearest port, Baltra, for due process.

Disappointments

Sadly, on many occasions our country's legal system has let us down. The GNP's mandate goes only as far as the administrative process — the capture and gathering of evidence. Almost invariably, district judges refuse to comprehend the gravity of environmental crimes, and they sympathize with fishers, even those flagrantly acting against the interests of the Galápagos community. Instead of prison terms and confiscation of vessels, ridiculously low fines are applied. Political expediency or curried favors at high level may play a part too. The most common scenarios rely on blurred evidence: captains drop their lines and claim they were only passing through the GMR after fishing in international waters, or that they were experiencing engine trouble and were taking shelter. In the case of the *Triunfo*, it took only two days before it sailed back to Costa Rica, having evidently lost no more than its illicit catch of the trip. *Tatiana II*'s crew were released on bail and within a few weeks this ship also sailed away, another case of some higher authority ruling in favor of the fishermen rather than stringently protecting nature.

Some years ago we brought in another Costa Rican vessel, which was held in harbor while the court case dragged on for months. We had the fuel tanks drained and the propellers chained, just in case. Then one morning the vessel was simply gone. Clearly some local divers had been paid off to cut the chains and smuggle fuel aboard under cover of darkness. By the time our air patrol tried to give chase, the poachers had slipped out of Galápagos waters.

We've had some successes, with due punishments carried out — prison terms and ships sold off at auction — but too often it seems the process is derailed and the poachers get away lightly. And now some foreign operators are becoming shrewd, keeping the mother vessel outside reserve waters and sending in only their fast tenders to gather up their illegal catch — we've caught a few, but they are swift and small, so very hard to spot and pursue. It can be really disheartening, but we must keep on trying, because the future of these rich waters depends on us.

ABOVE LEFT: A pile of dried fins confiscated by National Park guards at the airport represents harrowing numbers of dead sharks.
ABOVE RIGHT: The author, second from right, helps his park colleagues and marine biologists tally a haul of illegally caught dried sea cucumbers.

Photo courtesy Galápagos National Park Archive

Photo courtesy Galápagos National Park Archive

FAR LEFT: A fleet of captured speedboats, used to make rapid forays into the GMR by foreign vessels approaching from offshore waters, are lined up at the GNP dock in Puerto Ayora.
LEFT: Another poacher's load is dumped at sea.

Executive Director,
Charles Darwin Foundation,
Puerto Ayora, Santa Cruz,
Galápagos, Ecuador.
www.darwinfoundation.org

A Perspective on People and the Future

The Search for Harmony

Graham Watkins

Dr. Graham Watkins holds a degree in Zoology from St. Catherine's College, Oxford, and a Ph.D. in Ecology and Evolution from the University of Pennsylvania. His professional experience includes ecological research, collaborative wildlife and fisheries management, and sustainable enterprise development in aquaculture, fisheries and tourism in tropical South America, where he has worked for over 20 years, including the position of Director General of the Iwokrama International Center for Rain Forest Conservation and Development in Guyana. His involvement with Galápagos spans the guide-training course in 1987 and his recent tenure as Executive Director of the Charles Darwin Foundation (CDF) from 2005 to 2009.

THE GALÁPAGOS ISLANDS ARE a paragon of scientific research and conservation. The islands are also of interest to tourists, tourism businesses, fishers, the military, colonists and various governments. Since Charles Darwin's visit in 1835, the islands have been a focus for biophysical research, particularly in natural history, evolution, climate and geology [Larson, 2001]. However, because of the relatively small and comparatively new human presence in the islands, until recently they attracted attention from only a comparative smattering of anthropologists, social scientists and economists. But as the human population and both international and national interests in the islands have exploded in recent years, so they have come under the microscope of social and cultural researchers [Grenier, 2007; Ospina and Falconí, 2007]. As a microcosm of socio-economic issues, Galápagos has become an important prospective source for lessons to be learned in conflict management, and could potentially acquire even greater importance in the development of sustainable models for the world. The ideas outlined here stem from knowledge gained during my four-year tenure as executive director of the CDF.

Although, so far, socioeconomic studies have been rare, there is growing interest in analysis and discussion of the human issues in the islands. The *Economist* has featured several articles on the subject, including a 2008 evaluation of the viability of the predominant economic Galápagos paradigm based on a recent paper by J. Edward Taylor et al (2008). Tourism is the major determinant of change in the islands [Grenier, 2007; Taylor et al, 2008; Watkins and Cruz, 2007]. In 2006, an estimated USD$63 million of the $418 million Galápagos tourism business entered the local economy — equivalent to over 60% of the total island economy [Charles Darwin Foundation et al, 2008]. Over 161,000 visitors came to the islands in

BELOW LEFT: The tracks of a morning jogger and a marine iguana heading out to feed symbolize the crossroads of people and wildlife in Galápagos. BELOW RIGHT: As the shoreline of Puerto Ayora is gradually concreted over, marine iguanas refuse to cede territory, adapting to a strange new landscape.

LEFT: Continuous public education aimed at harmonizing human behavior with the needs of Galápagos wildlife allows a bevy of animals to frequent the local fishing docks fairly peaceably, perhaps glossing over deeper ecological clashes.
ABOVE: The ever-increasing needs of local people and tourist industry alike result in a constant flow of goods — in this case beer — arriving from the mainland, each load with its attendant risk of introducing more foreign pest species.
LEFT: A waved albatross sculpture created by local artists presides over a busy harbor, reminding its people of their special connection with such unique species.

2007 — an increase of 11% compared to 2006. Such growth results from three factors: a strong market, strong outside investment interests and strong local entrepreneurial spirit. Galápagos tourism data has demonstrated current annual growth of over 14% — outpacing growth rates of almost any other place in the world [Epler, 2007; Taylor et al, 2008; Taylor et al, 2006]. Visitation quadrupled between 1991 and 2007, and likewise hotels, restaurants and bars. In the five years from 2001 to 2006 the number of daily flights to the islands doubled from 15 to 30 [CDF et al, 2008].

Past industries in Galápagos have exhibited the same pattern of rapid expansion that we are seeing in tourism: whaling, fur sealing and the sea cucumber fisheries all grew rapidly before collapsing rather ignominiously. This 'boom and bust' economy is typical of islands where strong markets and outside investment drive rapid growth followed by resource collapse or market shifts.

Importantly, tourism is the basis for the many emerging secondary businesses in the areas of transportation, artisanal crafts, construction and local commerce. In 2006, employment was distributed in public administration and education (22%), transport and communications (17%), commerce (12%), agriculture (9%), construction (8%), hotels and restaurants (7%), community services (6%), manufacturing (5%), domestic services (4%) and fishing (4%). Furthermore, the percentages employed in transport, hotels and restaurants are directly linked to tourism, and many of the other employment opportunities are based on money likewise generated from tourism, showing that this is clearly the engine of growth in Galápagos. With its associated value chains, a major challenge in the islands is the management of the linked consequences of this rapid economic expansion [CDF et al, 2008; Epler, 2007; Taylor et al, 2008; Watkins and Cruz, 2007].

ABOVE: The bustling Saturday market in Puerto Ayora throbs with shopping activity, where large amounts of fresh goods — most of them flown in from mainland Ecuador — are sold, a dangerous portal for biological hitchhikers in spite of quarantine efforts by the public health authorities.
RIGHT: More and more traffic clogs the busy streets and short roads of Santa Cruz.

Economic studies also suggest that the distribution of the financial benefits from tourism is not structured to increase local equity [Taylor et al, 2008; Watkins and Cruz, 2007].

Larger outside operators frequently outcompete local small-business owners because they can more readily access capital, trained staff and markets, raising concerns that the present tourism model will lead to a continuing decline in the proportion of locally owned enterprises. There is also poor management of the urban consequences of financial flows from employment and tourism purchasing power down to the average household.

In addition, several economists have questioned the net financial flows from tourism to government, with some observing that the Government of Ecuador and the international community appear to be subsidizing highly lucrative tourism businesses. The overall contribution of tourism to the maintenance of the resource on which the industry depends is relatively low compared to what might be considered sustainable tourism enterprises [Epler, 1993; Epler, 2007]. Similarly, in fisheries and other businesses, government subsidies generate a real negative, or inverse, flow from the business to the state [CDF et al, 2008; Kerr et al, 2004; Watkins and Cruz, 2007].

The overall contribution from fisheries to the local economy (estimated in 2006 as less than $3 million) is lower than the contribution to the economy arising from research and conservation (estimated as $5.8 million in 2006) [Watkins and Cruz, 2007].

The ideal of economic growth, with trickle-down benefits and investment to limit or control consequent ecological damages, has proven not to be sustainable in Galápagos [Taylor et al, 2008]. As the economy continues to expand, social and ecological risks increase disproportionately — mainly through the arrival of invasive species due to massive transport of people and goods, plus increased pollution and additional pressures on natural resources, and habitat loss due to ramped up infrastructure. These risks are compounded by the high dependence on tourism associated with a shift toward high volume, low per capita value markets, driven by open competition with poorly focused legislative controls [Blanton, 2006; Plog, 1974]. This shift, and its potentially catastrophic consequences for the local environment, has been experienced in many tourism destinations around the world [Butler, 2006]; it is worrying that tourism product cycles have not been fully considered in the planning of the industry in Galápagos. Economic analysis and planning needs to focus on improving equity in financial flows, enhancing urban business planning, and ensuring the complete integration of economic, social and ecological issues into how we view the future of Galápagos.

All islands in the world are susceptible to outside influences from invasive species, pollution and the extraction of natural resources for international markets. As the economy of Galápagos has grown, social complexity has increased and the islands have become more 'open' to the rest of the world (see also chapter: Merlen). This process began in the 1600s, when an extractive economy first began to grow through the British privateers and later through British and American whalers and fur sealers. The difference now — besides the fact that back then there was no resident population, thus the economy was a transient one — is that recent growth and 'opening' has been exponential. This brings more and more players and economic interests into the Galápagos arena, including individuals seeking to capitalize on rarity markets for shark fins, sea cucumbers and sea lion penises, new educational, conservation and research organizations, and global real estate speculation, to name but a few.

As the economy, the local population and these varied interests have grown concurrently, it is not surprising that conflicts have also developed; the heterogeneity of the mainly migrant Galápagos 'community' compounds frictions. Substantial research and management effort has gone into how to resolve the disputes among the 'users' of the Galápagos and this, in turn, has contributed to the institutionalizing of numerous sector-based groups (associations, cooperatives and foundations) and the

CONSTITUTIONAL RIGHTS FOR NATURE

Following a referendum to approve a new constitution, in 2008 Ecuador became the first country in the world to enshrine the inalienable Rights of Nature in law. Five articles acknowledge rights possessed by Nature, referred to as the goddess Pachamama, whose name roughly translates to Mother Earth or Mother Universe, revered by indigenous Andean peoples. Some of the legally enforceable concepts include:

- **Granting Nature the right to exist, persist, maintain and regenerate its vital cycles, structure, functions and evolutionary processes.**
- **Promoting respect toward all elements that form an ecosystem.**
- **Confirming that the State will apply precautionary restrictions to all activities that may lead to the extinction of species, the destruction of ecosystems or the permanent alteration of natural cycles.**
- **Prohibiting the introduction of organisms and materials — organic or inorganic — that might definitively alter the national genetic patrimony, for the benefit of all people, communities and nationalities whose well-being depends on the environment and its natural wealth.**
- **Ensuring that environmental services cannot be appropriated; and that all use and exploitation of the environment will be regulated by the State.**

Source: Galápagos Conservation Trust

BELOW LEFT AND MIDDLE: Consequences of increasing human activity in Galápagos are notable in mounting coastal pollution (left), and marine life exploitation such as unsustainable harvesting of native and endemic molluscs (right) for 'cebiche de canchalagua,' a favorite seafood marinade served in local restaurants.

ABOVE: The beautiful *Chiton sulcatus* is one of the species targeted for consumption.

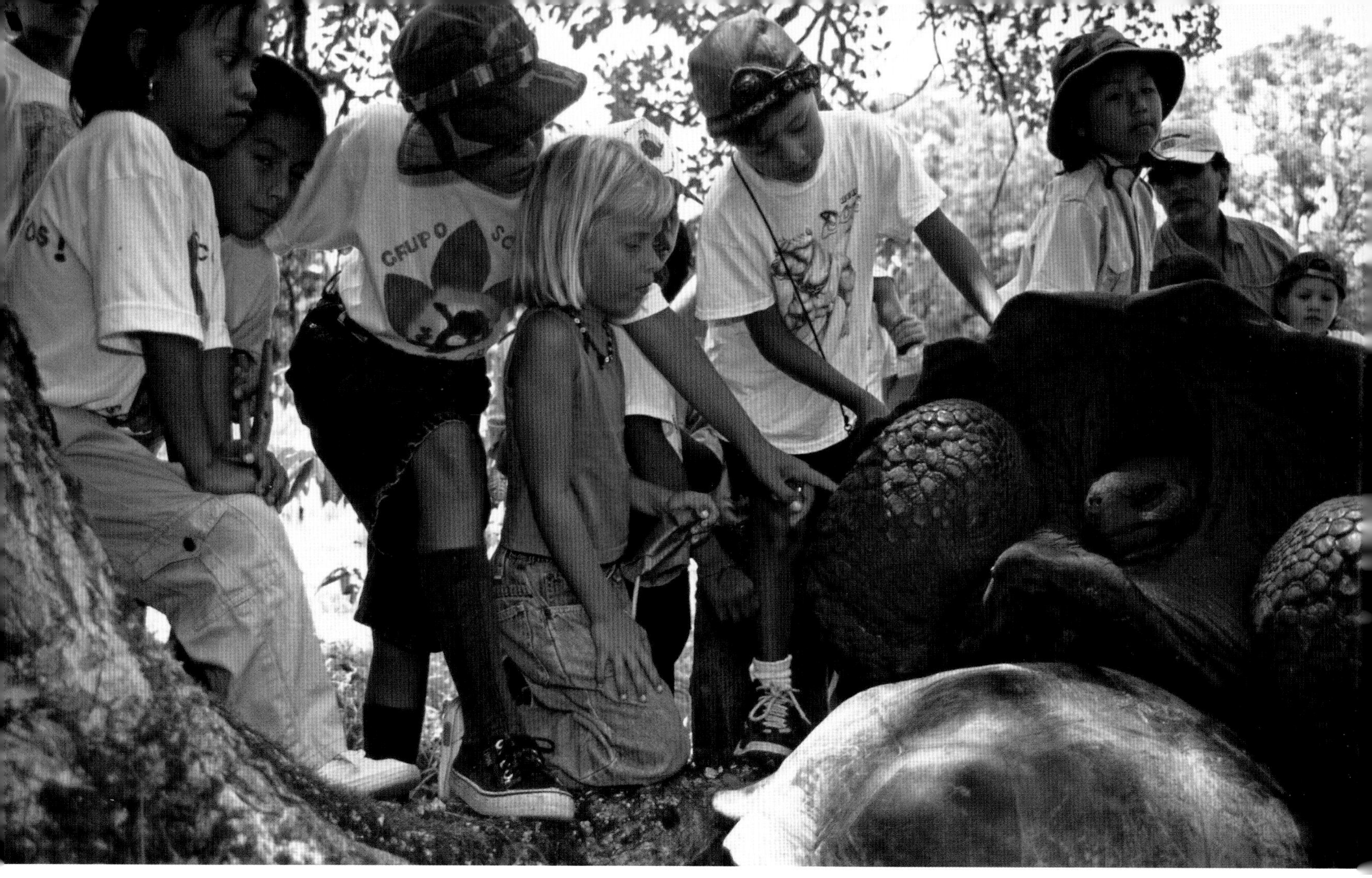

ABOVE: Schoolchildren on an environmental day-trip are fascinated by mating giant tortoises in the Santa Cruz highlands.
RIGHT: A local dance troupe performs shows with conservation themes for visiting tourists.
BELOW: Volunteers help replant *Scalesia* seedlings.

slow process of building a sense of community in the archipelago. The fishing, tourism, conservation, transport, agricultural, commercial, youth and professional sectors have all become increasingly organized, with over 180 civil society organizations of varying capacities now registered in the Galápagos. In addition, the INGALA Council (the National Galápagos Institute, a semi-autonomous central government institution in charge of planning and immigration control) and the Participatory Management Board for the Galápagos Marine Reserve both have established legal fora in which distinct stakeholder groups can resolve differences and establish agreement by consensus [Gobierno de la República del Ecuador: Presidencia de la República, 1998].

As of the early 1990s, a growing sense of a Galápagos 'community' has begun to emerge, a concept obviously very difficult to construct in a population made up mainly of recent immigrants and transient expatriates. Such a Galápagos community identity will be a key element in creating a sustainable future for the islands. Indeed, this community will need to focus on building an island culture and strengthening its social capital by developing conflict resolution mechanisms, building trust and improving internal networking. A strong social fabric is a critical element in the future of a sustainable society in the islands. Without this element, and an effective organization of existing and future interest groups, leadership and direction will always be lacking. In the absence of a shared Galápagos vision, it is more than likely that particular interests will gain space, leading to increasing conflict and dissatisfaction among minority or less powerful groups, and an increasing disintegration of social cohesion.

Some of the first steps toward establishing the required conditions for sustainability were taken with the approval of the 1998 Special Law for Galápagos. This law established local involvement in decision-

GECKO WARS

Invasive gecko species include the mourning gecko, *Lepidodactylus lugubris* (left), and a leaf-toed gecko, *Phyllodactylus tuberculosus.*

Galápagos geckoes consist of six endemic species in the genus *Phyllodactylus*, one of which is found on most central islands, and the other five distributed on more isolated outliers around the edge of the archipelago, with one species on the northern outpost of Wolf Island. In recent years, three additional geckoes have been introduced with cargo coming from mainland Ecuador. These have spread quickly on inhabited islands, particularly the invasive mourning gecko, *Lepidodactylus lugubris*, which breeds very rapidly and is now found in vast numbers in and around houses and other manmade structures. It has completely displaced the shyer natives in this environment, though in wilder habitat the latter seem to be holding their own.

Tui De Roy

The Wenman island endemic gecko, *Phyllodactylus gilberti.*

making, regulated immigration, limited growth, increased local equity, and mandated educational reform and mechanisms to reduce the ecological impacts of growth. Somewhat dormant for nearly a decade, work on critical elements, including reorganizing public administration and leadership, educational reform and integrated urban planning began in earnest during 2007 and 2008.

Culturally, Galápagos has undergone enormous change during its brief human history. Left untouched for nearly two centuries after their official 'discovery' by Tomás de Berlanga in 1535, Galápagos history is a fascinating tapestry involving privateers, whalers, fishers, colonists, prisoners, military and tourism entrepreneurs [Hickman, 1985], all crammed within just the last three centuries. At different times throughout this short encounter with humanity, the Galápagos Islands have been 'captured' by singular interests; in some sense, they have suffered as an international commons. The establishment of an Ecuadorian claim and presence in 1832 began a long process to regulate these commons. The arrival of Charles Darwin in 1835 and the subsequent stream of research work that followed the publication of *On the Origin of Species* in 1859 established a link second to none between Darwin and the Galápagos [Larson, 2001]. Within the islands, the early Galápagos 'pioneers' were the basis for a special culture of hardship and enterprise that has been diluted by later arrivals during the 'gold rush' economic development that began in the 1980s. Today, as the recent and older arrivals settle side by side, a unique and vibrant local culture is slowly growing. For the future, a sense of Galápagos identity will develop provided the arrival of new immigrants can be limited and this identity can be 'stamped' rapidly on any such new blood. Obviously, this identity will evolve over time, but will build on key aspects of local culture: entrepreneurship, a sense of adventure, an appreciation for solitude, the environment and service, and hopefully trust. It is possible that this cultural identity, when developed, could serve as a model for the rest of the world.

The magic of the Galápagos Islands extends well beyond its extraordinary natural treasures and

ABOVE FAR LEFT: Tourists and wildlife show their mutual interest.
BELOW: In the 1960s, cattle egrets colonized naturally as farmland expanded.

ALIEN INVADERS

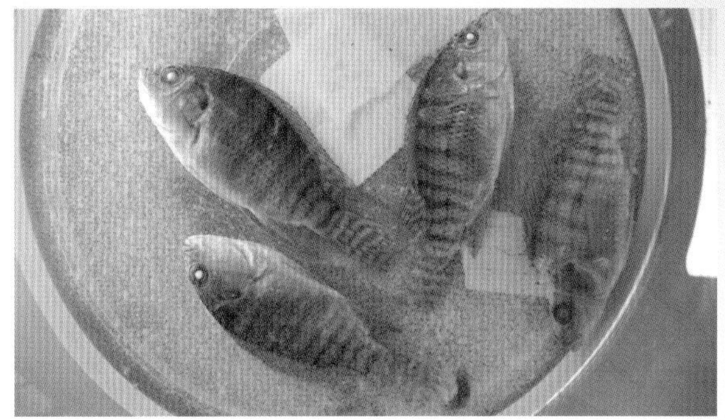

Introduced tree frogs (left) recently collected by the GNP in the Santa Cruz highlands, and young tilapia from El Junco Lake on San Cristóbal Island.

With more and more human traffic between Galápagos and mainland Ecuador, the rate at which new invasive species are being introduced continues to escalate. The majority are insidious insects and other microorganisms, but even alien vertebrates are making it to Galápagos undetected. In the late 1990s the tree frog *Scinax quinquefasciatus* appeared at Puerto Villamil on Isabela, though its means of arrival was not tracked. The 1997/98 El Niño event helped it spread and, being rather salt tolerant, by the time the rains stopped it was well established throughout the extensive system of brackish water pools around this village — a RAMSAR site classifying the area as a wetland of global significance. Within a few years it had reached the highlands of Santa Cruz (where the frogs, above left, were collected), although again how it jumped islands is not known. In 2006 came the discovery that tilapia had been intentionally introduced to El Junco Lake on San Cristóbal (where the young tilapia, above right, were found), though the culprit was never caught. Since neither amphibians nor freshwater fish existed naturally in Galápagos, the potential for such species dramatically altering the ecosystem is very high. Fortunately, the National Park was able to quickly eliminate the tilapia with the use of rotenone, a natural plant extract used by Amazon Indians to selectively catch river fish, as it does not affect other life forms. Frog control, however, remains unresolved.
Tui De Roy

includes the outstanding people who inhabit the archipelago. As the economy and population have grown in the islands, the social, economic and cultural nature of the archipelago has changed and become more complex.

In 2007, both the Government of Ecuador and UNESCO declared that the islands were at risk resulting from concerns about weak planning, rampant development, increased immigration, weak organizations and the need for educational reform. There is a growing call for detailed information on these aspects to couple with the biological knowledge that has been available over the years. Increasingly, we will need to integrate our understanding of both the natural and human sides of Galápagos; this holistic approach will provide us with critical lessons and help us more effectively plan for the future. Tomorrow's conservation and sustainable development will depend on simplifying and clarifying governance, fundamentally reforming and strengthening key organizations, further building social capital and ensuring sustainable enterprise development.

The failure to implement sweeping changes while continuing business as usual will result in a continually growing economy, increasing human population, ongoing social and cultural disintegration, and ultimately the demise of this extraordinary archipelago that has inspired so many. My first experience with the outstanding wildlife and landscapes of these islands in the late 1980s drives my interest in the social and cultural dynamics of Galápagos. A biologist by training, I have grown to understand that we need to comprehend the forces that drive the whole ecosystem, including people, if we are to manage such very special areas of the world. As part of this process, I hope we will be able to attract deeper interest in understanding the social, economic and cultural aspects of Galápagos as we strive to build a sustainable future.

RIGHT: A barn owl killed while preying on an introduced ship rat demonstrates the problems of increased human activities, including rising road kills of many species, from lizards to herons.

Reflections

'Noe reall Islands . . .', But Paradise

Godfrey Merlen

Isla Santa Cruz,
Galápagos,
Ecuador.
<merlenway@gmail.com>

Godfrey Merlen is a long-standing Galápagos resident and naturalist. His involvement with the Galápagos National Park Service (GNPS), where he concentrated his efforts in building a patrol and vigilance fleet, has given him deep insight into the issues that threaten the Islands. Working with visiting scientists, he has investigated the whale and dolphin populations around the archipelago, and in recent years has dedicated time to seeking solutions to the ever-present threat of introduced organisms that he sees as the principal menace to the fauna and flora of Galápagos.

WHEN I FIRST ARRIVED IN the port city of Guayaquil, Ecuador, in 1970 and cast my eyes to the west I was cautioned to search elsewhere for Paradise. It was almost as though the warnings given centuries ago to Spanish travelers, that there were 'but shadowes and noe reall Islands,' still endured in the dying years of the second millennium. Indeed, three decades ago, the Galápagos Islands still seemed lost in the sea of the setting sun, unloved, drifting in another time, wonderful to the imagination, feared by man. Biologically, to remain lost was their saving, and the title of Beebe's 1924 book, *World's End*, was eminently fitting. One thousand kilometers distant under a tropical sun, their essence is a mixture of baking lava and hair-clinging moisture from drifting fogs, the sea rank with seaweed smell.

Yet all distances are relative — and I have come to reflect on this, not in geographical terms but biological ones. It is true that, geographically, the islands are still fundamentally 1000 km (600 miles) from the continental coast, even accepting the 35 m (115-ft) tectonic shift of the Nazca Plate toward the east over the last 500 years since the archipelago's discovery. But in the present world what is much more relevant is their *biological* distance: how easy it is for organisms, including humans, to arrive and potentially establish themselves on these wild shores.

One thousand kilometres allows only waifs and strays to locate the islands by chance; zero kilometres removes all barriers to arrival. There is no question that it was the isolation of Galápagos that caused unique and fascinating plants and animals to evolve here. Cut off from their ancestral populations, their genetic quotient has been molded by adaptive processes epitomized in Darwin's staggering and enquiring remark, 'the first appearance of new beings on this earth.'

But there is a real and present danger in depending on isolation as a condition of life. Organisms become naive: tameness in the absence of predators is common, and immune systems relax without the pressure of constant bombardment by diseases kept away by distance. In other words, these curious and rare organisms have no resistance to the rough and tumble existence which is the norm on continents, where multitudinous species constantly interact and play out the survival game. Isolation *is* the defense of oceanic island species. Within it, their Paradise is secure; without it, their Paradise is lost.

Today, we are losing Paradise — so what has changed? I dare to suggest that the fundamental

BELOW: In a scene from the dawn of time, sunrise over the western shores of Fernandina reminds us that there is still so much that could be either lost or saved in Galápagos.

ABOVE: Only a short walk from the busy town of Puerto Ayora, the space and natural simplicity of Turtuga Bay's immaculate beach replenish the soul. BELOW LEFT AND RIGHT: In the free spirits of a mother sea lion playing with her pup, or the vulnerability of a swallow-tailed gull chick snuggling under the tail of its parent, lies the essence of Galápagos wildness.

reason is quite simple: the islands' biological distance is nowhere near 1000 km anymore. In terms of biological access, the distance is now a mere 100 km, or perhaps even less, effectively reduced to one-tenth of the original barrier of ocean, wind and time.

What could cause such a rapid and dangerous contraction to occur — a metaphorical decrease that has placed the blue-eyed flightless cormorant, the engaging Darwin's finches, the shiny-shelled tortoises and the bobbing daisy-headed *Scalesias* in the inescapable path of the 'invading hordes' from the east? It is humanity and its actions, our untidy lifestyle, and our huge propensity to travel, through our burgeoning technology, to the furthest places on Earth. The invention of sails and, much more particularly, the emergence of engines for ships and airplanes, give easy access to remote places such as Galápagos, and this is now pushing the extraordinary results of millions of years of evolution to the brink of annihilation. Man is *the* invasive species. The world over, few — if any — places remain free from his influence which, generally, has been systematically disastrous to natural ecosystems.

There is only one solution for Galápagos: to push back the ever-shrinking biological distance to its original 1000 km. I emphasize that this is *the* only answer. No play of words, no 'Well ...' or 'But ...' to interject, and no half-open doors to pacify human demands, will suffice. And it is definitely not by introducing new terminology, such as the recently coined notion of a human-Galápagos ecosystem, that we will win the day, for it is exactly by breaking the isolation of the islands that our species has created the most grave issues threatening the

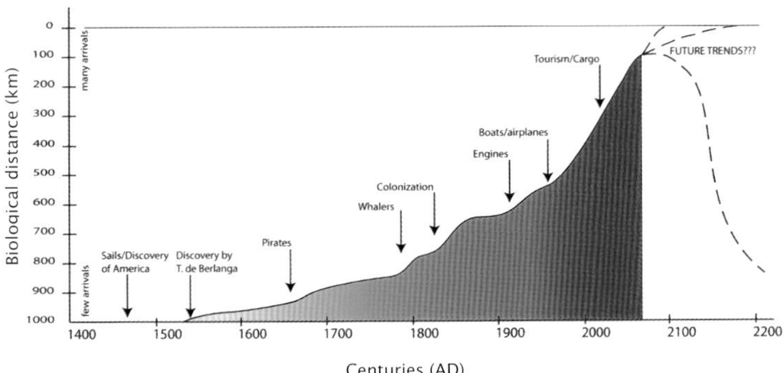

future of this insular environment. If we erroneously recognize human intervention to be a natural process and allow it to insinuate itself into the prehuman Galápagos ecosystem, we must also accept the ecological holocaust following in Man's wake as a natural consequence of this shift. In that scenario, the introduction of diseases, feral animals and commercial exploitation — and the eventual stamping out of all original biological balances — must all be seen as part of that 'natural' sequence too! Few are those who would readily accept such a notion.

So to achieve our objective, to re-create the ancient natural barriers and safeguards of yesteryear, how do we proceed? Moreover, is there evidence for this reduction of the biological distance, or is it merely theoretical? A few examples help to elucidate the present situation and demonstrate that small, yet important organisms, now easily cross the previously impossible distance, increasingly threatening the integrity of the islands' biota.

Take the botfly, *Philornis downsi*, which in its larval stage is an obligate small-bird parasite. It was first collected in 1964, and by 1997 was shown to be causing an alarming mortality of nestlings, laying its eggs in the host nests, where the larvae feed directly on the chicks (see also chapter: Parker). This fly has been found in the nests of 18 species of terrestrial birds, including 11 species of Darwin's finches, and recent studies have detected its presence on 13 islands. A terrifying 97% of nests examined contained evidence of *Philornis* fly larvae, with a staggering chick mortality rate of 76%. Birgit Fessl, a Charles Darwin Research Station scientist working with Darwin's finches, clearly sees the danger and emphasizes the 'extremely serious threat this parasite poses for the endemic passerine fauna of the Galápagos Islands' (2006).

Among many other insects that have arrived by such means, the fly was no doubt introduced from a mainland cargo vessel or perhaps, even 44 years ago, brought over accidentally by aircraft. For years it seems it was constrained, unnoticed, to just one or two islands. However, recent surveys show this insidious fly is regularly attracted to

DARWIN'S INSPIRATION IN THE BALANCE

San Cristobal mockingbird, *Mimus melanotis*

Española mockingbird, *Mimus macdonaldi*

Floreana mockingbird, *Mimus trifasciatus*

Galápagos mockingbird, *Mimus parvulus*

Contrary to popular belief, it was the four species of mockingbirds that first stirred Darwin's thinking on evolution. During his five weeks in Galápagos, he pondered their subtle differences and made copious notes on the matter — the finches came to his attention only after the 13 species were described by John Gould upon his return to England. But today, the mockingbirds, like the finches, face a growing barrage of alien threats, including the ravages of avian pox (see chapter Parker) and predation by cats. The Floreana mockingbird is gone from the island where Darwin collected his specimens, surviving only on tiny offshore islets (see chapter Steadman), and the San Cristóbal species is no longer abundant. Fortunately, no major threats have yet invaded Española, and the more widespread Galápagos mockingbird is still doing well on several islands. The challenge is to halt, or even reverse, the tide of invasive species.

Tui De Roy

A fledgling succumbs to the ravages of avian pox.
Photo: Godfrey Merlen

ABOVE FAR LEFT: A conceptual graph paints the manmade breakdown of the biological time barrier.
LEFT: A young visitor tries a giant tortoise on for size.

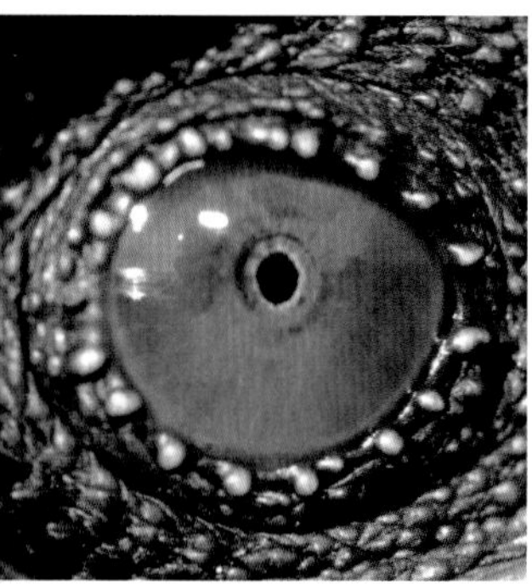

ABOVE: To look into the cobalt eye of a flightless cormorant is to peer into another world.
BELOW: The cycle of the seasons turns slowly on Seymour Island.

the lights of tour vessels that navigate extensively throughout the archipelago. This bears proof that even within the island group there exists an extremely efficient transport system for insects, that expanded with accelerating speed and efficacy as the passenger fleet grew (see also chapter: Watkins). Possibly by now *Philornis* may have reached the whole archipelago, including the two most isolated islands, Wolf and Darwin, where the unique blood-drinking vampire finches live. In an article published in *Noticias de Galápagos* 30 years ago, Robert Silberglied of Harvard University pointed out these dangers. We now reap the reward for ignoring this early warning!

Another graphic case is that of *Culex quinquefasciatus*, a mosquito species which arrived in Galápagos in 1985 (see also chapter: Parker). It is a known vector for avian pox, avian malaria and West Nile virus, the latter recently introduced into the New World and known to cause mortality in species occurring in Galápagos. This mosquito is now found on all the human-inhabited islands: Santa Cruz, San Cristóbal, Isabela and Floreana. Its distribution was thought to be related to human habitations, yet it has recently been identified from Caleta Iguana, Islas Marielas and La Muñeca on Isabela, all sites remote from human settlements, the first two being prime penguin breeding areas. Concurrently, recent research shows a strain of malaria to be present in some Galápagos penguins, with as yet unknown consequences.

With each new somber discovery many new questions are raised. What is the nature of this malaria strain? How dangerous is it for the penguins? Is *Culex* the vector? Urgent work is in progress, led by Patty Parker from the University of Missouri-St. Louis, to elucidate the present worrying situation.

Neither is looking ahead any more reassuring. Standing fresh water, essential for the breeding cycle of this mosquito, is normally in short supply along the generally arid coastlines, but the 'bigger picture' of a world undergoing global climate change suggests that more rain will fall. What are the implications? More freshwater ponds forming? An escalating population of disease-carrying insects? Reduced immune response among native species stressed by dwindling food supplies in a gradually heating world?

In real-world analogies, diseases transmitted by introduced mosquitoes to Hawaii are responsible for the extinction of many terrestrial birds, representing the principal cause of the disappearance of 50% of this island chain's avifauna. The mosquitoes arrived on boats. Likewise, the introduction of the brown tree snake to the island of Guam in the 1940s, suspected to be by cargo ship, has led to the extirpation of nine of 11 native forest birds, including five endemic species.

The problems span the animal and plant kingdoms. Galápagos now has far more introduced plant species than native ones. The latest figures show a ratio of 800 to 500! Some of these are highly invasive and are destroying native habitat and pushing endemic plants and birds to extinction. *Scalesia* forests are overrun by hill blackberry (*Rubus niveus*), likewise endemic *Miconia* by the quinine tree, *Cinchona* (see also chapters: McMullen; Hamann; Tye).

Unfortunately, the theory of reduced biological distance has become an inescapable fact. More airplanes and cargo vessels arrive each year, fueled by both an ever-increasing number of tourists and local communities growing accustomed to a wealthier lifestyle. It is this economic development that is placing the gravest risk on Galápagos biodiversity. What to do? First and foremost must come a clear recognition that it is man alone who is the prime invasive species and the root cause of these alarming changes. It is not a process directed by prehuman nature, whose evolutionary 'technology' achieved the astounding ecosystems of Galápagos over many millennia, and whose natural limitations preserved the isolation of the islands. Secondly, we have to know, and believe, that we possess the power to recreate this vital biological isolation. Our technology has got us into this dangerous situation, let our creativity now work for us to reverse direction. Let us rebuild an effective barrier equivalent to those thousand kilometers of open ocean.

For decades, aircraft have been recognized the world over as an extremely efficient transport mode for diseases and their insect vectors. The flights to Galápagos are no exception, and monitoring has shown that mosquitoes, flies and grasshoppers are common hitchhikers, easily surviving the 90-minute flight. Yet, once in the air, an aircraft is a sealed tube. Surely these insects could be eliminated before they escape into the vulnerable island environment. Careful inspection of all baggage is essential, and this is indeed being carried out by Agrocalidad-SICGAL, the Ministry of Agriculture Inspection and Quarantine System. But alone this is not sufficient. The fumigation of holds and cabins is also absolutely vital. But then again, who would want to be fumigated? Not I! Yet who wants to be responsible for the potential destruction of Galápagos penguins through the introduction of West Nile virus? Categorically not I!

So, recently, I decided that this would be a good place to start. Working closely with the Civil Aviation Authority, who control all movements of aircraft, and the technicians of TAME and Aerogal, the two commercial airlines flying to Galápagos, I found that

Photo courtesy Godfrey Merlen

ABOVE: A critically endangered Galápagos petrel chick prepares to fledge from its nesting burrow beneath the dripping canopy of native *Miconia robinsoniana*, thanks to intensive National Park control of rats and other aggressive species threatening the area, including invasive alien plants.

LEFT: The hallmark of evolutionary processes in Galápagos, a brood of Darwin's finches preparing to leave their nest confronts an uncertain future in the face of introduced diseases, parasites and a changing habitat.

ENDURING PUZZLE OF THE FERNANDINA TORTOISE

At the beginning of the 20th century, at a time when scientists around the world feared that most of the extraordinary animals of Galápagos would soon become extinct due to human inroads, the California Academy of Sciences mounted perhaps the most thorough collecting expeditions of all time. Exploring the interior of Fernandina Island in 1906, the intrepid expeditioner Rollo Beck found a single huge male saddleback tortoise, which he promptly dispatched as one more museum specimen. This was such a remarkable animal it was later described as *Geochelone elephantopus phantastica* — the 'fantastic tortoise.' But little did anyone realize then that no other tortoise would ever be seen on Fernandina, very odd considering this island escaped both the raids of the tortoise-gathering whalers and the invasion of introduced mammals. Did Mr. Beck's visit happen to have coincided with a time when the last survivor of a separate tortoise race was living out his final years in the wake of some cataclysmic but unrecorded volcanic upheaval? Or will genetic testing someday reveal that this majestic individual had in fact come from another island, perhaps a castaway Pinta tortoise (the next largest known saddleback) dropped by a passing ship long ago, having made it ashore and just grown and grown in his lonely idyll? The mystery endures.

Tui De Roy

Photos courtesy of Dr. Peter Pritchard, Chelonian Research Institute

ABOVE RIGHT: A new fumigation protocol for incoming aircraft is part of the ongoing efforts to slow the invasion by foreign species.

there really are allies where one might least expect them. A pleasant surprise indeed. The Executive Director of TAME was amazed to learn of the vampire finches of Wolf Island, which drink the blood of seabirds to survive. He was horrified to learn that the company's airplanes might actually be responsible for their demise, and understood for the first time that the shrinking biological distance is a product of efficient transport systems. As a result, he was willing to write into the airline's operational programs a total commitment to protocols that will help ward off this imminent danger. The owner of Aerogal offered her full support as well. As a result, all airplanes today are fumigated, a reminder to every passenger that the road to Paradise has to be paved with a willingness to suffer a little on the way. Also agreed with the aviation authorities is that no international flights may come directly to the islands, but must first be checked and fumigated in one of two of Ecuador's national airports, either Guayaquil or Quito.

But even these measures will not be enough by far. Of almost greater import, a full-blown sanitation plan for cargo ships must be drawn up as well. In 1970 the islands were serviced by an erratic cargo service run by two vessels carrying basic supplies to an impoverished community. Today, seven ships, each on tight three-week turnarounds, ferry goods of every imaginable description out from dirty docks in the heart of Guayaquil city. With a conservative estimate of 550 tonnes (600 tons) per trip, this amounts to over 3600 tonnes (4000 tons) of island-bound, potentially pestilent cargo per month!

Marine Pollution and Rats

To stem the tide, our vision of tomorrow must be a rigorous one. Cleanliness must be the byword, and vigilance a daily routine. All cargo, as well as the vessels themselves, must be guaranteed free of living organisms. All fresh foods free of soils and diseases. To achieve these high standards, quarantine inspections will have to be thorough, yet simple and practical. Vessels must comply with severe hygienic standards as well as with all International Maritime Organization (IMO) protocols for security and environmental standards, plus pre-embarkation and pre-disembarkation checks must be carried out on all cargoes. The procedure for boat sanitation needs to be refined and strictly applied by the captain and crew, using one page of clearly written guidelines posted in wheelhouse and galley as an aide mémoire. Fumigation will be carried out by certified companies. Compliance must be backed by law, and that law stringently enforced. Dock facilities in Guayaquil will have to be vastly improved, through cooperation with the municipality of this sprawling river city, to ensure that no wharf inhabitants, such as rats and snakes, can reach docked vessels. Embarking cargoes must arrive properly cleansed, and food packed in refrigerated containers. Agrocalidad-SICGAL will be the overall government agency responsible for the application of these vital services, and certificates issued as proof of enforcement.

In addition, all private yachts will also need to comply with standards that prevent invasive organisms sneaking in on encrusted hulls and other hidden nooks. No ornamental or culinary herbs will grace port lights or galley recesses. The reward for this vigilance? Paradise regained!

Can this dream be achieved? Can we truly push back the biological distance to 1000 kilometers? That will depend primarily on the will and dedication of 'the people' and the authorities through powerful educational programs. Money there is, but will that be all that's required? No. Success will ultimately depend on the need for us to recognize that this economic competitiveness (perhaps a rather poor reflection of biological fitness) is dragging down the planet's ecosystems. We need to challenge ourselves and our disposition if this is not to continue happening to Galápagos.

In the end, the islands can NEVER be considered free of risk. UNESCO cannot declare them as such even if the archipelago is removed from the World Heritage Sites in Danger list, the category they were assigned in 2007. Tourism will continue, even at the loss of biological integrity. The almost mythical allure of Galápagos will endure, attracting hundreds of thousands of visitors for decades to come. This is a place of unparalleled surprises, both for the mind and the senses. It has a raw beauty that takes one's breath away, causing the heart to beat a little harder. We have no right to the resources of Galápagos if we do not accept the total responsibility to reflect deeply on the actions of our own species and its effect upon the planet, and use our technological ability to ease the encroaching threats that beset Darwin's Islands.

ABOVE: A black rat escaping from the sinking cargo ship *Iguana* in 1988 in Puerto Ayora is a graphic example of how such foreign species became established. BELOW: A mockingbird singing to a misty volcano reminds us of our responsibility to preserve Galápagos.

Friends of Galápagos Around the World

FROM DARWIN'S DAY ONWARD, Galápagos has been hailed as a natural laboratory of evolution, a microcosm that can teach us how life functions on this planet, and a place from where to draw inspiration. But in the last half-century these islands have also come to represent our fervent hopes that we can indeed save a piece of the natural world as an icon precious to us all. This vision is personified by the network of charitable organizations around the world who work tirelessly to offer support for the conservation cause in Galápagos. This they achieve through a wide array of tools, including awareness-raising, technical assistance and, most critical of all, active fund-raising drives. While the government of Ecuador and tourism levies finance much of the Galápagos National Park's conservation work, the Charles Darwin Foundation's applied science and restoration efforts remain entirely dependent on private funding. In this fast-changing world, each year the challenges of preserving Galápagos become increasingly complex as well as crucial, with innumerable projects carried out by both institutions relying directly on help from the international community. This is where the Friends of Galápagos around the world step in. Few are the visitors who leave Galápagos without taking a little bit of the magic home with them in their hearts, and most pledge their support by becoming members of one of these organizations nearest them. To a large extent, the future of Galápagos rests on the shoulders of this worldwide family.

UNITED STATES
Galapagos Conservancy
Tel: +1 703 538 6833
Email: darwin@galapagos.org
Web: www.galapagos.org

CANADA
Charles Darwin Foundation of Canada
Tel: +1 416.964.4400
Email: garrett@lomltd.com

JAPAN
The Japanese Association for Galapagos
Tel/Fax: +81-3-6751-0321
Email: info@j-galapagos.org
Web: www.j-galapagos.org

NEW ZEALAND
Friends of Galapagos New Zealand
Email: info@galapagos.org.nz
Web: www.galapagos.org.nz

UNITED KINGDOM
Galapagos Conservation Trust
Tel: +44 (0)20 7629 5049
Email: gct@gct.org
Web: www.savegalapagos.org

FINLAND
Nordic Friends of Galapagos
Tel: +3358-50-5644279
Email: k.kumenius@kolumbus.fi
Web: www.galapagosnordic.org

THE NETHERLANDS
Stichting Vrienden van de Galapagos Eilanden
Tel: +31 313 421 940
Email: fin.galapagos@planet.nl
Web: www.galapagos.nl

GERMANY
Zoologische Gesellschaft, Frankfurt
Tel: +49 (0) 69-943446-0
Fax: +49 (0) 69-439348
Web: www.zgf.de

SWITZERLAND
Freunde der Galapagos Inseln
Tel: +41 (0)1 254 26 70
Email: galapagos@zoo.ch
Web: www.galapagos-ch.org

Acknowledgments

A BOOK OF THIS scope is a massively complex undertaking. Spanning 17 countries and five continents, every smallest piece of information had to be sourced from its reliable origin, reviewers' comments integrated, historic and specialized illustrations tracked down, new findings updated, and every fact and figure double-checked time and again. It is clear that such a project — resembling a multi-layered, constantly evolving, giant jigsaw puzzle — could never have come to fruition without the direct and extensive involvement of a very wide array of people.

From the first moment that I began to float the idea of this compendium, right through the year-long editing process, I was struck by one amazing fact: every person I approached agreed, without the slightest reservation, to help in any way he or she could. While far too many to mention by name in every instance, I will remain eternally grateful for the total support, and indeed unabashed enthusiasm, demonstrated by all, without any offer on my part of remuneration or reward other than a heartfelt 'thank you'.

First and foremost, I should mention the unconditional endorsement offered by the Galápagos National Park (GNP), the institution which, on behalf of the Ecuadorian government, is in charge of ensuring the wise use of the islands while maintaining their ecological integrity for posterity. Official recognition of this volume as a 50th anniversary publication imbued it with considerable credence. Enormous personal assistance was also received in the form of transport and access to the most sensitive areas of the park, as well as inside information and archival materials. For this, I am deeply grateful to the GNP director and heads of departments in the areas of sustainable development, information, tourism, marine and terrestrial protection, enforcement and patrol, and all their field staff. I likewise wish to thank the board, directors and staff of the Charles Darwin Foundation (CDF) for similarly supporting this book as a showcase of the science and conservation work at the heart of the organization's core mission.

Above all, my most profound appreciation goes directly to all 31 authors who have contributed freely of their time and their knowledge, producing superb essays, replete with fascinating yet little-known findings, and often infused with personal touches and humour. Your dedication, your patience and your good cheer have been truly inspiring. In displaying your extraordinary work to the public at large with such complete candour, I wish to thank you all for making this volume what it is.

Beyond the solid support of institutions and the individual contributors themselves, comes a very long list of people who lent invaluable assistance in more subtle ways. Among them are professional photographers and others who donated the use of precious images: Pete Oxford (peteoxfordphotos.com), Heidi Snell (Visual Escapes.smugmug.com), Jeffrey Mangel (Pro Delphinus, Peru), Dr. Peter Pritchard (Chelonian Research Institute), Dr. Beth Buckles (Cornell University), Jane Merkel (St Louis Zoo), Dr. Noah Whiteman (Harvard University), Judy O'Connor (Finders University, Australia), Freda Chapman (CDF), Mark Jones (Roving Tortoise Photos), Jacquie Grace (and Terri Maness), Noémi d'Ozouville, Frank Bungartz, Ole Hamman, Jack Grove, Christine Parent, Dave Anderson, Alan Tye, Project Isabela, GNP Archives. Equally important were the maps and diagrams drafted especially to illustrate points covered by several authors. For this, I wish to thank: James Ketchum (Biotelemetry Lab, UC-Davis), German Soler (Fundacion Malpelo), Cesar Peñaherrera (CDF), Christian Lavoie.

Many more people have helped in other special ways, including in no particular order: Sylvia Harcourt-Carrasco, Felipe Cruz, Victor Carrión, Washington Tapia, Gonzalo Banda, Sixto Naranjo, Edwin Naula, Fabian Oviedo, Edinson Cárdenas, Oscar Ramón, Eliecer Cruz, Luis Suarez, Rocio Cedeño, Monnik Desmeth, Timothy Silcott, Mark Gardner, Stuart Banks, Henri Herrera, Ivonne Guzman, Rachel Atkinson, Andrea Marin, Leon Baert, Veronica Toral, Robert Dowler.

Most of the authors also asked me to extend their personal thanks to their colleagues, collaborators, reviewers and supporters, among them James Gibbs, Lazaro Roque, Paulina Grove, Daniel Pauly, Pamela Mateson, Victoria Todd, Peter Vitousek. For my part, I would like to give special thanks to my editor and associate publisher, Tracey Borgfeldt, for her steadfastness and patience when my mood became frayed, and for her infinite attention to detail which sets this book apart from many others. And finally, my deepest thanks to Mark and Julie for all your work, and especially to Alan, for being who you are.

SPECIAL THANKS

To Monnik Desmeth of the Belgian Government Science Policy Office, BELSPO, for the generous subsidy that enabled the Charles Darwin Foundation to make copies of the book available in Galápagos, and to the regional office of Conservation International in Ecuador, in particular Luis Suarez, whose special grant provided for the translation of the text into Spanish.

Galápagos Vertebrate Checklist

RESIDENT REPTILES

Common name	Latin name
Green turtle	*Chelonia mydas*
San Cristóbal tortoise	*Geochelone chathamensis*
Española tortoise	*Geochelone hoodensis*
Pinzón tortoise	*Geochelone ephippium*
Pinta tortoise	*Geochelone abingdoni*
Floreana tortoise (extinct)	*Geochelone galapagoensis*
Santa Fe tortoise (extinct)	*Geochelone* sp.
Santa Cruz tortoise	*Geochelone porteri*
Santiago tortoise	*Geochelone darwini*
Wolf tortoise	*Geochelone becki*
Darwin tortoise	*Geochelone microphyes*
Alcedo tortoise	*Geochelone vandenburghi*
Sierra Negra tortoise	*Geochelone guntheri*
Cerro Azul tortoise	*Geochelone vicina*
Fernandina tortoise (extinct)	*Geochelone phantastica*
Galápagos land iguana	*Conolophus subcristatus*
Santa Fe land iguana	*Conolophus pallidus*
Pink iguana (new species)	*Conolophus* sp.
Galápagos marine iguana	*Amblyrhynchus cristatus*
Galápagos lava lizard	*Microlophus albemarlensis*
San Cristóbal lava lizard	*Microlophus bivittatus*
Española lava lizard	*Microlophus delanonis*
Floreana lava lizard	*Microlophus grayii*
Pinzón lava lizard	*Microlophus duncanensis*
Marchena lava lizard	*Microlophus habelii*
Pinta lava lizard	*Microlophus pacificus*
Galápagos gecko	*Phyllodactylus galapagensis*
San Cristóbal gecko	*Phyllodactylus leei*
San Cristóbal gecko	*Phyllodactylus darwini*
Santa Fe gecko	*Phyllodactylus barringtonensis*
Floreana gecko	*Phyllodactylus baurii*
Wolf Island gecko	*Phyllodactylus gilberti*
Galápagos snake	*Antillophis slevini*
Central Galápagos snake	*Antillophis steindachneri*
Southern Galápagos snake	*Alsophis biserialis*
Española snake	*Philodryas hoodensis*

RESIDENT MAMMALS

Common name	Latin name
Galápagos rice-rat	*Aegialomys galapagoensis*
Darwin's rice-rat (extinct)	*Nesoryzomys darwini*
Santa Cruz rice-rat (extinct)	*Nesoryzomys indefessus*
Santiago rice-rat	*Nesoryzomys swarthi*
Large Fernandina rice-rat	*Nesoryzomys narboroughi*
Small Fernandina rice-rat	*Nesoryzomys fernandinae*
Galápagos giant rat	*Megaoryzomys curioi*
Galápagos red bat	*Lasiurus borealis brachyotis*
Hoary bat	*Lasiurus cinereus*
Galápagos fur seal	*Arctocephalus galapagoensis*
Galápagos sea lion	*Zalophus wollebaeki*
Bottlenose dolphin	*Tursiops truncatus*
Common dolphin	*Delphinus delphis*
Cuvier's beaked whale	*Ziphius cavirostris*
Sperm whale	*Physeter macrocephalus*
Bryde's whale	*Balaenoptera edeni*

RESIDENT SEABIRDS AND SHORE BIRDS

Common name	Latin name
Galápagos penguin	*Spheniscus mendiculus*
Waved albatross	*Phoebastria irrorata*
Galápagos petrel	*Pterodroma phaeopygia*
Galápagos shearwater	*Puffinus subalaris*
Band-rumped storm petrel	*Oceanodroma castro*
White-vented storm petrel	*Oceanites gracilis galapagoensis*
Wedge-rumped storm petrel	*Oceanodroma tethys*
Magnificent frigatebird	*Fregata magnificens*
Great frigatebird	*Fregata minor*
Brown pelican	*Pelecanus occidentalis urinator*
Red-billed tropicbird	*Phaethon aethereus*
Flightless cormorant	*Phalacrocorax harrisi*
Nazca booby	*Sula granti*
Blue-footed booby	*Sula nebouxii excisa*
Red-footed booby	*Sula sula*
Swallow-tailed gull	*Creagrus furcatus*
Lava gull	*Larus fuliginosus*
Brown noddy	*Anous stolidus galapagensis*
Sooty tern	*Sterna fuscata*
Great blue heron	*Ardea herodias*
Cattle egret	*Bubulcus ibis*
Striated heron	*Butorides striatus sundevalli*
Yellow-crowned night heron	*Nyctanassa violacea pauper*
Caribbean flamingo	*Phoenicopterus ruber*
Oystercatcher	*Haematopus palliatus galapagoensis*
Black-necked stilt	*Himantopus mexicanus*

RESIDENT LAND BIRDS

Common name	Latin name
Galápagos hawk	*Buteo galapagoensis*
Short-eared owl	*Asio flammeus galapagoensis*
Barn owl	*Tyto alba*
Galápagos dove	*Zenaida galapagoensis*
Small ground finch	*Geospiza fuliginosa*
Medium ground finch	*Geospiza fortis*
Large ground finch	*Geospiza magnirostris*
Sharp-beaked ground finch	*Geospiza difficilis*
Cactus finch	*Geospiza scandens*
Large cactus finch	*Geospiza conirostris*
Small tree finch	*Camarhynchus parvulus*
Medium tree finch	*Camarhynchus pauper*
Large tree finch	*Camarhynchus psittacula*
Woodpecker finch	*Camarhynchus pallidus*
Mangrove finch	*Camarhynchus heliobates*
Vegetarian finch	*Platyspiza crassirostris*
Warbler finch	*Certhidea olivacea*

Yellow warbler	*Dendroica petechia aureolla*
Galápagos flycatcher	*Myiarchus magnirostris*
Vermilion flycatcher	*Pyrocephalus rubinus*
Galápagos mockingbird	*Mimus parvulus*
Española mockingbird	*Mimus macdonaldi*
San Cristóbal mockingbird	*Mimus melanotis*
Floreana mockingbird	*Mimus trifasciatus*
Dark-billed cuckoo	*Coccyzus melacoryphus*
Galápagos martin	*Progne modesta*
White-cheeked pintail	*Anas bahamensis*
Common gallinule	*Gallinula chloropus*
Galápagos rail	*Laterallus spilonotus*
Paint-billed crake	*Neocrex erythrops*

MIGRANT AND VAGRANT BIRDS

Common name	Latin name
Peregrine falcon	*Falco peregrinus*
Osprey	*Pandion haliaetus*
Belted kingfisher	*Ceryle alcyon*
Bobolink	*Dolichonyx oryzivorus*
Barn swallow	*Hirundo rustica*
Rose breasted grosbeak	*Pheucticus ludovicianus*
Summer tanager	*Piranga rubra*
Purple martin	*Progne subis*
Eastern kingbird	*Tyrannus tyrannus*
Common nighthawk	*Chordeiles minor*
Pied Billed grebe	*Podilymbus podiceps*
American coot	*Fulica americana*
Sora rail	*Porzana carolina*
Purple gallinule	*Porphyrio martinica*
Blue-winged teal	*Anas discors*
Brown booby	*Sula leucogaster*
Royal albatross	*Diomedea epomophora*
Wandering albatross	*Diomedea exulans*
Black-browed albatross	*Thalassarche melanophris*
Northem giant petrel	*Macronectes halli*
Pintado petrel	*Daption capense*
Flesh-footed shearwater	*Puffinus carneipes*
Pink-footed shearwater	*Puffinus creatopus*
Sooty shearwater	*Puffinus griseus*
Wedge-tailed shearwater	*Puffinus pacificus*
White-bellied storm petrel	*Fregetta grallaria*
Leach's storm petrel	*Oceanodroma leucorhoa*
White-faced storm petrel	*Pelagodroma marina*
Markham's storm petrel	*Oceanodroma markhami*
Black storm petrel	*Oceanodroma melania*
Laughing gull	*Larus atricilla*
Gray-headed gull	*Larus cirrocephalus*
Ring-billed gull	*Larus delawarensis*
Southern black-backed gull	*Larus dominicanus*
Frankin's gull	*Larus pipixcan*
Common tern	*Sterna hirundo*
Elegant tern	*Thalasseus elegans*
Royal tern	*Thalasseus maximus*
Long-tailed jaeger	*Stercorarius longicaudus*
Parasitic jaeger	*Stercorarius parasiticus*
Pomarine jaeger	*Stercorarius pomarinus*
Great skua	*Stercorarius skua*
Ruddy turnstone	*Arenaria interpres*
Spotted sandpiper	*Actitis macularia*
Least sandpiper	*Calidris minutilla*
Sanderling	*Calidris alba*
Willet	*Catoptrophorus semipalmatus*
Semipalmated plover	*Charadrius semipalmatus*
Wandering tattler	*Heteroscelus incanus*
Short-billed dowitcher	*Limnodromus griseus*
Whimbrel	*Numenius phaeopus*
Black-bellied plover	*Pluvialis squatarola*
Lesser yellowlegs	*Tringa flavipes*
Western sandpiper	*Calidris mauri*
Red-necked phalarope	*Phalaropus lobatus*
Wilson's phalarope	*Phalaropus tricolor*
Common egret	*Ardea alba*
Snowy egret	*Egretta thula*

VISITING MARINE REPTILES AND MAMMALS

Common name	Latin name
Leatherback turtle	*Dermochelys coriacea*
Hawksbill turtle	*Eretmochelys imbricata*
Olive ridley turtle	*Lepidochelys olivacea*
Yellow-bellied sea snake	*Pelamis platurus*
Striped dolphin	*Stenella coeruleoalba*
Risso's dolphin	*Grampus griseus*
Short-finned pilot whale	*Globicephala macrorhynchus*
False killer whale	*Pseudorca crassidens*
Melon-headed whale	*Peponocephala electra*
Orca	*Orcinus orca*
Hump-backed whale	*Megaptera novaengliae*
Blue whale	*Balaenoptera musculus*
Sei whale	*Balaenoptera borealis*
Minke whale	*Balaenoptera acutorostrata*

INTRODUCED FERAL VERTEBRATES

Common name	Latin name
Rock dove (Domestic pigeon)	*Columba livia*
Smooth-billed ani	*Crotophaga ani*
Domestic chicken	*Gallus gallus*
Pig	*Sus scrofa*
Dog	*Canis familiaris*
Cat	*Felis catus*
Donkey	*Equus asinus*
Horse	*Equus caballus*
House mouse	*Mus musculus*
Norwegian (brown) rat	*Rattus norvegicus*
Black (ship) rat	*Rattus rattus*
Cattle	*Bos taurus*
Goat	*Capra hircus*
Mourning gecko	*Lepidodactylus lugubris*
Peters' leaf-toed gecko	*Phyllodactylus reissii*
Yellowbelly (leaf-toed) gecko	*Phyllodactylus tuberculosus*
Shieldhead gecko	*Gonatodes caudiscutatus*
Nile tilapia	*Oreochromis niloticus*
Tree frog	*Scinax quinquefasciatus*

Further Reading

The Galápagos Islands have been the subject of innumerable books, from poetry to field guides, fiction to learned treatise. Below is a rounded list of titles picked from a choice of hundreds.

CONTEMPORARY SELECTION

Angemeyer, Johanna. *My Father's Island: A Galapagos Quest.* New York: Viking Adult, 1990.

Bassett, Carol Ann. *Galapagos at the Crossroads: Pirates, Biologists, Tourists, and Creationists Battle for Darwin's Cradle of Evolution.* Washington: National Geographic Books, 2009.

D'orso, Michael. *Plundering Paradise: The Hand of Man on the Galapagos Islands.* London: Harper Perennial, 2003.

Green, Jonathan R. *Galapagos: Ocean, Earth, Wind & Fire.* (2nd ed.) Quito: Imprenta Mariscal, 2006.

Hickman, John (editor). *Galapagos, Through Writer's Eyes.* London: Eland Books, 2009.

Hidrobo, Hugo. *Footsteps in Paradise.* Quito: Libri Mundi ediciones, 2005.

Larson, Edward J. *Evolution's Workshop: God and Science on the Galápagos Islands.* New York: Basic Books, 2002.

Nicholls, Henry. *Lonesome George: The life and loves of a conservation icon.* New York: Palgrave Macmillan, 2007.

Oxford, Pete and Renee Bish. *Galapagos: The Untamed Isles.* Qutio: Libri Mundi, 1999/New York: Imagine Publishing, 2009.

Stewart, Paul D. *Galapagos: The Islands That Changed the World.* New Haven: Yale University Press, 2007

Watkins, Graham and Pete Oxford. *Galapagos Both Sides of the Coin*, with introduction by Prince Philip. New York: Imagine Publishing, 2009.

Weiner, Jonathan. *The Beak of the Finch: A Story of Evolution in Our Time.* London: Vintage, 1995.

Woram, John. *Charles Darwin Slept Here.* New York: Rockville Press, 2005.

PRIME GUIDEBOOKS

Andrew, David. *Watching Wildlife Galápagos Islands.* Melbourne: Lonely Planet, 2005.

Boyce, Barry. *A Traveler's Guide to the Galapagos Islands.* (4th ed.) New Jersey: Hunter Publishing, 2004.

Castro, Isabel and Antonia Phillips. *A Guide to the Birds of the Galapagos Islands.* New Jersey, Princeton University Press, 1996.

Constant, Pierre. *Galapagos: A Natural History Guide.* (7th ed.) Hong Kong: Odyssey Illustrated Guides, 2006.

Constant, Pierre. *Marine Life of the Galapagos: Divers' Guide to the Fish, Whales, Dolphins and Marine Invertebrates.* (2nd ed.) Hong Kong: Odyssey Illustrated Guides, 2008.

Fitter, Julian, Daniel Fitter and David Hosking. *Wildlife of the Galapagos* (Traveller's Guide). London: Harper Collins, 2007.

Harris, Michael and Barry Kent MacKay. *The Collins Fieldguide to the Birds of the Galapagos.* London: Collins, 1982.

Hickman, Cleveland P. *A Field Guide to Marine Molluscs of Galapagos.* Lexington: Sugar Spring Press, 1999.

Hickman Jr, Cleveland P. and Todd L. Zimmerman. *A Field Guide to Crustaceans of Galapagos.* Lexington: Sugar Spring Press, 2000.

Hickman Jr, Cleveland P. *A Field Guide to Sea Stars & Other Echinoderms of Galapagos.* Lexington: Sugar Spring Press, 1998.

Hickman Jr, Cleveland P. *A Field Guide to Corals and Other Radiates of Galapagos.* Lexington: Sugar Spring Press, 2008.

Horwell, David and Pete Oxford. *Galapagos Wildlife: A Visitor's Guide* (2nd ed.). Chalfont St Peter: Bradt Travel Guide, 2005.

Humann, Paul. *Reef Fish Identification: Galapagos.* Jacksonville: New World Publications, 2003.

Jackson, Michael H. *Galapagos: A Natural History.* Calgary: University of Calgary Press, 1994.

Kavanagh, James. *Galapagos Wildlife: An Introduction to Familiar Species.* (Pocket Naturalist). Phoenix: Waterford Press, 2001.

Kricher, John. *Galapagos: A Natural History.* New Jersey: Princeton University Press, 2006.

McMullen, Conley K. *Flowering Plants of the Galapagos.* Ithaca: Cornell University Press, 1999.

Pons, Alain. *Galapagos Safari Companion.* Rickmansworth: Evans Mitchell Books, 2004.

Rosenberg, Steve and Ellen I. Sarbone. *The Diving Guide: Galapagos Islands.* Dunedin, Florida: Cruising Guide Publications, 2004.

Swash, Andy and Rob Still. *Birds, Mammals, and Reptiles of the Galapagos Islands: An Identification Guide*, (2nd ed.). New Haven: Yale University Press, 2006.

MAPS

Zagier, Sergio and John Woram. *Islas Galapagos Historical Chart.* Buenos Aires: Zagier & Urruty Publications, 2008.

Galapagos Islands Explorer Map. Chicester: Ocean Explorer Maps, n.d.

TREATISES AND MONOGRAPHS

Chester, Roy. *Furnace of Creation, Cradle of Destruction: A Journey to the Birthplace of Destruction.* New York: Amacom, 2008.

Grant, Peter R. *Ecology and Evolution of Darwin's Finches.* Introduction by Jonathan Weiner. New Jersey: Princeton Science Library, 1999.

Grove, Jack and Robert Lavenberg. *The Fishes of the Galapagos Islands.* Palo Alto: Stanford University Press, 1997.

McBirney, Alexander R. and Howel Williams. *Geology and Petrology of the Galapagos Islands.* Boulder: Geological Society of America,1969.

Wiggins, Ira L. and Duncan M. Porter. *Flora of the Galapagos Islands.* Palo Alto: Stanford University Press, 1971.

OLD CLASSICS

Beebe, William. *Galapagos: World's End.* London: Dover Publications, 1988.

Darwin, Charles and Janet Browne & Michael Neve. *The Voyage of the Beagle: Charles Darwin's Journal of Researches.* Abridged. London: Penguin Classics, 1989.

Darwin, Charles. *The Origin of Species.* London: Bantam Classics, 1999.

Lack, David. *Darwin's Finches: An essay on the general biological theory of evolution.* New York: Harper Torchbooks, 1961.

Nelson, Bryan. *Galapagos: Islands of Birds.* New York: William Morrow & Co., 1968.

Porter, Eliot. *Galapagos: The Flow of Wildness.* San Francisco: Sierra Club, 1968.

Steadman, David W., and Steven Zousmer and Lee M. Steadman. *Galapagos: Discovery on Darwin's Islands.* Washington DC: Smithsonian Institution, 1989.

Treherne, John. *The Galapagos Affair.* London: Pimlico Books, 1983.

Vonnegut, Kurt. *Galapagos: A Novel.* (Delta Fiction) New York: The Dial Press, 1999.

Wittmer, Margret. *Floreana: A Woman's Pilgrimage to the Galapagos.* Kingston: Moyer Bell, 1990.

OTHER BOOKS BY TUI DE ROY

Signed copies can be mail-ordered by writing to: books@rovingtortoise.co.nz

Galapagos: Wild Portraits. (8th ed.) Tui De Roy & Mark Jones. New York: Abbeville, 2009.

Galapagos: Islands Born of Fire (10th anniversary edition). Auckland: David Bateman Ltd., 2007.

Galapagos: Inseln aus Feuer Geboren. Steinfurt: Tecklenborg, 2000.

Spectacular Galapagos. New York: Universe Books, 1999.

Galapagos: Islands Lost in Time. Tui De Roy Moore, with foreword by Peter Matthiessen. New York: Penguin Books, 1980.

Albatross: Their World, Their Ways by Tui De Roy, Mark Jones and Julian Fitter. Auckland: Bateman/Toronto: Firefly Books, 2008.

New Zealand: A Natural History by Tui De Roy and Mark Jones. Auckland: Bateman/Toronto: Firefly Books, 2006.

The Andes: As the Condor Flies. Auckland: Bateman/Toronto: Firefly Books, 2005.

Dawn to Dusk in the Galapagos: Flightless Birds, Swimming Lizards, and Other Fascinating Creatures. Tui De Roy and Rita Gelman. New York: Little Brown, 1991.

Wild Ice: Antarctic Journeys. Colin Monteath, Tui De Roy, Mark Jones Ron Naveen. Washington DC: Smithsonian Institution Press, 1990.

Index